Fundamentals
of
engineering drawing

Fundamentals of

Warren J. Luzadder, p.e.
PURDUE UNIVERSITY

Seventh edition

engineering drawing

for design

product development,

and numerical control

PRENTICE-HALL, INC., ENGLEWOOD CLIFFS, NEW JERSEY 07632

Library of Congress Cataloging in Publication Data

Luzadder, Warren Jacob.
    Fundamentals of engineering drawing for design,
product development, and numerical control.

    First-5th ed. published under title: Fundamentals of
engineering drawing for technical students and
professional draftsmen; 6th ed.: Fundamentals of
engineering drawing for design, communication, and
numerical control.
    Bibliography: p. 600
    Includes index.
    1. Mechanical drawing. I. Title.
T353.L88   1977   604'.2'4   76-49886
ISBN 0-13-338368-7

Fundamentals of engineering drawing
for design, product development, and numerical control

SEVENTH EDITION

by Warren J. Luzadder, P.E.

Current printing (last digit):

10 9 8 7 6 5 4 3 2 1

other books by the author:

Innovative Design with an Introduction to Design Graphics, Prentice-Hall, Inc. 1975.
Basic Graphics for Design, Analysis, Communications and the Computer,
    2nd ed., Prentice-Hall, Inc., 1968.
Fundamentos De Dibujo Para Ingenieros, 2nd ed., Compania Editoria
    Continental, 1967.
Graphics for Engineers, Prentice Hall of India Private Limited, 1964.
Technical Drafting Essentials, 2nd ed., Prentice-Hall, Inc., 1956.
Problems in Engineering Drawing, 7th ed., Prentice-Hall, Inc., 1977.
Engineering Graphics Problems for Design, Analysis and Communications,
    Prentice-Hall, Inc., 1968.
Problems in Drafting Fundamentals, Prentice-Hall, Inc., 1956.
Purdue University Engineering Drawing Films, with J. Rising et al.

PRENTICE-HALL INTERNATIONAL, INC., *London*
PRENTICE-HALL OF AUSTRALIA, PTY. LTD., *Sydney*
PRENTICE-HALL OF CANADA, LTD., *Toronto*
PRENTICE-HALL OF INDIA PRIVATE LIMITED, *New Delhi*
PRENTICE-HALL OF JAPAN, INC., *Tokyo*
PRENTICE-HALL OF SOUTHEAST ASIA PTE. LTD., *Singapore*
WHITEHALL BOOKS LIMITED, WELLINGTON, *New Zealand*

To my wife Frances Jeannette,
who was always willing to do ''just a little more typing.''

Contents

11

Pictorial presentation *211*

Part **III** Design

12

The design process and graphics *243*

Part **IV** Graphics for design and communication

13

Dimensions, notes, limits, and geometric tolerances *275*

14

Threads and standard machine elements *307*

20

Graphical mathematics   *459*

Part VII   Design and communication drawing in specialized fields

21

Design of machine elements—gears, cams, and linkages   *473*

22

Electronic drawings   *485*

23

Welding drawings   *501*

24

Piping drawings and models   *509*

25

Structural drawings   *521*

26

Topographic and engineering map drawings   *541*

Appendix

Preface

In addition to meeting present-day course requirements, a worthwhile text should anticipate trends and include new material that may profitably be presented to students over the several years that must pass before another edition can be justified. Anticipating trends is sometimes difficult, but the author has made an attempt to do so by expanding and up-dating computer graphics and by giving the measurement values for most of the illustrations and problems in millimeters, centimeters, kilometers, kilograms, and so forth. When our largest corporations, such as General Motors, Ford, IBM, TRW, McDonnell Douglas, Honeywell, Rockwell International, John Deere and Company, and International Harvester are all converting to metric, it is time for authors and educators to turn to the SI metric system (Système International d'Unités). The conversion period may be a long one, covering a decade or more. During this time, we could have widely different rates of progress in industry, with some companies, such as John Deere and Company and International Harvester, changing over quickly and totally to metric (as they have already done), while other companies remain in limbo for several years, depending upon a dual-dimensioning system to meet their needs. The long period of time that will be needed for the American National Standards Institute (ANSI) to change the many standards will act as a brake to slow down the rapidity of our movement toward a totally metric U.S. However, with the change to metric now well under way, a total changeover seems to be inevitable. At the time that this preface is being written, a special committee set up by ANSI to study and produce an Optimum Metric Fastener System has released OMFS Recommendations 2, 3, 4, and 6, which relate to metric screw threads. Hopefully, this will lead to a new and complete standard for ANSI–OMFS Metric Screw Threads sometime in the near future. The completion of this task and the release of a new standard for threads, however, will still leave much to be done before new sets of standards for fasteners can be made available to users. The instructor using this text is advised to point out to the students that many of those companies that claim to have gone totally metric still specify threads, fasteners, and standard parts using the inch system of the existing standards. In addition, their drawings include an inch conversion chart for the use of suppliers who are not equipped to handle metric dimensions. Metric equivalents, conversion multipliers, and tables for metric drills and metric threads have been added to the Appendix.

It has been reported that ANSI committee Y14.5 is at work on a revision of the standard to incorporate metric dimensioning. If so, it is possible for this revision to be approved and published as early as 1978. In the revision of this text, the author has been guided largely by the metric design and drafting standards furnished by General Motors, John Deere and Company, and International Harvester. International Harvester furnished an almost complete standard, while GM furnished material needed for metric screw threads and surface quality. The thoughtfulness and kindness of the several administrators and standards engineers is deeply appreciated.

This text has not been made totally metric because ANSI standards were not available to make this possible. Also, since the English inch, foot, mile, acre, pound, and so forth will be with us for many years to come, it was thought to be wise to retain some drawings that show dimensions in inches and to prepare some of the illustrations showing how dimensions are to be given using both systems. In general, this was done in the chapter covering dimensioning. Elsewhere, illustrations and problems show all metric or they have been dual-dimensioned. Where problems show inch measurements, the instructor may either require the part to be dimensioned in inches or he may ask the students to use millimeters. If a conversion to metric is made by the students, they will acquire a better understanding of the metric system. The author has had the good fortune to have been forced to face the metric system in 1960 with a Spanish translation of an earlier edition.

For convenience this text has been arranged in seven main parts: Part I, Basic Graphical Techniques; Part II, Spatial Graphics: Shape Description and Spatial Relationships; Part III, Design; Part IV, Graphics for Design and Communication; Part V, The Computer; Part VI, Graphic Methods for Engineering Communication, Design, and Computation; and Part VII, Design and Communication Drawing in Specialized Fields.

Part I covers the use of instruments, lettering, and engineering geometry. Part II provides essential information on basic graphics, spatial geometry, and pictorial representation. Part III presents the procedures of the design process and endeavors to show the interrelationship between design and manual drafting that includes both sketching and instrumental drawing. This chapter is new for this edition. Part IV presents information needed to prepare engineering drawings. The chapters in this part are: Chapter 13, Size Description: Dimensions and Specifications; Chapter 14, Threads and Standard Machine Elements; Chapter 15, Shop Processes, Shop Terms, and Tool Drawings; and Chapter 16, Production Drawings: Preparation and Duplication. Part V (Chapters 17 and 18) covers the field of computer-aided design, automated drafting, and numerically controlled machine tools. At all institutions, particularly where computer-aided design systems are available, Chapter 17 should be assigned for reading and discussion. In addition, since practical knowledge of these systems can only be gained by either working experience or by on-the-scene observations, instructors of graphics and design should schedule plotter time and CRT (cathode-ray tube) console time for demonstrations. If this is not possible, several excellent motion pictures are available (see the Bibliography) that may be shown in lieu of actual demonstrations.

Part VI presents the graphic methods commonly employed for communication, design, and computation. Finally, Part VII provides the information needed to prepare drawings in specialized fields so that problems may be assigned to broaden overall knowledge of the field of graphics, particularly in an area of individual and class interest.

To bring this text abreast of new technological developments, a number of leading industrial organizations have generously assisted the author by supplying appropriate illustrations that were needed in developing specific subjects. Every commercial illustration supplied by American industry has been identified using a courtesy line. The author deeply appreciates the kindness and generosity of these many companies and the busy people in their employment who found the time to select these drawings and photographs that appear in almost every chapter.

The author is grateful to Professors K. E. Botkin, R. L. Airgood, R. P. Thompson, and other members of the graphics staff at Purdue University for their many valuable suggestions in regard to the content and organization of this seventh edition.

Special appreciation must be expressed for the contributions of Professors W. L. Baldwin, R. H. Hammond, Byard Houck, and J. F. Zimmerly of the Ford Motor Company. Professor Baldwin of Purdue University developed the material on linkages. Professors Hammond and Houck contributed new material to the chapter on computer-aided design. Professor Houck is an authority on the TRIDM program that has been prepared at North Carolina State University at Raleigh, since he played a large role in its development. Mr. Zimmerly supplied the author with new information and material in the area of computer graphics that has been included in Chapter 17. Not to be forgotten is the fact that there are many persons, some known and others unknown, who have made valuable contributions to this classroom text. The author's indebtedness to these persons is hereby reaffirmed.

Last, I would like to acknowledge two gentlemen for their outstanding work on this text. The first, Mark A. Binn, senior book designer, did an excellent job on the cover and interior designs. The second, Ken Wisman, production editor, did superb work handling all aspects of production. My sincere thanks to both.

W. J. L.
Purdue University

Fundamentals
of
engineering drawing

ORSIII OV5-8 Satellite-friction experiment. Working drawings prepared in the drafting room convey the ideas of the designer to the men who must fabricate the parts and assemble the system. Many drawings were needed for the satellite shown above. Communication drawings, both detail and assembly, are discussed in Chapters 12 and 16. (Note drawing as a backdrop.) (*Courtesy TRW Systems Group*)

1

Introduction

(1.1)
Brief History of Drawing. For upward of twenty thousand years, a drawing has been the principal means for the portrayal of ideas through the use of lines. Its beginnings, however, are still further back in time, for our early ancestors undoubtedly explained their ideas by marking in the dust on the floors of their caves. It is deeply rooted in our instincts and, in a sense it is our one universal language even today, when some of our drawings are prepared by computers and plotters (Fig. 17.17). The earliest records of man are graphic, depicting people, deer, buffalo, and other animals of the time on the rock walls of caves. These drawings were to satisfy an elemental need for expression, long before the development of writing. However, drawing gradually freed itself from this early usage when writing was developed and it then came to be used primarily by artists and engineering designers as a means of setting forth ideas for the construction of finished works such as pyramids, war chariots, buildings, and simple mechanisms useful to man. Most of the very early drawings that still exist were made on parchment, which was very durable. Later, during the twelfth century, paper was developed in Europe and came into general use for drawings.

Only a few of the earliest drawings for fortresses, buildings, and simple mechanisms are in existence today. Those that have come down to us have been largely pictorial in nature, and they exist as carvings and paintings on walls of structures or have been woven into tapestry. One of the earliest representations shows the use

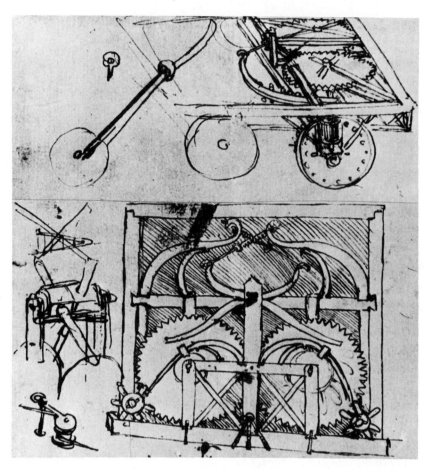

Fig. 1.1 Idea sketch prepared by Leonardo da Vinci (1452–1519). The da Vinci "automobile" was to have been powered by two giant springs and steered by the tiller, at the left in the picture, attached to the small wheel. (*From Collection of Fine Arts Department, International Business Machines Corporation*)

of the wheel about 3200 B.C. in Mesopotamia. The drawing presents a wheelbarrow-like structure being used by a man to transport his wife or child. The pictorial is primitive without any depth of perspective. Also, a ground-plan type of drawing for a fortress exists on a tablet prepared about 4000 B.C.

At the beginning of the Christian era, Roman architects had become skillful in preparing drawings of buildings that were to be constructed. They used straightedges and compasses to lay out the elevation and plan views and were able to prepare well-executed perspectives. However, the theory of projection of views upon imaginary planes of projection was not developed as a means of representation until sometime during the Renaissance period. Even though it is probable that Leonardo da Vinci was aware of the theory of multiview drawing (Fig. 1.5), his training as an artist prevailed and he recorded his ideas and designs for war machines

and mechanical constructions by preparing perspective sketches and drawings such as the one shown in Fig. 1.1. No multiview drawings prepared by da Vinci have been found. He knew the value of a pictorial drawing, and it is interesting to note that even in this day of space travel, we prepare pictorial drawings to supplement our other design drawings (see the pictorial drawing of Skylab in Fig. 1.2).

1.2

Interrelationship of Engineering Graphics (Drawing) and Design. Since design graphics and design are interrelated in the total design process, those persons preparing the design layouts, the detailers, and the production engineer who has been assigned to the project must work closely at times with the project group leader as a part of the total design team. In general, all persons assigned to a project, both designers and those who support the design effort at any stage, should have full

Skylab

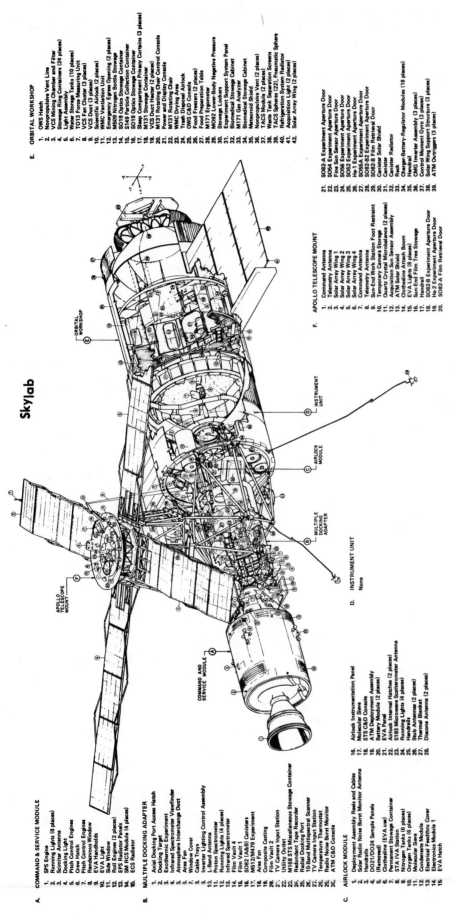

A. COMMAND & SERVICE MODULE
1. SPS Engine
2. Running Lights (8 places)
3. Scimitar Antenna
4. Docking Light
5. Pitch Control Engines
6. Crew Hatch
7. Pitch Control Engines
8. Rendezvous Window
9. EVA Handholds
10. EVA Light
11. Side Window
12. Cable Window
13. Roll Engines (2 places)
14. EPS Radiator Panels
15. SM RCS Module (4 places)
16. ECS Radiator

B. MULTIPLE DOCKING ADAPTER
1. Axial Docking Port Access Hatch
2. Docking Target
3. Exothermic Experiment
4. Infrared Spectrometer Viewfinder
5. Atmosphere Interchange Duct
6. Area Fan
7. Window Cover
8. Cable Trays
9. Inverter Lighting Control Assembly
10. L-Band Antenna
11. Proton Spectrometer
12. Running Lights (4 places)
13. Infrared Spectrometer
14. Film Vault 4
15. Film Vault 1
16. S082 (A&B) Canisters
17. M512/M479 Experiment
18. Area Fan
19. Composite Casting
20. Temperature Thermostat
21. TV Camera Input Station
22. Utility Outlet
23. Redundant Tape Recorder
24. M188 STS Miscellaneous Stowage Container
25. Radial Docking Port
26. TD-Band Multispectral Scanner
27. TV Camera Input Station
28. Radio Noise Burst Monitor
29. ATM C&D Console

C. AIRLOCK MODULE
1. Deployment Assembly Reels and Cables
2. Solar Radio Noise Burst Monitor Antenna
3. Handrails
4. DO21/DO24 Sample Panels
5. (Removed)
6. Clothesline (EVA use)
7. Permanent Stowage Container
8. ST4 EVA Station
9. Nitrogen Tanks (6 places)
10. Oxygen Tanks (6 places)
11. Molecular Sieve
12. Condensate Module
13. Electrical Feedthru Cover
14. Electronics Module 1
15. EVA Hatch

16. Airlock Instrumentation Panel
17. Molecular Sieve
18. STS C&D Console
19. ATM Deployment Assembly
20. Battery Module (2 places)
21. EVA Panel
22. Airlock Internal Hatches (2 places)
23. S193 Microwave Scatterometer Antenna
24. Running Lights (4 places)
25. Handrails
26. Stub Antennas (2 places)
27. Thermal Blanket
28. Discone Antenna (2 places)

D. INSTRUMENT UNIT
None

E. ORBITAL WORKSHOP
1. OWS Hatch
2. Nonpropulsive Vent Line
3. VCS Mixing Chamber and Filter
4. Stowage Ring Containers (24 places)
5. Light Assembly
6. Water Storage Tanks (10 places)
7. TO13 Force Measuring Unit
8. VCS Fan Cluster (3 places)
9. VCS Duct (3 places)
10. Scientific Airlock (2 places)
11. WMC Ventilation Unit
12. Emergency Egress Opening (2 places)
13. M509 Nitrogen Bottle Stowage
14. SO19 Optics Stowage Container
15. S149 Particle Collection Container
16. SO19 Optics Stowage Container
17. Sleep Compartment Privacy Curtains (3 places)
18. M131 Stowage Container
19. VCS Duct Heater (2 places)
20. M131 Rotating Chair Control Console
21. Power and Display Console
22. M131 Rotating Chair
23. WMC Drying Area
24. Trash Disposal Airlock
25. OWS C&D Console
26. Food Freezers (2 places)
27. Food Preparation Table
28. M171 Ergometer
29. MO92 Lower-Body Negative Pressure
30. Stowage Lockers
31. Experiment Support System Panel
32. Biomedical Stowage Cabinet
33. M171 Gas Analyzer
34. Biomedical Stowage Cabinet
35. Meteoroid Shield
36. Nonpropulsive Vent (2 places)
37. TACS Module (2 places)
38. Waste Tank Separation Screens
39. Refrigeration System Radiator
40. TACS Spheres (22), Pneumatic Sphere
41. Acquisition Light (2 places)
42. Solar Array Wing (2 places)

F. APOLLO TELESCOPE MOUNT
1. Command Antenna
2. Telemetry Antenna
3. Solar Array Wing 1
4. Solar Array Wing 2
5. Solar Array Wing 3
6. Solar Array Wing 4
7. Command Antenna
8. Telemetry Antenna
9. Sun-End Work Station Foot Restraint
10. Temporary Camera Storage
11. Quartz Crystal Microbalance (2 places)
12. Acquisition Sun Sensor Assembly
13. ATM Solar Shield
14. Clothesline Attach Boom
15. EVA Lights (8 places)
16. Sun-End Film Tree Stowage
17. Handrail
18. SO82-B Experiment Aperture Door
19. He-2 Experiment Aperture Door
20. SO82-A Film Retrieval Door

21. SO82-A Experiment Aperture Door
22. SO54 Experiment Aperture Door
23. Fine Sun Sensor Aperture Door
24. SO56 Experiment Aperture Door
25. SO52 Experiment Aperture Door
26. He-1 Experiment Aperture Door
27. SO55A Experiment Aperture Door
28. SO82-B2 Experiment Aperture Door
29. SO82-B Film Retrieval Door
30. Canister Solar Shield
31. Canister
32. Canister Radiator
33. Rack
34. Charger-Battery-Regulator Modules (18 places)
35. Handrail
36. Control Moment Gyro (3 places)
37. CMG Inverter Assembly (3 places)
38. Solar Wing Support Structure (3 places)
39. ATM Outriggers (3 places)

Fig. 1.2 Skylab—manned orbital scientific space station. The Skylab was designed to expand our knowledge of manned earth-orbital operations and to accomplish carefully selected scientific, technological, and medical investigations. (*Courtesy National Aeronautics and Space Administration*)

knowledge of engineering graphics. A design engineer himself, if he is to be successful as a designer, should have thorough training in this area. At the very least, he should be capable of preparing well-executed freehand design sketches and have a working knowledge of all of the forms of graphical expression that have been presented in this text.

Persons in the design support area, who may be expected to solve some of the design problems that arise graphically, prepare design layouts and models, and, finally, prepare the detail and assembly drawings needed for use in the production shops, must all have some basic education in the classroom and then acquire added experience where they are employed in order to become acquainted with company standards and practices. Some of these present-day "design-room professionals," in addition to their usual assignments, as mentioned, may also be expected to have some knowledge of production methods, particularly concerning numerically controlled (NC) machines, and have an understanding of computers (Chapters 17 and 18). In addition, specific design tasks may require the use of digitizers and plotters. (See the left-hand page facing the beginning of Chapter 2.)

1.3

Engineering Graphics Today. Even though the attention of a designer may be said to be centered mainly on the problems that arise in design and development (Fig. 1.3), he must have a complete working knowledge of communication drawing, for it is often his responsibility to direct the preparation of the final drawings. These follow the preliminary design drawings and instructive sketches that he and his aids have prepared in accordance with the basic principles underlying the preparation of working drawings.

For a full and complete exchange of ideas with others, the engineering technologist must be proficient in the three means of communication that are at his disposal: (1) English, both written and oral; (2) symbols, as used in the basic sciences; and (3) engineering drawing.

As a member of the engineering team, the design draftsman must be fully capable of preparing the final drawings that will convey the information about size and shape needed to fabricate the parts and assemble the structure. This must be done in accordance with company practices and with the practices recommended in the publications of the American National Standards Institute. It is expected that the design draftsman keep up to date with the changes in standards that are constantly being made by standards engineers employed by his company and by the committees working under the sponsorship of the American National Standards Institute.

Engineering technicians, assigned to production areas or to the engineering department to aid the engineers, must have considerable knowledge of engineering drawing. Those closely assisting a design engineer or technologist may be called on to solve problems graphically, and to make working sketches relating

Fig. 1.3   Linear induction motor test vehicle on a run at DOT's testing facility at Pueblo, Colorado. (*Courtesy U.S. Department of Transportation*)

to mechanisms, electrical circuits, and structural systems.

1.4

Organization of the Text. The purpose of this text is to present the grammar and composition of drawing so that those students in engineering colleges and technical institutes who conscientiously study the basic principles will eventually be able to prepare satisfactory industrial drawings and, after some practical experience, be capable of directing the work of others. To facilitate study, the subject matter has been separated into its various component parts: engineering geometry, multiview drawing, dimensioning, pictorial drawing, sketching, design, and so forth. Later chapters discuss the preparation of working drawings, both detail and assembly, computer-aided design, the preparation of topographic drawings, and the construction of charts and graphs. The major portion of the material presented leads up to the preparation of machine drawings, which the prospective members of some of the other branches of engineering technology think is not of interest to them. Since the methods used in the preparation of machine drawings, however, are the same methods used in the preparation of drawings in other fields, a thorough understanding of machine drawing assures a good foundation for later study in some specialized field, such as structural drawing. For those interested in specific types of drawing, some material has been presented with the assumption that the student already possesses a working knowledge of projection and dimensioning.

Proficiency in applying the principles of orthographic projection leads to easy graphical methods of solving space problems such as the determination of the clearance distance between a wheel and a fender or the true angle between a turbine blade route and the engine axis (Chapter 9).

1.5

Graphics and the Computer. The graphic language has become a means for an intimate and continuous conversational style of interchange between man and computer in the process of creative design. At present, computer systems have been developed that are in daily use and that interact with a human partner using the designer's own graphic language (Fig. 1.4). In bringing the computational power

Fig. 1.4 An engineer working at a graphics display unit to find solutions to problems that have arisen in the design. With the electronic "light pen" the designer can revise the image and change the parameters. An engineer may work with diagrams, drawings, or printed letters and numbers directly on the face of the tube. This direct man–computer interaction leads to quick answers and permits concepts to be easily evaluated and then accepted or rejected. (*Courtesy International Business Machine Corporation*)

of the digital computer to bear on graphical design, special image processing systems have been built that allow the computer to both read and generate drawings. Chapter 17 covers computer graphics in considerable detail.

1.6

The Present and the Future. This is a period of revolutionary change in the field of communication technology. At the present time, we are integrating new methods with the stylized methods of the past, using each method where it would seem to be best suited for the production of technical communications. We will continue to draw graphical representations manually and print out dimensions, notes, lists of materials, and so forth for the foreseeable future. However, plotters used on-line with computers or off-line directed by tapes will continue to gain wider utilization. Plotters are used in particular by aircraft and aerospace companies that produce products having complex contours. They are used also to a limited extent for the preparation of ordinary drawings of machine parts, for printed circuitry, for structural steel drawings, for highway route plans, and for determining the tool paths of milling (NC) machine tapes (Chapters 17 and 18).

Since nothing has happened to alter the

fact that we communicate with others best in the graphic language, one advance that has come about in the use of computers has been the development of computer graphics—that is, the development of the hardware and the software that enable the computer to accept, understand, analyze, and produce engineering design data in graphic form. Persons who are responsible for the preparation of technical representations must become familiar with programming and with the use of both the computer and the plotter. These are their newest design and drafting tools. Although plotters are somewhat sophisticated as well as expensive, they save the designer and draftsman hours of tedious manual labor.

No new developments appear on the immediate horizon that will relieve the designer of the task of thinking out his design or the draftsman of the responsibility of applying the knowledge of his trade. Furthermore, recent technological breakthroughs have served to extend, rather than supplant, the direct application of the principles of engineering drawing by designers and draftsmen. The technical field needs the graphic language in all of its many forms.

As technology advances to meet the rising expectations of people all around the world, more and more drawings and other forms of graphical representation will be required. Most of these, as might be expected, will continue to be prepared manually.

Automation has already entered the field of communication technology and therefore must be accepted by those persons now in the field and by those about to enter. The only question that either a student or a technically trained man now at work can ask himself is whether or not he can adjust to new knowledge, new methods, and different requirements during his career.

1.7

Role of the Computer and Plotter in the Drawing Room. Although words alone may not relieve entirely a person's concern about the continued importance of drafting, an understanding of the true role of the computer and plotter in the design room will answer some, if not all, of the questions that may arise in one's mind. Those who have been employed, where computers have taken over a number of graphic functions, welcome their use because they have learned that computers can accomplish those repetitive tasks, such as wire routing, schematics, and repetitious mechanical drafting, that can be very boring and time consuming to the draftsman. This leaves a large number of drawings in the field of mechanical design still to be done by draftsmen and detailers at the drawing board (see the page facing Chapter 2). In this category are drawings of piece parts that must be prepared manually. In the aircraft industry, where computers, digitizers, and plotters are readily available, manually prepared drawings now constitute at least 50% of the total drawing output. Each of these drawings done on the board is so unique and different as to require human intelligence for its preparation, together with indepth knowledge of graphics and shop practices. It should also be noted that the design layout, which defines the function and shape of these parts being drawn, was of necessity prepared manually in the preliminary design stage of the design project. Design layouts will continue to be drawn manually until the time comes when we are willing to forgo the idea of producing better products and commit ourselves to the continual production of products and systems already designed. Knowing the restlessness and competitive nature of humankind, this author predicts that this day is far in the future.

It should be noted, however, that at this point in time there are a few types of mechanical drawings that can be prepared other than manually through the use of a digitizer or a cathode ray tube coupled to a design-assist system (Chapter 17), the output in this case being a magnetic tape for a numerically controlled machine (Chapter 18).

Those who may be interested in learning more about the role of the computer in design should read Chapters 17 and 18. These two chapters present detailed information on computer-aided design and numerically controlled machine tools. In reading these chapters, one should come to the conclusion that there is a need to know graphics and that the computer will never make the teaching of graphics obsolete at any level of our educational system.

1.8

Major U.S. Corporations Go Metric. Although the metric system was legally approved for the United States by an Act of Congress more than 100 years ago (1866), it has not been widely used up to the present

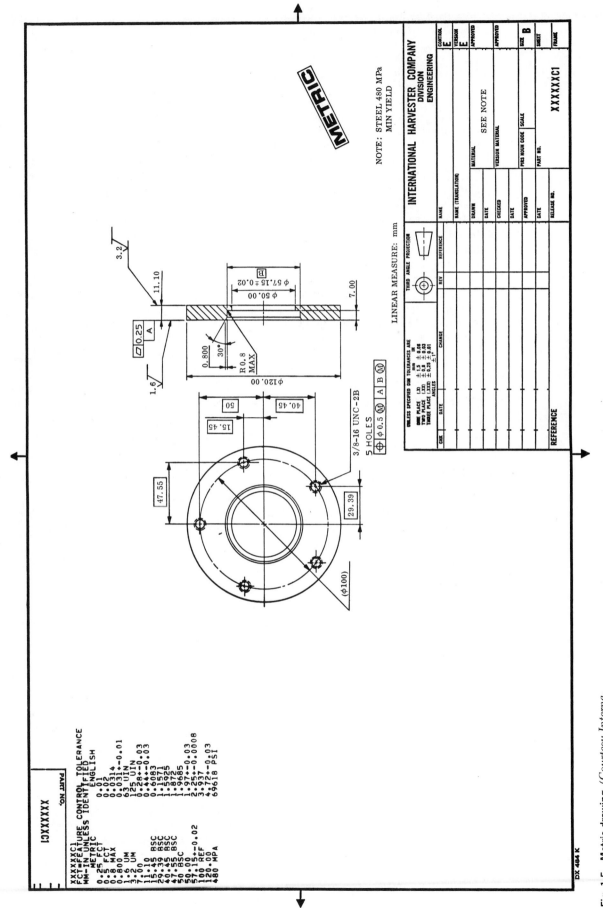

Fig. 1.5 Metric drawing. (*Courtesy International Harvester Company*)

time because its usage was never made mandatory. Now, we find ourselves living alone in isolation on a nonmetric island in a world where all the major nations use the metric system. At least this statement was valid until a few years ago, when a number of our large U.S. multinational corporations started to convert to SI (Système International d'Unités) units, the primary reason for making a change at the time being more profitable international trade. These companies, with their worldwide operations, were finding it increasingly difficult to do business in countries committed to the metric system, with its different units of weights and measures. In addition, all these companies import metric machinery from their foreign subsidiaries. This makes it necessary to supply spare parts, tools, and manuals for the needed interface with U.S.-designed equipment. The change to metric in the industrial field is now well under way and large manufacturing companies, such as General Motors, Ford, IBM, 3M, International Harvester, John Deere and Company, Caterpillar Tractor, Honeywell, McDonnell Douglas, Rockwell International, and TRW, are all converting to metric.

Total conversion will not come about immediately for much of American industry. For this reason, and because it will be a number of years before the ANSI standards will be converted to metric, this text may be said to be only about 75% metric. It must be noted that companies who claim to be totally metric give thread specifications in the inch system, as shown in Fig. 1.5. This practice will be followed until a new ANSI standard for threads has been developed and approved. A similar situation exists for fasteners, bearings, keys, pipes, drills, and cylindrical fits. However, the student will find that the Appendix of this text contains numerous conversion tables, along with tables for metric drills and metric threads. At this time a special committee set up to study and produce an Optimum Metric Fastener System has released OMFS Recommendations 2, 3, 4, and 6, which relate to screw threads. Eventually there should be a complete standard for ANSI-OMFS metric screw threads. Recently, ANSI committee Y14.5 reported that the committee was at work on a new dimensioning standard covering metric that could be approved and published as early as 1978. In the revision of this text, the author has been guided largely by the metric design and drafting standards furnished by General Motors, John Deere and Company, and International Harvester.

Educational Value of Graphics. Engineering drawing offers students an insight into the methods of attacking engineering problems. Its lessons teach the principles of accuracy, exactness, and positiveness with regard to the information necessary for the production of a nonexisting structure. Finally, it develops the engineering imagination that is so essential to the creation of successful design (Fig. 1.2).

Part I

Basic graphical techniques

A typical design room with designers and draftsmen working at their drawing boards.
Note the computer in the foreground. At this time, computers and plotters can be
seen in most drawing rooms. Many designs are worked out using a computer, and
some of the drawings are prepared by a plotter. All engineers, technologists, and technicians should be familiar with their capabilities. (*Courtesy International Business
Machines Corporation*)

2

Instrumental drawing— practices and techniques

A □ Drawing equipment and use of instruments

2.1
Introduction. The instruments and materials needed for making ordinary engineering drawings are shown in Fig. 2.1. The instruments in the plush-lined case should be particularly well made, for with inferior ones it is often difficult to produce accurate drawings of professional quality.

2.2
List of Equipment and Materials. The following list is a practical selection of equipment and materials necessary for making pencil drawings and ink tracings.

1. Case of drawing instruments
2. Drawing board
3. T-square
4. 45° triangle*
5. 10–in. 30°–60° triangle
6. French curve
7. Scales (Secs. 2.19–2.20)
8. Drawing pencils
9. Pencil pointer (file or sandpaper pad)
10. Thumbtacks, brad machine, or Scotch tape
11. Pencil eraser
12. Cleaning eraser
13. Erasing shield
14. Dusting brush

*A 6-in. 45° Braddock lettering triangle, which may be used as either a triangle or a lettering instrument, may be substituted for this item.

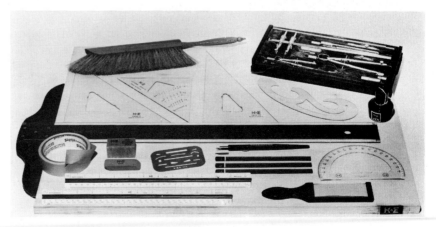

Fig. 2.1 Essential drafting equipment.

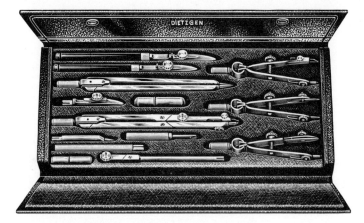

Fig. 2.2 A standard set of drawing instruments.

Fig. 2.3 Purdue Riefler set.

15. Protractor
16. Lettering pens
17. Black drawing ink
18. Pad of sketching paper (plain or ruled)
19. Drawing paper
20. Tracing paper
21. Tracing cloth or film (Mylar)

To these may be added the following useful items:

22. Piece of soapstone
23. Tack lifter
24. Slide rule

2.3

Set of Instruments. A standard set of drawing instruments in a velvet-lined case and a large-bow set, which is capable of fulfilling the needs of most engineers and draftsmen, are shown in Figs. 2.2 and 2.3, respectively.

Large-bow sets are preferred by many persons, especially in the aircraft and automotive fields. A complete set might include the following large bow (compass), beam compass with an extension bar for drawing very large circles, dividers, small bow, and a slip-handle ruling pen. Since the large bow (Fig. 2.3) is particularly suited for drawing circles of very small sizes up to 10 in. in diameter, its range covers that of both the bow instruments and the compass (without the extension bar) of the standard set (Fig. 2.2).

The large bow is preferred by its advocates because its sturdy construction permits the draftsman to exert the pressure necessary to secure black, opaque lines on pencil drawings that are to be used for making prints.

2.4

Protractor. The protractor (Fig. 2.4) is used for measuring and laying off angles.

2.5

Special Instruments and Templates. A few of the many special instruments and templates that are convenient for drawing are shown in Figs. 2.5–2.11.

Because of the time consumed in cutting back the wood to repoint an ordinary drawing pencil, some draftsmen favor the use of artist's automatic pencils (Fig. 2.5). Separate leads for these pencils may be purchased in any of the 17 degrees of hardness available for regular drawing pencils.

The flexible curves shown in Fig. 2.6, because of their limitless variations, are extremely convenient. The type shown in (a) is a lead bar enclosed in rubber. The more desirable one shown in (b) has a steel ruling edge attached to a spring with a lead core.

The drop pen (pencil) (Fig. 2.7) is designed for repeated drawing of circles of

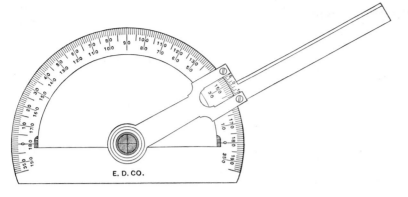

Fig. 2.4 Protractor.

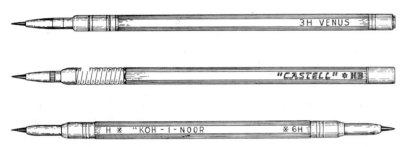

Fig. 2.5 Artist's pencils.

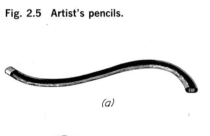

(a)

(b)

Fig. 2.6 Flexible curves.

Fig. 2.7 Drop pen (pencil).

small diameter, such as the circles representing rivet heads.

The electric erasing machine (Fig. 2.8) saves valuable drafting time of those persons who prepare reproductions in ink and who find it necessary to make frequent erasures.

Several of a number of special ruling pens that are available are shown in Fig. 2.9. The contour pen (*a*) is used freehand for tracing contours on a map. The inking leg is free to swivel in the handle. The railroad pen (*b*) draws parallel lines that may represent roads or railroad tracks. The fountain ruling pen (*c*) is useful when much inking is to be done. It does not require frequent filling. The border pen (*d*) draws a line of fixed width. It is desirable to have a set of these pens available, since

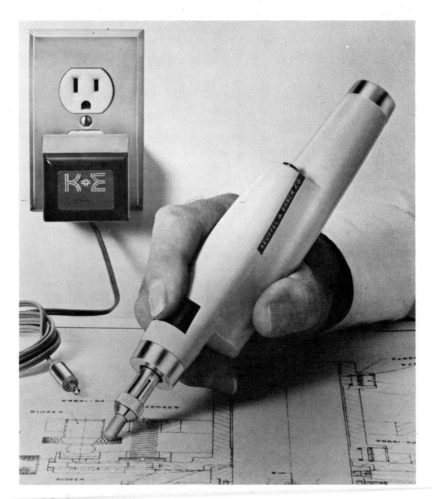

Fig. 2.8 Erasing machine.

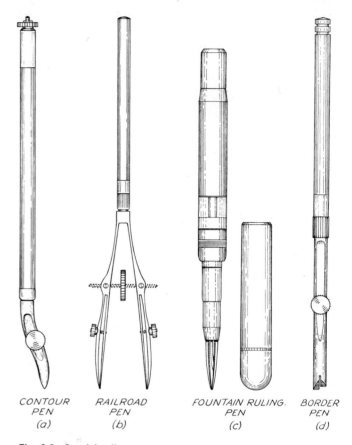

CONTOUR PEN (a) RAILROAD PEN (b) FOUNTAIN RULING PEN (c) BORDER PEN (d)

Fig. 2.9 Special ruling pens.

Fig. 2.10 Proportional dividers.

a separate pen is required for each different line width.

Proportional dividers (Fig. 2.10) are used to reproduce distances to a reduced or an enlarged scale.

The use of templates (Fig. 2.11) can save valuable time in the drawing of standard figures and symbols on plans and drawings.

2.6
Drafting Machine. The drafting machine (Fig. 2.12) is designed to combine the functions of the T-square, triangles, scale, and protractor. Drafting machines are used extensively in commercial drafting rooms because it has been estimated that their use leads to a 25–50% saving in time.

2.7
Tracing Paper, Tracing Cloth, and Mylar.
White lightweight tracing paper, on which pencil drawings can be made and from which blueprints can be produced, is used in most commercial drafting rooms in order to keep labor costs at a minimum.

The two general types of cloth available are ink cloth and pencil cloth. The cloth used for ink is clear and transparent, dull on one side and glossy on the other. Pencil cloth is a white cloth with a surface specially prepared to take pencil marks readily.

The matte drawing surface of a plastic film, commonly called *Mylar*, is excellent for both pencil and ink work. The film has high dimensional stability and the printing qualities are very good. Erasures leave no ghost marks, as happens with cloth and vellum. Some users think of it as being virtually indestructible, since it is resistant to cracking and almost impossible to tear. It is rapidly replacing cloth and in some cases designers prefer it in place of vellum. It may be purchased in rolls or standard size sheets.

2.8
Pencils. The student and professional man should be equipped with a selection of good, well-sharpened pencils with leads of various degrees of hardness, such as 9H, 8H, 7H, and 6H (hard); 5H and 4H (medium hard); 3H and 2H (medium); and H and F (medium soft).

The grade of pencil to be used for various purposes depends on the type of line

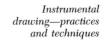

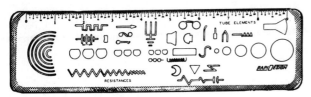

ELECTRO SYMBOL TEMPLATE

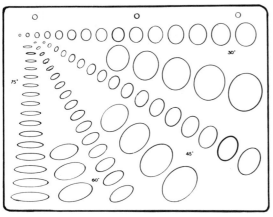

ELLIPSES

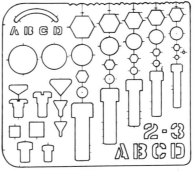

TOOLING TEMPLATE

TILT-HEX DRAFTING TEMPLATE

Fig. 2.11 Special templates. (*Courtesy Frederick Post Co.*)

desired, the kind of paper employed, and the humidity, which affects the surface of the paper. Standards for line quality usually will govern the selection. As a minimum, however, the student should have available a 6H pencil for the light construction lines in layout work where accuracy is required, a 4H for repenciling light finished lines (dimension lines, center lines, and invisible object lines), a 2H for visible object lines, and an F or an H for all lettering and freehand work.

2.9

Pointing the Pencil. Many persons prefer the conical point for general use (Fig. 2.13), while others find the wedge point more suitable for straight-line work, as it requires less sharpening and makes a denser line (Fig. 2.14).

When sharpening a pencil, the wood should be cut away (on the unlettered end) with a knife or a pencil sharpener equipped with draftsman's cutters. About .38 in. (10 mm) of the lead should be exposed and should form a cut, including the wood, about 1.5 in. (38 mm) long. The lead then should be shaped to a conical point on the pointer (file or sandpaper pad). This is done by holding the file stationary in the left hand and drawing the lead toward the handle while rotating the pencil against the movement (Fig. 2.13). All strokes should be made in the same manner, a new grip being taken each time so that each stroke starts with the pencil in the same rotated position as at the end of the preceding stroke.

Fig. 2.12 Drafting machine. A draftsman discusses part of the design of a jet engine with his supervisor. (*Courtesy Allison Division, General Motors Corporation*)

2.10

Drawing Pencil Lines. Pencil lines should be sharp and uniform along their entire length and sufficiently distinct to fulfill their ultimate purposes. Construction

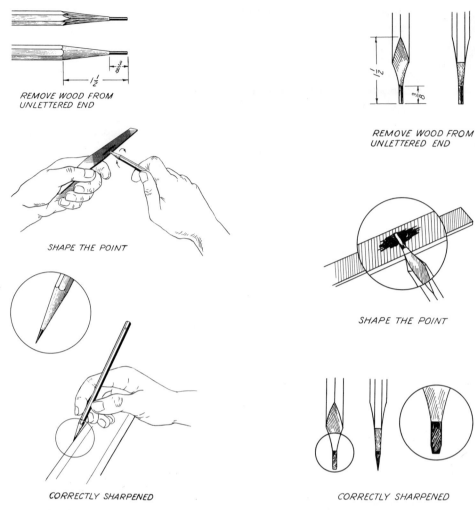

REMOVE WOOD FROM
UNLETTERED END

SHAPE THE POINT

CORRECTLY SHARPENED

Fig. 2.13 Conical point.

REMOVE WOOD FROM
UNLETTERED END

SHAPE THE POINT

CORRECTLY SHARPENED

Fig. 2.14 Wedge point.

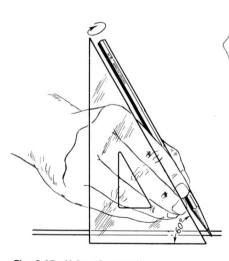

Fig. 2.15 Using the pencil.

lines (preliminary lines) should be drawn *very* lightly so that they may be easily erased. Finished lines should be made boldly and distinctly, so that there will be definite contrast between visible and invisible object lines and auxiliary lines, such as dimension lines, center lines, and section lines. To give this contrast, which is necessary for clearness and ease in reading, object lines should be of medium width and very black, invisible lines black and not so wide, and auxiliary lines dark and thin.

When drawing a line, the pencil should be inclined slightly (about 60°) in the direction in which the line is being drawn (Fig. 2.15). The pencil should be "pulled" (never pushed) at the same inclination for the full length of the line. If it is rotated (twirled) slowly between the fingers as the line is drawn, a symmetrical point will be maintained and a straight uniform line will be ensured.

2.11
Placing and Fastening the Paper. For accuracy and ease in manipulating the T-square, the drawing paper should be located well up on the board and near the left-hand edge. The lower edge of the sheet (if plain) or the lower border line (if printed) should be aligned along the working edge of the T-square before the sheet is fastened down at all four corners with thumbtacks, Scotch tape, or staples.

2.12
T-square. The T-square is used primarily for drawing horizontal lines and for guiding the triangles when drawing vertical and inclined lines. It is manipulated by sliding the guiding edge (inner face) of the head along the left edge of the board [Fig. 2.16(*a*)] until the blade is in the required position. The left hand then should be shifted to a position near the center of the blade to hold it in place and to prevent its

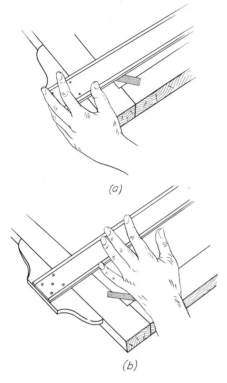

(a)

(b)

Fig. 2.16 Manipulating the T-square.

deflection while drawing the line. Experienced draftsmen hold the T-square, as shown in Fig. 2.16(*b*), with the fingers pressing on the blade and the thumb on the paper. Small adjustments may be made with the hand in this position by sliding the blade with the fingers.

Horizontal lines are drawn from left to right along the upper edge of the T-square

(Fig. 2.17). (*Exception:* Left-handed persons should use the T-square head at the right side of the board and draw from right to left.) While drawing the line, the ruling hand should slide along the blade on the little finger.

2.13

Triangles. The 45° and the 30° × 60° triangles (Fig. 2.18) are the ones commonly used for ordinary work. A triangle may be checked for nicks by sliding the thumbnail along the ruling edges, as shown in Fig. 2.19.

2.14

Vertical Lines. Vertical lines are drawn upward along the vertical leg of a triangle whose other (horizontal) leg is supported and guided by the T-square blade. The blade is held in position with the palm and thumb of the left hand, and the triangle is adjusted and held by the fingers, as shown in Fig. 2.20. In the case of a right-handed person, the triangle should be to the right of the line to be drawn.

Either the 30° × 60° or the 45° triangle may be used since both triangles have a right angle. However, the 30° × 60° is generally preferred because it usually has a longer perpendicular leg.

2.15

Inclined Lines. Triangles also are used for drawing inclined lines. Lines that make angles of 30°, 45°, or 60° with the horizontal may be drawn with the 30° × 60° or the 45° triangle in combination with

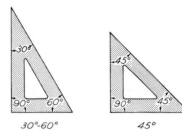

Fig. 2.18 Triangles.

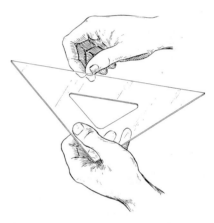

Fig. 2.19 Testing a triangle for nicks.

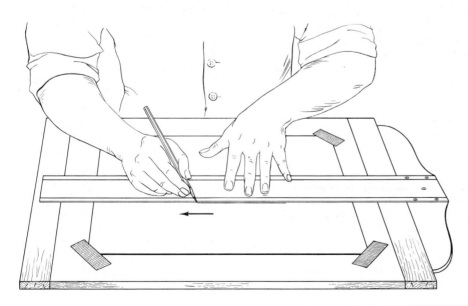

Fig. 2.17 Drawing horizontal lines.

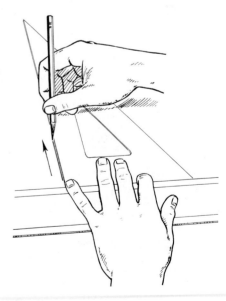

Fig. 2.20 Drawing vertical lines.

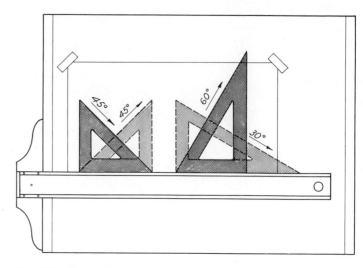

Fig. 2.21 Inclined lines.

the T-square, as shown in Fig. 2.21. If the two triangles are combined, lines that make 15° or a multiple of 15° may be drawn with the horizontal. Several possible arrangements and the angles that result are shown in Fig. 2.22.

The triangles used singly or in combination offer a useful method for dividing a

circle into 4, 6, 8, 12, or 24 equal parts (Fig. 2.23). For angles other than those divisible by 15, a protractor must be used.

2.16
Parallel Lines. The triangles are used in combination to draw a line parallel to a given line. To draw such a line, place a ruling edge of a triangle, supported by a T-square or another triangle, along the given line; then slip the triangle, as shown in Fig. 2.24, to the required position and draw the parallel line along the same ruling edge that previously coincided with the given line.

2.17
Perpendicular Lines. Either the sliding triangle method [Fig. 2.25(*a*)] or the revolved triangle method [Fig. 2.25(*b*)] may be used to draw a line perpendicular to a given line. When using the sliding triangle method, adjust to the given line a side of a triangle that is adjacent to the right angle. Guide the side opposite the right angle with a second triangle, as shown in Fig. 2.25(*a*); then slide the first triangle along the guiding triangle until it is in the required position for drawing the perpendicular along the other edge adjacent to the right angle.

Although the revolved triangle method [Fig. 2.25(*b*)] is not so quickly done, it is widely used. To draw a perpendicular using this method, align along the given line the hypotenuse of a triangle, one leg of which is guided by the T-square or another triangle; then hold the guiding member in position and revolve the triangle about the right angle until the other leg is against the guiding edge. The new

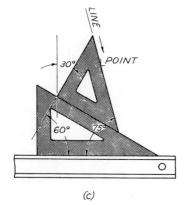

(a)

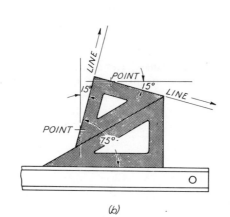

(b)

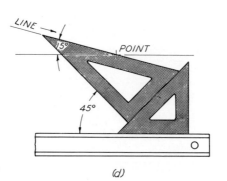

(c)   *(d)*

Fig. 2.22 Drawing inclined lines with triangles.

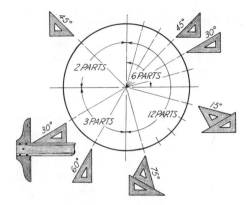

Fig. 2.23 To divide a circle into 4, 6, 8, 12, or 24 equal parts.

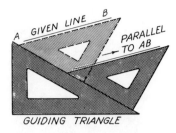

Fig. 2.24 To draw a line parallel to a given line.

position of the hypotenuse will be perpendicular to its previous location along the given line and, when moved to the required position, may be used as a ruling edge for the desired perpendicular.

2.18
Inclined Lines Making 15°, 30°, 45°, 60°, or 75° with an Oblique Line. A line making an angle with an oblique line equal to any angle of a triangle may be drawn with the triangles. The two methods previously discussed for drawing perpendicular lines are applicable with slight modifications. To draw an oblique line using the revolved triangle method [Fig. 2.26(a)], adjust along the given line the edge that is opposite the required angle; then revolve the triangle about the required angle, slide it into position, and draw the required line along the side opposite the required angle.

To use the sliding triangle method [Fig. 2.26(b)], adjust to the given line one of the edges adjacent to the required angle, and guide the side opposite the required angle with a straight edge; then slide the triangle into position and draw the required line along the other adjacent side.

To draw a line making 75° with a given line, place the triangles together so that the sum of a pair of adjacent angles equals 75°, and adjust one side of the angle thus formed to the given line; then slide the triangle, whose leg forms the other side of the angle, across the given line into position, and draw the required line, as shown in Fig. 2.27(a).

To draw a line making 15° with a given line, select any two angles whose difference is 15°. Adjust to the given line a side adjacent to one of these angles, and guide the side adjacent with a straightedge. Remove the first triangle and substitute the

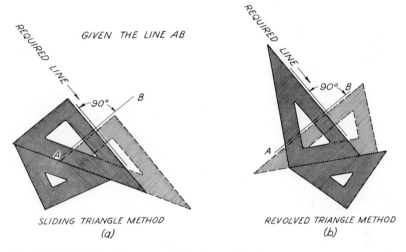

Fig. 2.25 To draw a line perpendicular to another line.

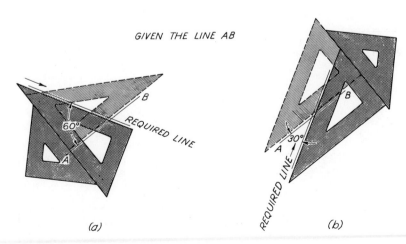

Fig. 2.26 To draw lines making 30°, 45°, or 60° with a given line.

GIVEN THE LINE AB

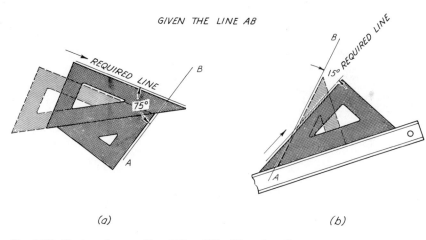

(a)   (b)

Fig. 2.27 To draw lines making 15° or 75° with a given line.

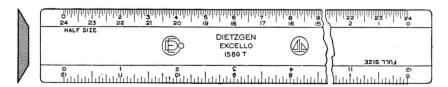

Fig. 2.28 Mechanical engineers' scale, full-divided.

Fig. 2.29 Engineers' decimal scale.

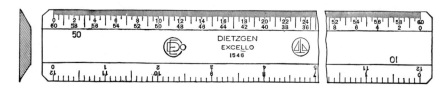

Fig. 2.30 Civil engineers' scale.

other so that one adjacent side of the angle to be subtracted is along the guiding edge, as shown in Fig. 2.27(*b*); then slide it into position and draw along the other adjacent side.

**2.19
Scales—Inches and Feet.** A number of kinds of scales are available for varied types of engineering design. For convenience, however, all scales may be classified according to their use as mechanical engineers' scales (both fractional and decimal), civil engineers' scales, architects' scales, or metric scales.

The mechanical engineers' scales are generally of the full-divided type, graduated proportionally to give reductions based on inches. On one form (Fig. 2.28) the principal units are divided into the common fractions of an inch (4, 8, 16, and 32 parts). The scales are indicated on the stick as eighth size ($1\frac{1}{2}$ in. = 1 ft), quarter size (3 in. = 1 ft), half size (6 in. = 1 ft), and full size.

Decimal scales are more widely used in industrial drafting rooms. The full size decimal scale shown in Fig. 2.29, which has the principal units (inches) divided into fiftieths, is particularly suited for use with the two-place decimal system. The half-size, three-eighths size, and quarter-size scales (Fig. 2.32) have the principal units divided into tenths.

The civil engineers' (chain) scales (Fig. 2.30) are full-divided and are graduated in decimal parts, usually 10, 20, 30, 40, 50, 60, 80, and 100 divisions to the inch.

Architects' scales (Fig. 2.31) differ from mechanical engineers' scales in that the divisions represent a foot, and the end units are divided into inches, half-inches, quarter-inches, and so forth (6, 12, 24, 48, or 96 parts). The usual scales are $\frac{1}{8}$ in. = 1 ft, $\frac{1}{4}$ in. = 1 ft, $\frac{3}{8}$ in. = 1 ft, $\frac{1}{2}$ in. = 1 ft, 1 in. = 1 ft, $1\frac{1}{2}$ in. = 1 ft, and 3 in = 1 ft.

The sole purpose of the scale is to reproduce the dimensions of an object full-size on a drawing or to reduce or enlarge them to some regular proportion, such as eighth-size, quarter-size, half-size, or double-size. The scales of reduction most frequently used are as follows:

Fractional

Mechanical engineers' scales

Full-size	(1″ = 1″)
Half-size	($\frac{1}{2}$″ = 1″)
Quarter-size	($\frac{1}{4}$″ = 1″)
Eighth-size	($\frac{1}{8}$″ = 1″)

Architects' or mechanical engineers' scales

Full-size	$(12'' = 1'\text{-}0)$
Half-size	$(6'' = 1'\text{-}0)$
Quarter-size	$(3'' = 1'\text{-}0)$
Eighth-size	$(1\frac{1}{2}'' = 1'\text{-}0)$

$1'' = 1'\text{-}0$	$\frac{1}{4}'' = 1'\text{-}0$
$\frac{3}{4}'' = 1'\text{-}0$	$\frac{3}{16}'' = 1'\text{-}0$
$\frac{1}{2}'' = 1'\text{-}0$	$\frac{1}{8}'' = 1'\text{-}0$
$\frac{3}{8}'' = 1'\text{-}0$	$\frac{3}{32}'' = 1'\text{-}0$

Decimal

Mechanical engineers' scales

Full-size	$(1.00'' = 1.00'')$
Half-size	$(0.50'' = 1.00'')$
Three-eighths-size	$(0.375'' = 1.00'')$
Quarter-size	$(0.25'' = 1.00'')$

Civil engineers' scales

10 scale:	$1'' = 1'$;	$1'' = 10'$;
	$1'' = 100'$;	$1'' = 1000'$
20 scale:	$1'' = 2'$;	$1'' = 20'$;
	$1'' = 200'$;	$1'' = 2000'$
30 scale:	$1'' = 3'$;	$1'' = 30'$;
	$1'' = 300'$;	$1'' = 3000'$
40 scale:	$1'' = 4'$;	$1'' = 40'$;
	$1'' = 400'$;	$1'' = 4000'$
50 scale:	$1'' = 5'$;	$1'' = 50'$;
	$1'' = 500'$;	$1'' = 5000'$
60 scale:		$1'' = 60'$; etc.
80 scale:		$1'' = 80'$; etc.

The first four scales, full-size, half-size, quarter-size, and eighth-size, are the ones most frequently selected for drawing machine parts, although other scales can be used. Since objects drawn by structural draftsmen and architects vary from small to very large, scales from full-size to $\frac{3}{32}$ in. = 1 ft ($\frac{1}{128}$-size) are commonly encountered. For maps, the civil engineers' decimal scales having 10, 20, 30, 40, 50, 60, and 80 divisions to the inch are used to represent 10, 20, 30 ft, and so forth, to the inch.

On a machine drawing, it is considered good practice to omit the inch ('') marks in a scale specification. For example, a scale may be specified as: FULL SIZE, 1.00 = 1.00, or 1 = 1; HALF SIZE, .50 = 1.00, or $\frac{1}{2}$ = 1; and so forth.

The decimal scales shown in Fig. 2.32 have been approved by the American National Standards Institute for making machine drawings when the decimal system is used.

It is essential that a draftsman always think and speak of each dimension as full-size when scaling measurements, because the dimension figures given on the finished drawing indicate full-size measurements of the finished piece, regardless of the scale used.

The reading of an open-divided scale is illustrated in Fig. 2.33 with the eighth-size ($1\frac{1}{2}$ in. = 1 ft) scale shown. The dimen-

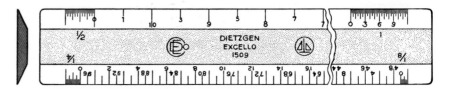

Fig. 2.31 Architects' scale, open-divided.

sion can be read directly as 21 in., the 9 in. being read in the divided segment to the left of the cipher. Each long open division represents 12 in. (1 ft).

To lay off a measurement, using a scale starting at the left of the stick, align the scale in the direction of the measurement with the zero of the scale being used toward the left. After it has been adjusted to the correct location, make short marks opposite the divisions on the scale that establish the desired distance (Fig. 2.34). For ordinary work most draftsmen use the same pencil used for the layout. When extreme accuracy is necessary, however, it is better practice to use a pricker and make slight indentations (not holes) at the required points. If a regular pricker is not available, the dividers may be opened to approximately 60° and the point of one leg used as a substitute.

To ensure accuracy, place the eye directly over the division to be marked, hold the marking instrument perpendicular to the paper directly in front of the scale

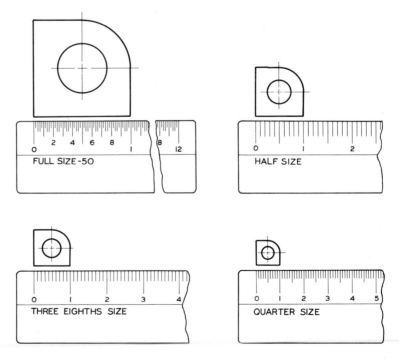

Fig. 2.32 Decimal scales.

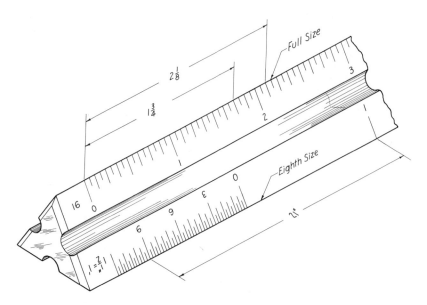

Fig. 2.33 Reading a scale.

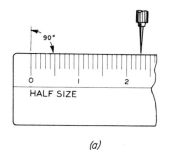

(a)

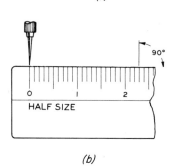

(b)

**Fig. 2.34 To lay off a measurement—
decimal scale.**

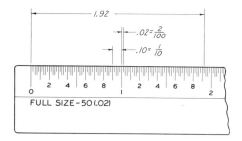

Fig. 2.35 Reading the decimal scale.

division, and mark the point. Always check the location of the point before removing the scale. If a slight indentation is made, it will be covered by the finished line; if a short mark is made and it is *very* light, it will be unnoticeable on the finished drawing.

To set off a measurement (say 2.30 in.) to half-size, the scale indicated either as half-size (Fig. 2.34) or 20 should be used. If the measurement is to be made from left to right, place the zero (0) division mark on the given line, and make an indentation (or mark) opposite the 2.30 division point [Fig. 2.34(*a*)]. The distance from the line to the point represents 2.30, although it is actually 1.15 in. To set off the same measurement from right to left, place the 2.30 mark on the given line, and make an indentation opposite the zero division mark [Fig. 2.34(*b*)].

The reading of the full-size decimal scale is illustrated in Fig. 2.35. The largest division indicated in the illustration represents one inch, which is subdivided into tenths and fiftieths (.02 in.). In Fig. 2.32, the largest divisions on the half size, three eighths size, and quarter size decimal scales represent one inch.

**2.20
Metric Scales.** The metric scale (Figs. 2.36 and 2.37) is used in those countries where the meter is the standard of linear measurement. As of this time, the International System of Units (Systéme International d'Unités) (SI) for the measurement of length, measurement of surface, measurement of volume, measurement of weight, and so forth, is used as a world standard in every industrialized country except the United States. We stand alone

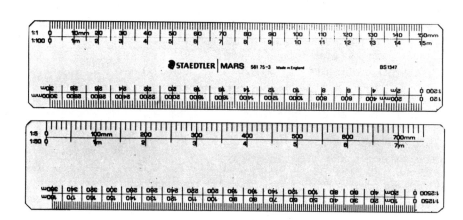

Fig. 2.36 Flat metric scale (front and reverse sides). *(Courtesy J. S. Staedtler, Inc.)*

with our inches, gallons, and pounds in spite of the fact that the metric system was legalized for use in this country by an Act of Congress more than one hundred years ago (1866). The metric system is legal in all states. Even though the cost of conversion to the metric system will be enormous, there can be little doubt that a changeover will be made in the near future. As a practical matter, it is probable that American industry will shift gradually to the metric system without prodding by an Act of Congress since even now the automobile industry has adopted and is using a dual-dimensioning system by which dimensions are given on drawings in both inches and millimeters. In the agricultural-machinery field both International Harvester and John Deere have gone *all metric*. Companies in other areas will probably make this step in the near future. It is for this reason that dimensional values for more than half of the problems and illustrations have been given in metric units in this text. If desired, problems given using the English system may be prepared showing metric values by utilizing the tables and other information given in the Appendix.

2.21

Compass or Large Bow. The compass or large bow is used for drawing circles and circle arcs. For drawing pencil circles, the style of point illustrated in Fig. 2.38(*c*) should be used because it gives more accurate results and is easier to maintain than most other styles. This style of point is formed by first sharpening the outside of the lead on a file (Fig. 2.39) or sandpaper pad to a long flat bevel approximately $\frac{1}{4}$ in. long [Fig. 2.38(*a*)] and then finishing it [Fig. 2.38(*b*)] with a slight rocking motion to reduce the width of the point. Although a hard lead (4H–6H) will maintain a point longer without resharpening, it gives a finished object line that is too light in color. Soft lead (F or H) gives a darker line but quickly loses its edge and, on larger circles, gives a thicker line at the end than at the beginning. Some draftsmen have found that a medium-grade (2H–3H) lead is a satisfactory compromise for ordinary working drawings. For design drawings, layout work, and graphical solutions, however, a harder lead will give better results.

The needle point should have the shouldered end out and should be adjusted approximately $\frac{3}{8}$ in. beyond the end of the split sleeve [Fig. 2.38(*a*)].

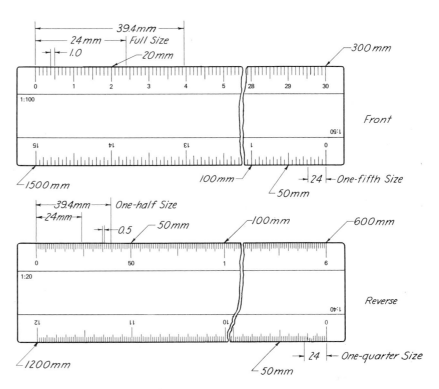

Fig. 2.37   Reading metric scales.

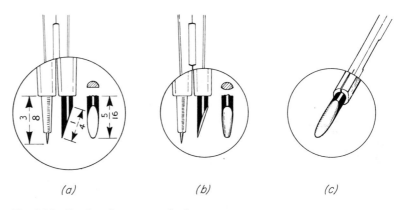

(a)   (b)   (c)

Fig. 2.38   Shaping the compass lead.

2.22

Using the Compass or Large Bow. To draw a circle, it is first necessary to draw two intersecting center lines at right angles and mark off the radius. The pivot point should be guided accurately into position at the center. After the pencil point has been adjusted to the radius mark, the circle is drawn in a clockwise direction, as shown in Fig. 2.40. While drawing the circle, the instrument should be inclined slightly forward. If the pencil line is not dark enough, it may be drawn around again.

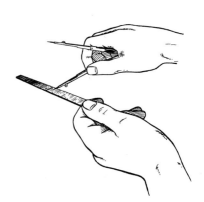

Fig. 2.39   Sharpening the compass lead.

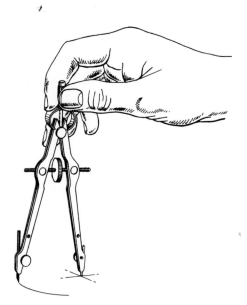

Fig. 2.40 Using the large bow (Vemco).

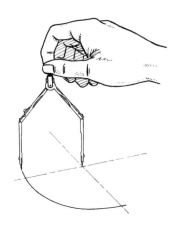

Fig. 2.41 Using the compass (legs bent).

When using a compass for a radius larger than 2 in., the legs should be bent at the knee joints to stand approximately perpendicular to the paper (Fig. 2.41). It is particularly important that this adjustment be made when drawing ink circles, otherwise both nibs will not touch the paper. For circles whose radii exceed 5 in., the lengthening bar should be used to increase the capacity.

The beam compass is manipulated by steadying the instrument at the pivot leg with one hand while rotating the marking leg with the other (Fig. 2.42)

2.23
Dividers. The dividers are used principally for dividing curved and straight lines into any number of equal parts and for transferring measurements. If the instrument is held with one leg between the forefinger and second finger, and the other leg between the thumb and third finger, as illus-

trated in Fig. 2.43, an adjustment may be made quickly and easily with one hand. The second and third fingers are used to "open out" the legs, and the thumb and forefinger to close them. This method of adjusting may seem awkward to the beginner at first, but with practice absolute control can be developed.

2.24
Use of the Dividers. The trial method is used to divide a line into a given number of equal parts (Fig. 2.44). To divide a line into a desired number of equal parts, open the dividers until the distance between the points is estimated to be equal to the length of a division, and step off the line *lightly*. If the last prick mark misses the end point, increase or decrease the setting by an amount estimated to be equal to the error divided by the number of divisions, before lifting the dividers from the paper. Step off the line again. Repeat this procedure until the dividers are correctly set, then space the line again and indent the division points. When stepping off a line, the dividers are rotated alternately in an opposite direction on either side of the line, each half-revolution, as shown in Fig. 2.44.

Although the dividers are used to transfer a distance on a drawing, they should never be used to transfer a measurement from the scale, as the method is slow and inaccurate and results in serious damage to the graduation marks. Care should be taken to avoid pricking large unsightly holes with the divider points. It is the common practice of many expert draftsmen to draw a small freehand circle around a very light indentation to establish its location.

2.25
Use of the Bow Instruments. The bow pen and bow pencil are convenient for

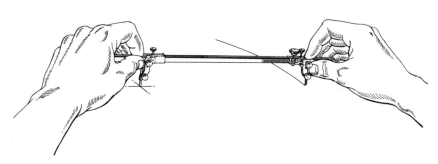

Fig. 2.42 Drawing large circles (Vemco beam compass).

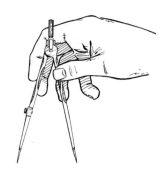

Fig. 2.43 To adjust the large dividers.

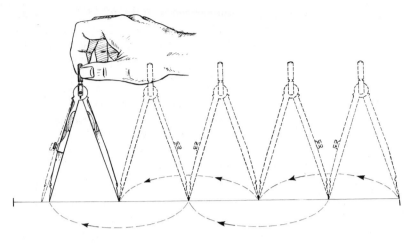

Fig. 2.44 Using the dividers.

drawing circles having a radius of 1 in. or less (Fig. 2.45). The needle points should be adjusted slightly longer than the marking points, as in the case of the compass.

Small adjustments are made by the fingers of the hand holding the instrument, with the pivot point in position at the center of the required circle or arc.

2.26
Use of the French Curve. A French curve is used for drawing irregular curves that are not circle arcs. After sufficient points have been located, the French curve is applied so that a portion of its ruling edge passes through at least three points, as shown in Fig. 2.46. It should be so placed that the increasing curvature of the section of the ruling edge being used follows the direction of that part of the curve which is changing most rapidly. To ensure that the finished curve will be free of humps and sharp breaks, the first line drawn should start and stop short of the first and last points to which the French curve has been fitted. Then the curve is adjusted in a new position with the ruling edge coinciding with a section of the line previously drawn. Each successive segment should stop short of the last point matched by the curve. In Fig. 2.46, the curve fits the three points, A, 1, and 2. A line is drawn from between point A and point 1 to between point 1 and point 2. Then the curve is shifted, as shown, to fit again points 1 and 2 with an additional point 3, and the line is extended to between point 2 and point 3.

Some people sketch a smooth continuous curve through the points in pencil before drawing the mechanical line. This procedure makes the task of drawing the curve less difficult, since it is easier to adjust the ruling edge to segments of the freehand curve than to the points.

2.27
Use of the Erasing Shield and Eraser. An erasure is made on a drawing by placing an opening in the erasing shield over the work to be erased and rubbing with a pencil eraser (never an ink eraser) until it is removed (Fig. 2.47). Excessive pressure should not be applied to the eraser because, although the lines will disappear more quickly, the surface of the paper is likely to be permanently damaged. The

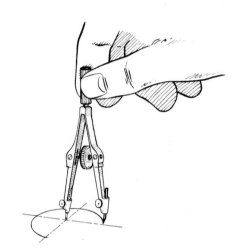

Fig. 2.45 Using the bow pencil.

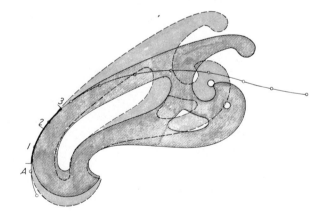

Fig. 2.46 Using the irregular curve.

Fig. 2.47 Using the erasing shield.

fingers holding the erasing shield should rest partly on the drawing paper to prevent the shield from slipping.

2.28

Use of the Ruling Pen. The ruling pen is used to ink mechanical lines. It is always guided by the working edge of a T-square, triangle, or French curve and is never used freehand.

When ruling a line, the pen should be in a vertical plane and inclined slightly (approximately 60°) in the direction of the movement. It is held by the thumb and forefinger, as illustrated in Fig. 2.48, with the blade against the second finger and the adjusting screw on the outside away from the ruling edge. The third and fourth fingers slide along the T-square blade and help control the pen. Short lines are drawn with a hand movement, long lines with a free arm movement that finishes with a finger movement. While drawing, the angle of inclination and speed must remain constant to obtain a line of uniform width and straightness. Particular attention should be given to the position of the pen, as practically all faulty lines are due to incorrect inclination or to leaning the pen so that the point is too close to the straightedge or too far away from it. The correct position of the pen for drawing a satisfactory line is illustrated in Fig. 2.48.

If the pen is held so that it leans outward, as shown in Fig. 2.49(a), the point

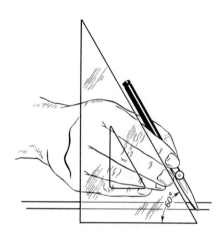

Fig. 2.48 Holding the pen.

will be against the straightedge, and ink will run under and cause a blot; or, if it leans inward, as in Fig. 2.49(b), the outer nib will not touch the paper and the line will be ragged.

Unnecessary pressure against the straightedge changes the distance between the nibs, which in turn may either reduce the width of the line along its entire length or cause its width to vary, as in Fig. 2.49(d).

It will not take any beginner long to discover that care must be taken when removing a T-square or triangle away from a wet ink line [Fig. 2.49(c)].

The ruling pen is filled by inserting the quill or dropper device of the stopper between the nibs. Care must be taken, while filling, to see that there is enough ink to finish the line and that none of the ink from the filler gets on the outside of the blades. No more than $\frac{1}{4}$ in. should ever be put in; there is a danger of blotting if the pen is used with a greater amount.

The special ink-bottle holder and filler shown in Fig. 2.50 is a convenience that saves time when filling a pen, because only one hand is needed to raise the stopper and insert the ink.

The width of a line is determined by the distance between the nibs, which is regulated by the adjusting screw. When setting the pen, a series of test lines should be drawn with a straightedge on a small piece of the same kind of paper or cloth to establish the setting to the desired width of the line. The draftsman's line gauge, shown in Fig. 2.51, is convenient for testing the widths of trial lines as illustrated. If the first trial line is not of the desired width, another and another should be drawn until the final one agrees with the selected width as given on the gauge.

If the ink refuses to flow when the pen is touched to the paper, either the ink has thickened or the opening between the nibs has become clogged. To start it flowing, draftsmen often touch the point to the back of a finger or pinch the blades to-

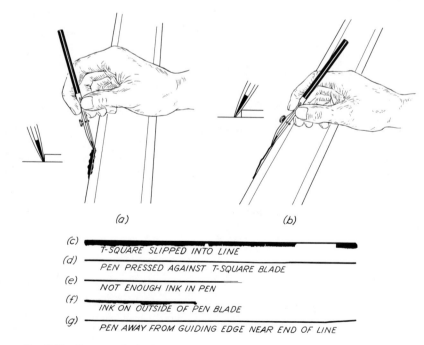

(a)   (b)

(c) T-SQUARE SLIPPED INTO LINE
(d) PEN PRESSED AGAINST T-SQUARE BLADE
(e) NOT ENOUGH INK IN PEN
(f) INK ON OUTSIDE OF PEN BLADE
(g) PEN AWAY FROM GUIDING EDGE NEAR END OF LINE

Fig. 2.49 Common faults in handling a ruling pen.

Fig. 2.50 Special ink-bottle holder and filler.

gether. Whenever this fails to produce an immediate flow, the pen should be cleaned and refilled.

A dirty pen in which the ink has been allowed to thicken will not draw any better than a dull one. To avoid changing the setting of the pen when cleaning it, fold the pen wiper twice at 90° and draw the corner of the fold between the ends of the blades (Fig. 2.52).

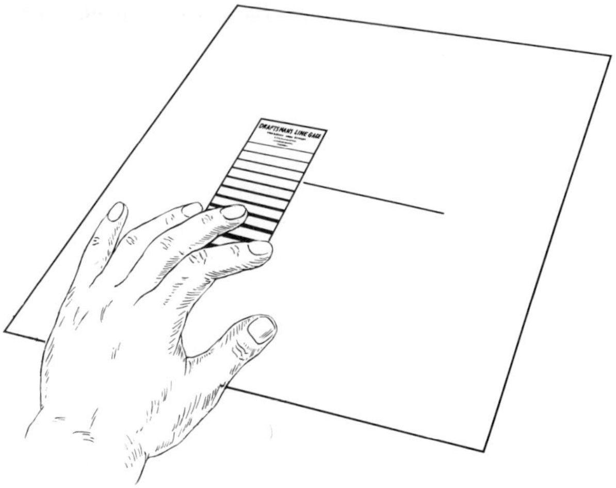

2.29

Tracing. Often, when it is necessary to make duplicate copies (blueprints) of important drawings for a machine or structure, the original pencil drawings are traced in ink on a tracing medium, usually tracing cloth. Contrary to the practice of old-time draftsmen and the intention of the early manufacturers of this medium, the dull side is now almost universally used for the inking surface instead of the slick side because it produces less light glare, will take both pencil and ink lines better, and will withstand more erasing. The fact that the dull side will take pencil lines is important because, on some occasions, in order to save time, drawings are made directly on the cloth and then traced. On completion of the tracing, all pencil lines including the guide lines and slope lines for the lettering may be removed by wiping the surface of the cloth with a rag moistened with a small amount of gasoline, benzene, or cleaning fluid.

When the tracing cloth has been fastened down over the drawing, a small quantity of tracing cloth powder may be sprinkled over the surface to make it take the ink evenly and smoothly. After it has been well rubbed in, the excess must be thoroughly removed by wiping with a clean cloth, for even a small amount of loose powder left on the surface can cause clogging of the pen. Powder is also used by some persons over a spot where an erasure has been made; but a better practice is to use a piece of soapstone, which will put a smooth, slick finish over the damaged area. In applying the soapstone, rub the spot and then wipe a finger over it a few times. Following this treatment, the erased area will take ink almost as well as the original surface.

Since ink lines are made much wider than pencil lines, in order to get a good contrast on a blueprint, they should be carefully centered over the pencil lines when tracing. The center of an ink line should fall directly on the pencil line, as shown correctly in Fig. 2.53. For ink work

Fig. 2.51 Testing a trial line on a "try sheet."

it might be said that ink lines are tangent when their center lines touch. In this same illustration, note the poor junctures obtained when ink lines are not centered so that their center lines are tangent.

Figure 2.54(a) shows the filled-in corner effect that frequently appears, to the disgust of the draftsman, when an ink line is drawn from or to another previously drawn line that is still wet.

When a working drawing is traced in ink on either paper or cloth, the lines should be "inked" in a definite order. Otherwise, the necessity of waiting for the ink to dry after every few lines not only wastes time but often results in a line here and there being left out. Furthermore, hit-and-miss inking may produce lines of unequal

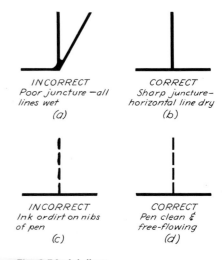

Fig. 2.52 Cleaning the ruling pen.

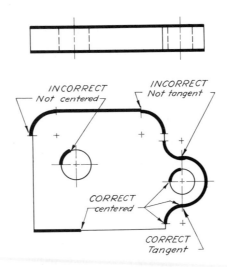

Fig. 2.53 Inking over pencil lines.

Fig. 2.54 Ink lines.

width. It is therefore recommended that the student make a conscientious attempt to follow the order of inking suggested in this chapter.

After the paper or cloth has been fastened down over the drawing, and before the inking is begun, each tangent point should be marked and all centers should be indented.

Order of inking

1. Curved lines
   a. Circles and circle arcs (small circles first) in the order of 1, 2, and 3 below.
      (1) Visible
      (2) Invisible
      (3) Circular center lines and dimension lines
   b. Irregular curves
      (1) Visible
      (2) Invisible
2. Straight lines
   a. Visible
      (1) Horizontal, from the top of the sheet down
      (2) Vertical, from the left side of the sheet to the right
      (3) Inclined, from the left to the right
   b. Invisible
      (1) Horizontal
      (2) Vertical
      (3) Inclined
   c. Auxiliary (center, extension, dimension lines, etc.)
      (1) Horizontal
      (2) Vertical
      (3) Inclined
      (4) Section lines
3. Arrowheads and dimension figures
4. Notes and titles
5. Border

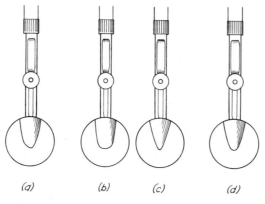

(a)   *(b)*   *(c)*   *(d)*

Fig. 2.55 Worn, incorrectly shaped, and correctly shaped blades.

2.30

Sharpening a Ruling Pen. Ruling pens, whether they have been used continuously or intermittently, eventually show signs of wear. When a pen's nibs have lost their elliptical shape and have become so dull that the ink spreads under their tips, fine lines cannot be drawn until the pen has been sharpened. The way to detect this condition is to examine the tips. If bright spots may be seen on the tips, the pen is too dull for satisfactory work. An example of a worn point is illustrated in Fig. 2.55(*a*).

Since even new pens are seldom sharpened properly and often require retouching, every draftsman should be able to reshape and sharpen his own pen.

Incorrectly sharpened points are illustrated in Fig. 2.55(*b*) and (*c*). Points shaped as shown in either (*b*) or (*c*) are aggravating, because if a point is rounded, as in (*b*), the ink flows too freely; if it is too pointed, as in (*c*), the ink cups up and the flow is difficult to start. Only a pen that is correctly shaped and sharpened, as shown in Fig. 2.55(*d*), will give the results that should be expected.

Although blades should be sharpened to a thin edge, care should be taken not to make them sharp enough to cut the surface of the paper. A pen should never be sharpened, even the slightest amount, on the inside of the blades.

A fine-grained Arkansas oilstone is the best all-purpose stone for sharpening a ruling pen. The first step in sharpening is to equalize the length of the nibs and correct their shape. This may be done by bringing the blades together so that they barely touch and then drawing them lightly back and forth across the stone while swinging the pen through an arc of approximately 120° each stroke (Fig. 2.56). During this operation, it is essential that the pen be held in a vertical plane and that an even pressure be maintained against the stone. When an inspection, under a magnifying glass, reveals that the nibs have been restored to their correct shape, the blades should be opened and each nib sharpened all around the outside to a thin edge. A blade is sharpened by holding the pen as shown in Fig. 2.57 and sliding it back and forth across the stone with a slight rolling motion to preserve its original convex shape. The pen should be examined from time to time so that the sharpening can be stopped as soon as the bright point disappears.

Finally, a test should be made by filling

Fig. 2.56 Shaping the nibs.

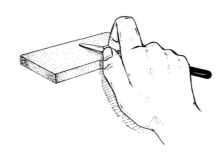

Fig. 2.57 Sharpening the blades.

the pen and drawing a series of lines of various weights on a piece of tracing paper or tracing cloth. If the instrument is capable of drawing satisfactory lines of any weight, particularly very fine lines, it has been correctly sharpened.

2.31
Conventional Line Symbols. Symbolic lines of various weights are used in making technical drawings. The recommendations of the American National Standards Institute, as given in ANSI Y14.2–1973, are the following:

Two widths of lines—thick and thin—are recommended for use on drawings (Fig. 2.58). Pencil lines in general should be in proportion to the ink lines except that the thicker pencil lines will be necessarily thinner than the corresponding ink lines but as thick as practicable for pencil work. Exact thicknesses may vary according to the size and type of drawing. For example, where lines are close together, the lines may be slightly thinner.

Ink lines on drawings prepared for catalogs and books may be drawn using three widths—thick, medium, and thin. This provides greater contrast between types of lines and gives a better appearance.

The lines illustrated in Fig. 2.58 are shown full-size. When symbolic lines are used on a pencil drawing they should not vary in color. For example, center lines, extension lines, dimension lines, and section lines should differ from object lines only in width. The resulting contrast makes a drawing easier to read. All lines, except construction lines, should be very dark and bright to give the drawing the "snap" that is needed for good appearance. If the drawing is on tracing paper the lead must be "packed on" so that a satisfactory print can be obtained. Con-

struction lines should be drawn *very* fine so as to be unnoticeable on the finished drawing. The lengths of the dashes and spaces shown in Figs. 5.29 and 7.21 are recommended for the hidden lines, center lines, and cutting-plane lines on average-size drawings.

B □ Freehand technical lettering

2.32
Introduction. To impart to the workers in the shops all the necessary information for the complete construction of a machine or structure, the shape description, which is conveyed graphically by the views, must be accompanied by size descriptions and instructive specifications in the form of figured dimensions and notes (Fig. 2.59).

All dimensions and notes should be lettered freehand in a plain, legible style that can be rapidly executed. Poor lettering detracts from the appearance of a drawing and often impairs its usefulness, regardless of the quality of the line work.

2.33
Single-stroke Letters (Reinhardt). Single-stroke letters are now used universally for technical drawings. This style is suitable for most purposes because it possesses the qualifications necessary for legibility and speed. On commercial drawings it appears in slightly modified forms, however, since each person finally develops a style that reflects his own individuality.

The expression *single-stroke* means that the width of the straight and curved lines that form the letters are the same width as the stroke of the pen or pencil.

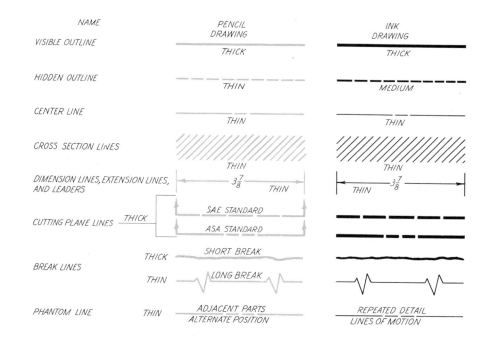

NAME	PENCIL DRAWING	INK DRAWING
VISIBLE OUTLINE	THICK	THICK
HIDDEN OUTLINE	THIN	MEDIUM
CENTER LINE	THIN	THIN
CROSS SECTION LINES	THIN	THIN
DIMENSION LINES, EXTENSION LINES, AND LEADERS	$3\frac{7}{8}$ THIN	THIN $3\frac{7}{8}$
CUTTING PLANE LINES THICK	SAE STANDARD / ASA STANDARD	
BREAK LINES THICK / THIN	SHORT BREAK / LONG BREAK	
PHANTOM LINE THIN	ADJACENT PARTS ALTERNATE POSITION	REPEATED DETAIL LINES OF MOTION

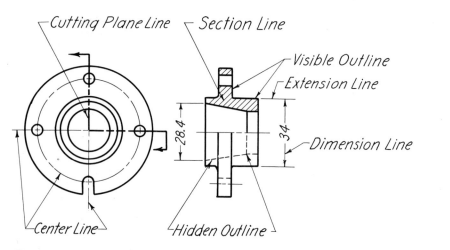

—Short Break Line

—Long Break Line

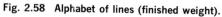

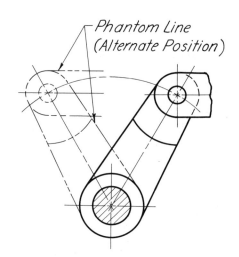

—Cutting Plane Line —Section Line

—Visible Outline

—Extension Line

28.4 34

—Dimension Line

—Center Line— —Hidden Outline—

—Phantom Line (Alternate Position)

Fig. 2.58 Alphabet of lines (finished weight).

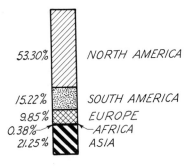

WORLD PRODUCTION
OF
CRUDE PETROLEUM
(a) Chart

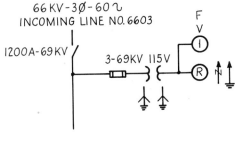

CLAMP WASHER
C.I   I REQ'D
(b) Machine Drawing

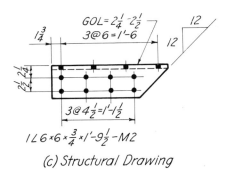

(c) Structural Drawing

66 KV-3∅-60〜
INCOMING LINE NO. 6603

(d) Electrical Diagram
From ASA Y32.2-1954

Fig. 2.59   Technical drawings.

2.34

General Proportions of Letters. Although there is no fixed standard for the proportions of the letters, certain definite rules must be observed in their design if one wishes to have his lettering appear neat and pleasing. The recognized characteristics of each letter should be carefully studied and then thoroughly learned through practice.

It is advisable for the beginner, instead of relying on his untrained eye for proportions, to follow the fixed proportions given in this chapter. Otherwise, his lettering will most likely be displeasing to the trained eye of the professional. Later, after he has thoroughly mastered the art of lettering, his individuality will be revealed naturally by slight variations in the shapes and proportions of some of the letters.

It is often desirable to increase or decrease the width of letters in order to make a word or group of words fill a certain space. Letters narrower than normal letters of the same height are called *compressed letters;* those that are wider are called *extended letters* (Fig. 2.60).

2.35

Lettering Pencils and Pens. Pencil lettering is usually done with a medium-soft pencil. Since the degree of hardness of the lead required to produce a dark opaque line will vary with the type of paper used, a pencil should be selected only after drawing a few trial lines. In order to obtain satisfactory lines, the pencil should be sharpened to a long conical point and then rounded slightly on a piece of scratch paper. To keep the point symmetrical while lettering, the pencil should be rotated a partial revolution before each new letter is started.

The choice of a type of pen point for lettering depends largely upon the personal preference and characteristics of the individual. The beginner can learn only from experience which of the many types available are best suited to him.

A pen that makes a heavy stroke should be used for bold letters in titles, and so forth, while a light-stroke pen is required for the lighter letters in figures and notes.

A very flexible point should never be used for lettering. Such a point is apt to

Fig. 2.60   Compressed and extended letters.

shade the downward stem strokes as well as the downward portions of curved strokes. A good point has enough resistance to normal pressure to permit the drawing of curved and stem strokes of uniform width.

There are ordinary steel pen points with special ink-holding devices that make them especially suitable for lettering. The *Henry tank* pen, for example, shown in Fig. 2.61, has an ink reservoir that holds the ink above the point so that it feeds down the slit in an even flow. This device further assists in maintaining a uniform line by preventing the point from spreading.

Four of the many special pens designed for single-stroke letters are illustrated in Fig. 2.62. The *Barch–Payzant pen* (a) is available in graded sizes from No. 000 (very coarse) to No. 8 (very fine). The very fine size is suitable for lettering $\frac{1}{8}$–$\frac{3}{16}$ in. high on technical drawings. The *Edco* (b) has a patented holder into which any one of a graded set of lettering nibs (ranging in sizes from No. 0 to No. 6) may be screwed. The tubular construction of the point makes it possible to draw uniform lines regardless of the direction of the stroke. Also of tubular construction is the *Leroy* (c). The *Speedball* (d) may be obtained in many graded sizes.

2.36
Devices for Drawing Guide Lines and Slope Lines. Devices for drawing guide lines (Fig. 2.63) are available in a variety of forms. The two most popular are the *Braddock lettering triangle* (Fig. 2.64), and the *Ames lettering instrument* (Fig. 2.65).

The Braddock lettering triangle is provided with sets of grouped countersunk holes that may be used to draw guide lines by inserting a sharp-pointed pencil (4H or 6H) into the holes and sliding the triangle

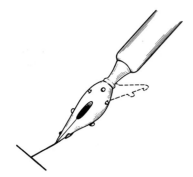

Fig. 2.61 Henry tank pen.

back and forth along the guiding edge of a T-square or a triangle supported by a T-square (Fig. 2.64). The holes are grouped to give guide lines for capitals and lower-case letters. The numbers below each set indicate the height of the capitals in thirty-seconds of an inch. For example, the No. 3 set is for capitals $\frac{3}{32}$ in. high, the No. 4 set is for capitals $\frac{1}{8}$ in. high, the No. 5 is for capitals $\frac{5}{32}$ in. high, and so on.

2.37
Uniformity in Lettering. Uniformity in height, inclination, spacing, and strength of line is essential for good lettering (Fig. 2.66). Professional appearance depends as much on uniformity as on the correctness of the proportion and shape of the individual letters. Uniformity in height and inclination is assured by the use of guide lines and slope lines and uniformity in weight and color, by the skillful use of the pencil and proper control of the pressure of its point on the paper. The ability to space letters correctly becomes easy after continued thoughtful practice.

2.38
Composition. In combining letters into words, the spaces for the various combinations of letters are arranged so that the areas appear to be equal (Fig. 2.67). For standard lettering, this area should be about equal to one-half the area of the letter M. If the adjacent sides are stems, this area is obtained by making the distance between the letters slightly greater than one-half the height of a letter, and a smaller amount, depending on the contours, for other combinations. Examples of good and poor composition are shown in Fig. 2.67.

The space between words should be equal to or greater than the height of a letter but not more than twice the height. The space between sentences should be

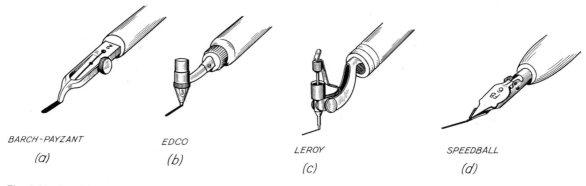

BARCH-PAYZANT
(a)

EDCO
(b)

LEROY
(c)

SPEEDBALL
(d)

Fig. 2.62 Special lettering pens.

somewhat greater. The distance between lines of lettering may vary from one-half the height of the capitals to $1\frac{1}{2}$ times their height.

2.39

Stability. If the areas of the upper and lower portions of certain letters are made equal, an optical illusion will cause them to appear to be unstable and top-heavy. To overcome this effect, the upper portions of the letters B, E, F, H, K, S, X, and Z and the figures 2, 3, and 8 must be reduced slightly in size.

An associated form of illusion is the phenomenon that a horizontal line drawn across a rectangle at the vertical center will appear to be below the center. Since the letters B, E, F, and H are particularly subject to this illusion, their central horizontal strokes must be drawn slightly above the vertical center in order to give them a more balanced and pleasing appearance.

The letters K, S, X, Z and the figures 2, 3, and 8 are stabilized by making the width of the upper portion less than the width of the lower portion.

2.40

Technique of Freehand Lettering. Any prospective engineer or technologist can learn to letter if he practices intelligently and is persistent in his desire to improve. The necessary muscular control, which must accompany the knowledge of lettering, can be developed only through constant repetition.

Pencil letters should be formed with strokes that are dark and sharp, never with strokes that are gray and indistinct. Beginners should avoid the tendency to form letters by sketching, as strokes made in this manner vary in color and width.

When lettering with ink, the results obtained depend largely on the manner in which the pen is used. Many beginners complain that the execution of good freehand lettering is impossible with an ordinary pen point, although their own incorrect habits may have resulted in the inability to make strokes of uniform width. This lack of uniformity may be due to one of four causes: (1) excessive pressure on the pen point; (2) an accumulation of lint, dirt, or dried ink on the point; (3) tilting the point while forming a stroke; or (4) fresh ink on the point. The latter cause requires some explanation, since very few persons know the proper method of "inking" the pen. The pen should be

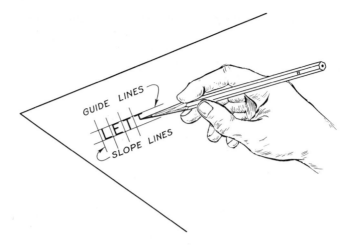

Fig. 2.63 Guide lines and slope lines.

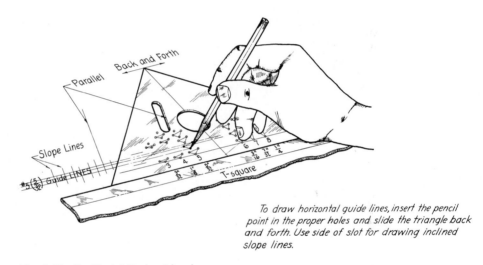

To draw horizontal guide lines, insert the pencil point in the proper holes and slide the triangle back and forth. Use side of slot for drawing inclined slope lines.

Fig. 2.64 Braddock lettering triangle.

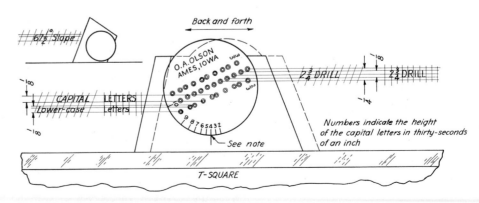

Fig. 2.65 Ames lettering instrument.

UNIFORMITY IN HEIGHT, INCLINATION, AND STRENGTH OF LINE IS ESSENTIAL FOR GOOD LETTERING

Fig. 2.66 Uniformity in lettering.

POOR

GOOD

Fig. 2.67 Letter areas.

Fig. 2.68 "Inking" the pen.

INK ON POINT

Fig. 2.69

Fig. 2.70 Holding the pen.

wiped thoroughly clean, and the ink should be deposited on the underside over the slot, well above the point (Fig. 2.68). When the pen is filled in this manner, the ink feeds down the slit in an even flow, making possible the drawing of uniform curved and straight lines. If ink is placed on the point or allowed to run to the point, an excessive amount of ink will be deposited on the first letters made, and the width of the strokes will be somewhat wider than the strokes of, say, the sixth or seventh letter (Fig. 2.69).

When lettering, the pen is held as shown in Fig. 2.70. It should rest so loosely between the fingers that it can be slid up and down with the other hand.

The thin film of oil on a new point must be removed by wiping before the point is used.

2.41
Inclined and Vertical Capital Letters. The letters shown in Figs. 2.71 and 2.72 have been arranged in related groups. In laying out the characters, the number of widths has been reduced to the smallest number consistent with good appearance; similarities of shape have been emphasized and minute differences have been eliminated. Each letter is drawn to a large size on a cross-section grid that is 2 units wider, to facilitate the study of its characteristic shape and proportions. Arrows with numbers indicate the order and direction of the strokes. The curves of the inclined capital letters are portions of ellipses, while the curves of the vertical letters are parts of circles.

The I, T, L, E, and F

The letter I is the basic or stem stroke. The horizontal stroke of the T is drawn first, and the stem starts at the exact center of the bar. The L is 5 units wide, but it is often desirable to reduce this width when an L is used in combination with such letters as A and T. It should be observed that the letter L consists of the first two strokes of the E. The middle bar of the E is $3\frac{1}{2}$ units long and is placed slightly above the center for stability. The top bar is $\frac{1}{2}$ unit shorter than the bottom bar. The letter F is the E with the bottom bar omitted.

The H and N

Stroke 3 of the H should be slightly above the center, for stability. The outside parallel strokes of the N are drawn first to permit an accurate estimate of its width. The inclined stroke should intersect these accurately at their extremities.

The Z and X

The top of the Z should be 1 unit narrower than the bottom, for stability. In the smaller sizes, this letter may be formed without lifting the pen. The X is similar to the Z in that the top is made 1 unit narrower than the bottom. The inclined strokes cross slightly above center.

The A, V, M, and W

The horizontal bar of the A is located up from the bottom a distance equal to one-third the height of the letter. The V is the letter A inverted without the crossbar, and is the same width. The letters M and W are the widest letters of the alphabet. The outside strokes of the M are drawn first, so that its width may be judged accurately. The inside strokes of this letter meet at the center of the base. The W is formed by two modified V's. Alternate strokes are parallel.

The K and Y

The top of the letter K should be made 1 unit narrower than the bottom, for stability. Stroke 2 intersects the stem one-third up from the bottom. Stroke 3 is approxi-

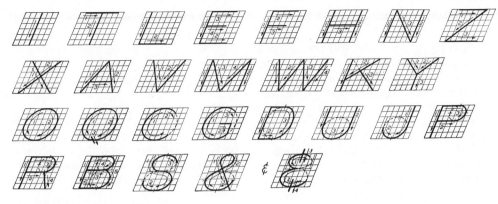

Fig. 2.71 Inclined capital letters.

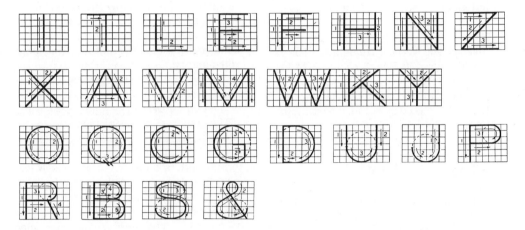

Fig. 2.72 Vertical capital letters.

mately perpendicular to stroke 2, and, if extended, would touch the stem at the top. The strokes of the Y meet at the center of the enclosing parallelogram or square.

The O, Q, C, and G

Stroke 1 of the letter O starts just to the right of the top and continues to the left around the side to a point beyond the bottom. Thus stroke 1 forms more than half of the ellipse or circle. The Q is the letter O with the added kern, which is a straight line located near the bottom tangent point. The C is based on the O, but since it is not a complete ellipse or circle, it is narrower than either the O or Q. The top extends 1 unit down and the bottom 1 unit up on the right side. G is similar to C. The horizontal portion of stroke 2 starts at the center.

The D, U, and J

The first two strokes of the D form an incomplete letter L. Stroke 3 starts as a horizontal line. The bottom third of the U is one-half of an ellipse or circle. J is similar to the letter U.

The P, R, and B

The middle horizontal bar of the P is located at the center of stroke 1. The curved portion of stroke 3 is one-half of a perfect ellipse or circle. The R is constructed similarly to the P. The tail joins at the point of tangency of the curve and middle bar. To stabilize the letter B, the top is made $\frac{1}{2}$ unit narrower than the bottom and the middle bar is placed slightly above the center. The curves are halves of ellipses or circles.

The S and &

The upper and lower portions of the S are perfect ellipses with one-quarter removed. The top ellipse should be made $\frac{1}{2}$ unit narrower than the lower one, for stability. In the smaller sizes this letter may be made with one or two strokes, depending upon its size. The true ampersand is made with three strokes. Professionals, however, usually represent an ampersand with a character formed by using portions of the upper and lower ellipses of the numeral 8 with the addition of two short bars (Fig. 2.71).

Although many favor the inclined letters, recent surveys indicate that vertical letters are used more generally.

2.42
Inclined and Vertical Numerals. The numerals shown in Figs. 2.73 and 2.74 have been arranged in related groups in accordance with the common characteristics that can be recognized in their construction.

The 1, 7, and 4

The stem stroke of the 4 is located 1 unit in from the right side. The bar is $1\frac{1}{2}$ units above the base. The stem of the 7 terminates at the center of the base.

The 0, 6, and 9

The cipher, which is 1 unit narrower than the letter O, is the basic form for this group. In the figure 6, the right side of the large ellipse ends 1 unit down from the top, and the left side ends at the center of the base. The small loop is slightly more than three-fourths of an ellipse. The 9 is the 6 inverted.

The 8, 3, and 2

Each of these figures is related to the letter S, and the same rule of stability should be observed in their construction. The top portion of the figure 8 is shorter and $\frac{1}{2}$ unit narrower than the lower portion. Each loop is a perfect ellipse. The figure 3 is the 8 with the lower-left quarter of the upper loop and the upper-left quarter of the lower loop omitted. The 2 is simply three-quarters of the upper loop of

the 8 and the upper-left quarter of the lower loop of the 8 with straight lines added.

The 5

This figure is a modification of the related groups previously described. The top is $\frac{1}{2}$ unit narrower than the bottom, for stability. The curve is a segment of a perfect ellipse, ending one unit up from the bottom.

2.43
Single-stroke Lowercase Letters. Single-stroke lowercase letters, either vertical or inclined, are commonly used on map drawings, topographic drawings, structural drawings, and in survey field books. They are particularly suitable for long notes and statements because, first, they can be executed much faster than capitals and, second, words and statements formed with them can be read more easily.

The construction of inclined lowercase letters (Fig. 2.75) is based on the straight line and the ellipse. This basic principle of forming letters is followed more closely for lowercase letters than for capitals. The body portions are two-thirds the height of the related capitals. As shown in Fig. 2.76, ascenders extend to the cap line, and descenders descend to the drop line. For lowercase letters based on a capital letter 6 units high, the waistline is 2 units down from the top and the drop line 2 units below the base line.

The order of stroke, direction of stroke, and formation of the letters follow the same principles as for the capitals. The letters are presented in family groups having related characteristics, to enable the beginner to understand their construction. The vertical lowercase letters, illustrated in Fig. 2.77, are constructed in the same manner as inclined letters.

The i, l, k, and t

All letters of this group are formed by straight lines of standard slope. The i is drawn 4 units high, and the dot is placed halfway between the waistline and cap line. Stroke 2 of the k starts at the waistline and intersects stroke 1 at a point 2 units above the base. Stroke 3, extended should intersect stroke 1 at the top. The t is 5 units high, and the crossbar is on the waistline.

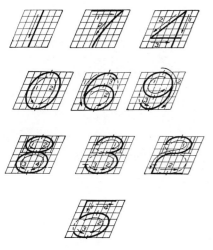

Fig. 2.73 Inclined numerals.

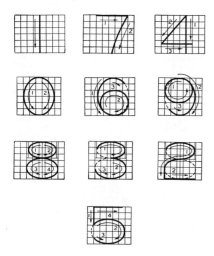

Fig. 2.74 Vertical numerals.

The v, w, x, and z

All of these letters are similar to the capitals. Alternate strokes of the w are parallel. The width of the top of both the x and the z is made $\frac{1}{2}$ unit less than the width across the bottom, for stability.

The o, a, b, d, p, and q

The bodies of the letters in this group are formed by the letter o, and they differ only in the position and length of the stem stroke. The o is made with two strokes, and the first stroke should form more than half of the character.

The g

The g is related to the letters o and y. Stroke 3 starts at the waistline and ends slightly beyond the point of tangency of the curve with the drop line.

The c and e

The c is a modified letter o. It is not a complete form, and, therefore, its width is less than its height. Stroke 1 ends 1 unit up on the right side, stroke 2 ends 1 unit down. The e is similarly constructed, except for the fact that stroke 2 continues as a curve and finishes as a horizontal line that terminates at the middle of the back.

The h, n, r, and m

The curve of the h is the upper portion of the letter o. Stroke 2 starts 2 units above the bottom of the stem and finishes parallel to stroke 1. The n differs from the h in that the stem stroke extends only from the waistline to the base line. The r is a portion of the letter n, stroke 2 ending 1 unit down from the top. The m consists of two modified letter n's. The straight portions of strokes 2 and 3 are parallel to stroke 1.

The u and y

The letter u is an inverted n, and the curve is a portion of the letter o. It should be noted that stroke 2 extends to the base line. The y is a partial combination of the letters u and g.

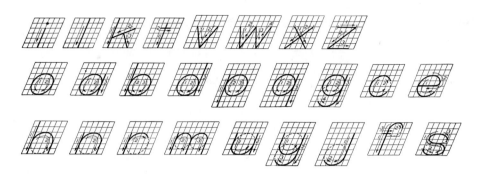

Fig. 2.75 Inclined lowercase letters.

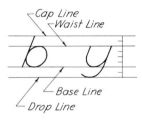

Fig. 2.76

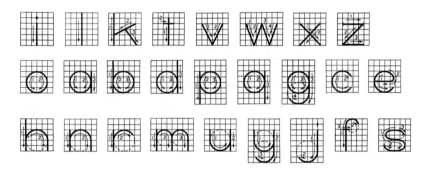

Fig. 2.77 Vertical lowercase letters.

The j and f

The portion of the j above the base line is the letter i. The curve is the same as that which forms the tail of the g. The curved portion of the letter f is $2\frac{1}{2}$ units wide, stroke 1 starting slightly to the right of the point of tangency with the cap line.

The s

The lowercase s is almost identical to the capital S.

Fig. 2.78

HANDLE PIN
C.R.S. I REQ'D.

Fig. 2.79 Use of large and small caps.

APPROVED BY CUSTOMER	ROSS-SHOER CO.
ADAMS MFG. CO. AURORA, ILL.	
ASSEMBLY STANDARD DUTY FLANGE UNIT	

SCALE *FULL SIZE*	DATE *JAN. 3, 1959*
DRAWN BY *CARL KING*	CHECKED BY *J.B.JONES*
TRACED BY *K. BODINE*	APPROVED BY *W.L. KOE*
ORDER NO. *M-L-24631*	DRG. *5-19-9612*

Fig. 2.80 Machine-drawing title block.

2.44

Fractions. The height of the figures in the numerator and denominator is equal to three-fourths the height of the whole number, and the total height of the fraction is twice the height of the whole number. The division bar should be horizontal and centered between the fraction numerals, as shown in Fig. 2.78. It should be noted that the sloping center line of the fraction bisects both the numerator and denominator and is parallel to the sloping center line of the whole number.

2.45

Large Caps and Small Caps in Combination. Many commercial draftsmen use a combination of large caps and small caps in forming words, as illustrated in Fig. 2.79. When this style is used, the height of the small caps should be approximately three-fifths the height of the first capital letter of the word.

2.46

Titles. Every drawing, sketch, graph, chart, or diagram has some form of descriptive title to impart certain necessary information and to identify it. On machine drawings, where speed and legibility are prime requirements, titles are usually single-stroke. On display drawings, maps, and so on, which call for an artistic effect, the titles are usually composed of "built-up" ornate letters.

Figure 2.80 shows a title block that might be used on a machine drawing. It should be noted that the important items are made more prominent by the use of larger letters formed with heavier lines. Less important data, such as the scale, date, drafting information, and so on, are given less prominence.

To be pleasing in appearance, a title should be symmetrical about a vertical

Fig. 2.81 Leroy lettering device.

center line and should have some simple geometric form. An easy way to ensure the symmetry of a title is first to count the letter and word spaces, then, working to the left and right from the middle space or letter on the vertical center line, to sketch the title lightly in pencil, before lettering it in finished form. An alternative method is to letter a trial line along the edge of a piece of scrap paper and place it in a balanced position just above the location of the line to be lettered.

2.47

Mechanical Lettering Devices and Templates. Although mechanical lettering devices produce letters that may appear stiff to an expert, they are used in many drafting rooms for the simple reason that they enable even the unskilled to do satisfactory lettering with ink. The average draftsman rightly prefers stiff uniformity to wavy lines and irregular shapes. One of the oldest instruments of this sort on the market is the *Wrico* outfit. With a satisfactory set of Wrico pens and templates, letters ranging in size from $\frac{3}{32}$ to $\frac{1}{2}$ in. in height may be executed. The letters are formed by a stylographic pen that is guided around the sides of openings in a template made of transparent pyralin.

The *Leroy* device, shown in Fig. 2.81, is possibly even more efficient than the Wrico, for it does not require the sliding of a template to complete a letter.

2.48

Commercial Gothic (Fig. 2.82). Gothic letters are comparatively easy to construct. They are drawn first in outline, in skeleton form, and then filled in with a ballpoint pen. This style, from which the single-stroke engineering letters are derived, is used to some extent for titles on machine drawings, but more frequently it appears on maps, charts, and graphs that are prepared for display purposes.

2.49

Modern Roman. The modern roman letters shown in Fig. 2.83 are used extensively by engineers for names and titles on maps. Students should be familiar with this alphabet, for it appears in all modern publications. As in the case of the gothic style, these letters must be drawn first in outline, in skeleton form, and then filled in with a ballpoint or wide-line pen. The straight lines are usually drawn with a ruling pen. The curved lines, except for the serifs, may be formed either mechanically or freehand. When attempting to construct this type of letter it is wise to bear in mind the following facts: (1) All vertical strokes are heavy except those forming the letters M, N, and U. (2) All horizontal strokes are light. (3) The width of the heavy strokes may vary from one-eighth to one-sixth the height of the letter.

Roman letters may be drawn either extended or compressed, depending on the area to be covered or the space restrictions determined by other lettering, such as on maps. If a draftsman using the roman-style letter would have his work look professional, he must pay particular attention to detail. Care must be taken to keep the proportions of a letter appearing more than once in a name or title identical.

ABCDEFG
HIJKLMN
OPQRSTU
VWXYZ &
1234567
890

SCALE

Fig. 2.82 Gothic letters and numerals.

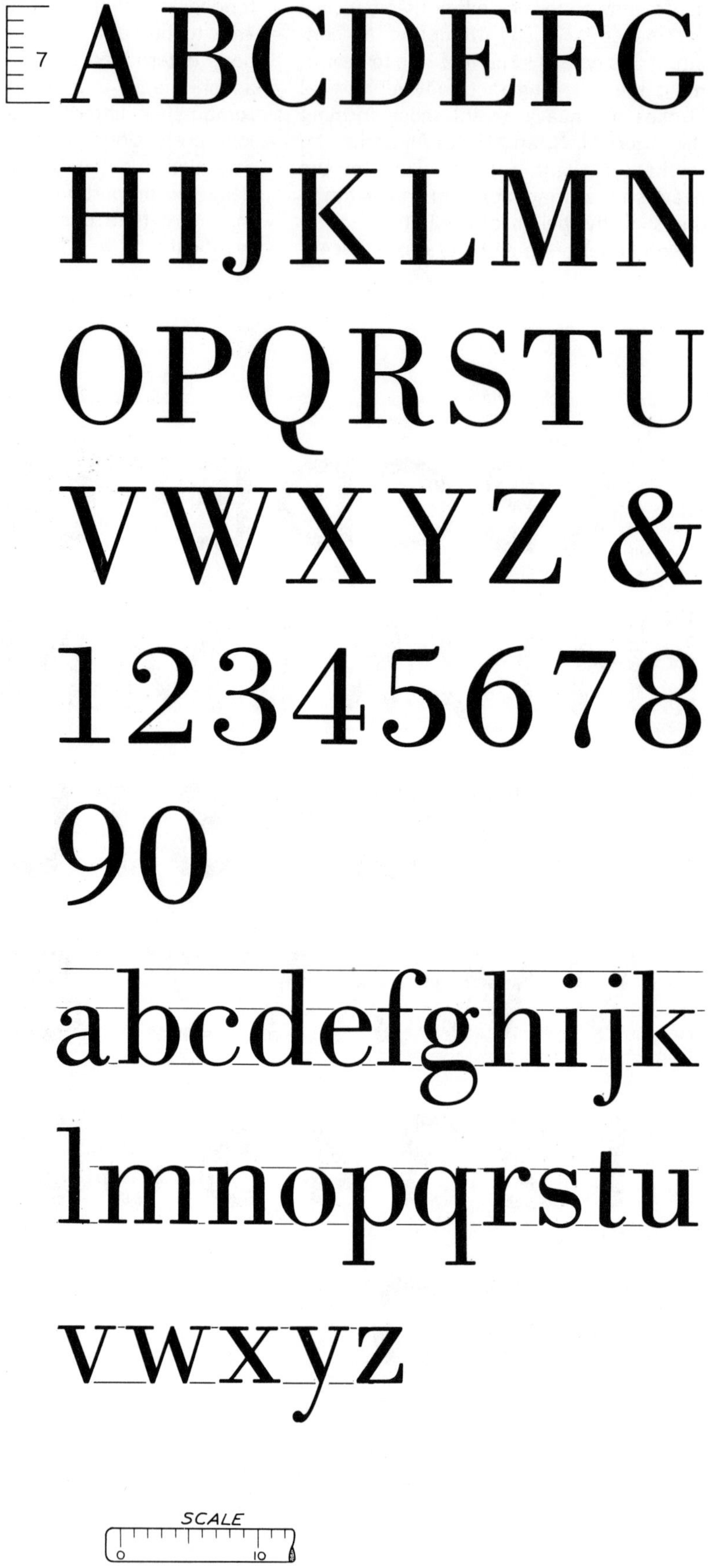

ABCDEFG
HIJKLMN
OPQRSTU
VWXYZ &
12345678
90

abcdefghijk
lmnopqrstu
vwxyz

SCALE

Fig. 2.83 Modern roman letters and numerals.

Exercises in instrumental drawing

The following elementary exercises have been designed to offer experience in the use of the drafting instruments. The designs ·should be drawn *lightly* with a hard pencil. After making certain that all constructions shown on a drawing are correct, the lines forming the designs should be heavied with a medium-hard pencil. The light construction lines need not be erased if the drawing is relatively clean.

1. (Fig. 2.84). On a sheet of drawing paper reproduce the line formations shown. If the principal border lines have not been printed on the sheet, they may be drawn first so that the large $5\frac{1}{2} \times 8\frac{1}{4}$-in. rectangle can be balanced horizontally and vertically within the border. To draw the inclined lines, first draw the indicated measuring lines through the lettered points at the correct angle, and mark off $\frac{1}{4}$-in. distances. These division points establish the locations of the required lines of the formation. The six squares of the formation are equal in size.

2. (Fig. 2.85). Reproduce the line formations shown, following the instructions given for Exercise 1.

3. (Fig. 2.86). This exercise is designed to give the student practice with the bow pencil and compass by drawing some simple geometric figures. The line work within each large circle

may be reproduced with only the knowledge that the diameter is $3\frac{1}{4}$ in. (82 mm). All circles and circle arcs are to be made finished-weight when they are first drawn, since retracing often produces a double line. Do not "overrun" the straight lines or stop them too short.

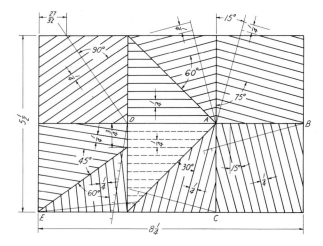

Fig. 2.85

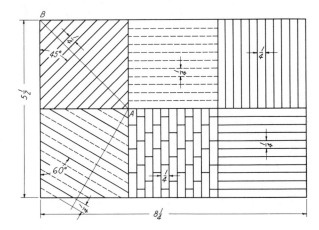

Fig. 2.84

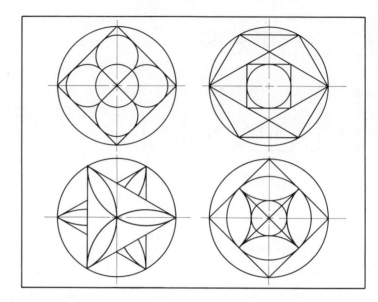

Fig. 2.86

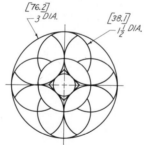

─ 3.00 DIA.

Fig. 2.87

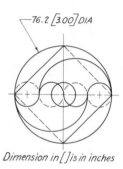

─ 76.2 [3.00] DIA

Dimension in [ ] is in inches

Fig. 2.88

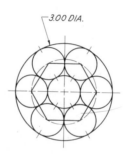

[76.2]
─ 3 DIA.

[38.1]
1½ DIA.

Dimensions in [ ] are in millimeters

Fig. 2.89

4–6. (Figs. 2.87–2.89). Reproduce the following designs according to the instructions given for problem 3, making the dashes of the arcs approximately ⅛ in. (3 mm) long.

7–9. (Figs. 2.90–2.92). Reproduce the line work within each square, using the dimensions given. (The dimensions shown, however, are for the student's use only and should not appear on the finished drawing.) Arcs should be made

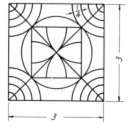

Fig. 2.90

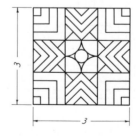

Fig. 2.91

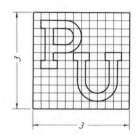

Fig. 2.92

finished-weight when first drawn. The straight
lines of each design may be drawn with a hard
pencil and later heavied with a softer pencil. Do
not erase the construction lines.

10–13. (Figs. 2.93–2.96). Reproduce the geometric shapes.

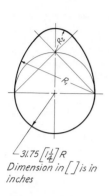

Dimension in [ ] is in
inches

Fig. 2.93   Oval.

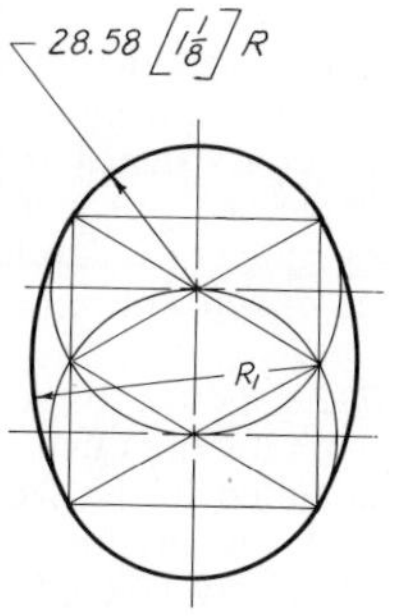

Dimension in [ ] is in
inches

Fig. 2.94   Ellipse (approximate).

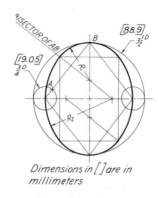

Dimensions in [ ] are in
millimeters

Fig. 2.95   Ellipse (approximate).

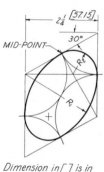

Dimension in [ ] is in
millimeters

Fig. 2.96   Ellipse (pictorial).

Exercises

It should be noted that while these exercises are offered to give the student practice in letter forms and word composition, they also contain statements of important principles of drawing, shop notes, and titles with which every engineer should be familiar. Each lettering exercise should be submitted to the instructor for severe criticism before the student proceeds to the next. Section 2.34 should be reread before starting the first exercise.

1. Letter the statement given in Fig. 2.97 in $5/32$-in. capital letters using an appropriate pencil that is suited to the type of paper being used and that will produce uniform opaque lines. The necessary guide lines should be drawn with a hard pencil.

2, 3. Letter the statements given in Figs. 2.98 and 2.99 in $1/8$-in. capital letters using an appropriate pencil that is suited to the type of paper being used and that will produce uniform opaque lines. The necessary guide lines should be drawn with a hard pencil.

4. Letter the statement given in Fig. 2.100 in $5/32$-in. capital letters using an appropriate pen-

cil that is suited to the type of paper being used and that will produce uniform opaque lines. The necessary guide lines should be drawn with a hard pencil.

5, 6. Letter the statements given in Figs. 2.101 and 2.102 in $1/8$-in. capital letters using an appropriate pencil that is suited to the type of paper being used and that will produce uniform opaque lines. The necessary guide lines should be drawn with a hard pencil.

7–13. Letter the following statements in $1/8$-in. capital letters using an appropriate pencil that is suited to the type of paper being used and that will produce uniform opaque lines. The necessary guide lines should be drawn with a hard pencil.

7. A GOOD STUDENT REALIZES THE IMPORTANCE OF NEAT AND ATTRACTIVE LETTERING.

8. THE POSITION OF THE VIEWS OF AN ORTHOGRAPHIC DRAWING MUST BE IN STRICT ACCORDANCE WITH THE UNIVERSALLY RECOGNIZED ARRANGEMENT ILLUSTRATED IN FIG. 5.6.

POOR LETTERING DETRACTS FROM THE APPEARANCE OF A DRAWING

Fig. 2.97

IN LEARNING TO LETTER, CERTAIN DEFINITE RULES OF FORM & DESIGN MUST BE OBSERVED

Fig. 2.98

WHEN LETTERING WITH INK, THE INK SHOULD BE WELL ABOVE THE TIP OF THE POINT

Fig. 2.99

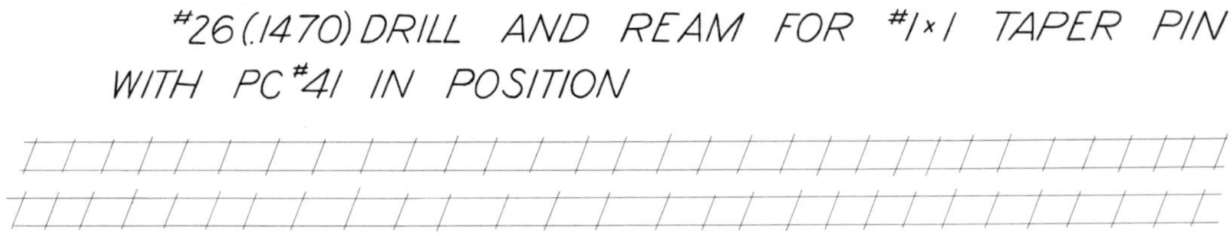

#26 (.1470) DRILL AND REAM FOR #1 x 1 TAPER PIN WITH PC #41 IN POSITION

Fig. 2.100 Lettered statement. Draw guide lines as shown.

9. THE VIEWS OF AN ORTHOGRAPHIC DRAW-ING SHOULD SHOW THE THREE DIMENSIONS: WIDTH, DEPTH, AND HEIGHT.

10. AN INVISIBLE LINE SHOULD START WITH A SPACE WHEN IT FORMS AN EXTENSION OF A SOLID LINE.

11. THE FRONT VIEW OF AN ORTHOGRAPHIC DRAWING SHOULD BE THE VIEW THAT SHOWS THE CHARACTERISTIC SHAPE OF THE OB-JECT.

12. AN AUXILIARY VIEW SHOWS THE TRUE SIZE AND SHAPE OF AN INCLINED SURFACE.

13. A SECTIONAL VIEW SHOWS THE INTERIOR CONSTRUCTION OF AN OBJECT.

14. Draw horizontal and inclined guide lines and letter the following detail titles. Use $\frac{5}{32}$-in. capitals for the part names and $\frac{1}{8}$-in. capitals for the remainder of the titles.

BASE
C.I.   1 REQ'D

BUSHING
BRO.   1 REQ'D

SPINDLE
C.R.S.   1 REQ'D

15–19. Using a hard pencil, draw two or more sets of horizontal and inclined guide lines for $\frac{5}{32}$-in. letters; then letter the following exercises in lowercase letters.

15. The front and top views are always in line vertically.

16. The front and side views are in line hori-zontally.

17. The depth of the top view is the same as the depth of the side view.

18. If a line is perpendicular to a plane of pro-jection, its projection will be a point.

19. If a line is parallel to a plane of projection, its projection on the plane is exactly the same length as the true length of the line.

S.A.E. 1020 – COLD DRAWN STEEL BAR
1-12 UNF-2B   1-8 UNC-2A   1-5 SQUARE
BREAK ALL SHARP CORNERS UNLESS OTHERWISE SPECIFIED

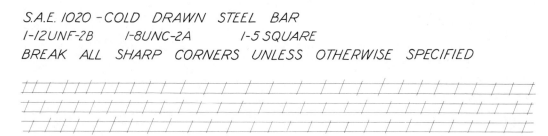

Fig. 2.101   Lettered statement. Draw guide lines as shown.

NECK 3 WIDE ×1.5 DEEP   3 ×45° CHAMFER   SPHERE R15 ±0.1

8.8 DRILL – M10×1.5 – 6H   16 DRILL – C'BORE 23 D×16 DEEP –2 HOLES

Fig. 2.102   Lettered statement. Draw guide lines as shown.

The design of highway interchanges involves the application of the geometry of circle arcs. The up-and-down curves of the concrete roadway are parabolic. In the mechanical engineering field, geometry is used in the design of an endless number of machine parts as well as for gears, cams, and linkages. (*Courtesy Department of Public Works, State of California*)

3

Engineering geometry

3.1

Introduction. The simplified geometrical constructions presented in this chapter are those with which an engineer should be familiar, for they occur frequently in engineering drawing. The methods are applications of the principles found in textbooks on plane geometry. The constructions have been modified to take advantage of time-saving methods made possible by the use of drawing instruments.

Since a study of the subject of plane geometry should be a prerequisite for a course in engineering drawing, the mathematical proofs have been omitted intentionally. Geometric terms applying to lines, surfaces, and solids, however, are given in Figs. 3.60 and 3.61 for the purpose of review.

3.2

To Bisect a Straight Line (Fig. 3.1).

(*a*) With A and B as centers, strike the intersecting arcs as shown using any radius greater than one-half of AB. A straight line through points C and D bisects AB.

(*b*) Draw either 60° or 45° lines through E and F. Through their intersection draw the perpendicular GH that will bisect EF.

The use of the dividers to divide or bisect a line by the trial method is explained in Sec. 2.24.

3.3

To Trisect a Straight Line (Fig. 3.2). Given the line AB. Draw the lines AO and OB making 30° with AB. Similarly, draw CO and OD making 60° with AB. AC equals CD equals DB.

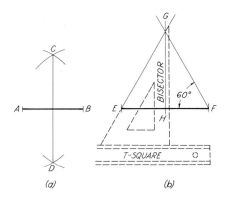

Fig. 3.1 To bisect a straight line.

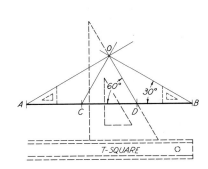

Fig. 3.2 To trisect a straight line.

47

48

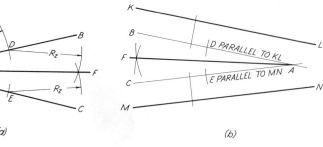

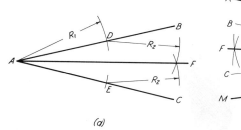

Fig. 3.3 To bisect an angle.

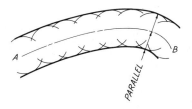

Fig. 3.4 To draw parallel curved lines.

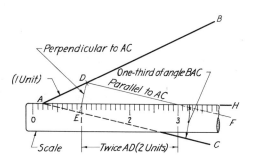

Fig. 3.5 To trisect an angle.

3.4
To Bisect an Angle (Fig. 3.3)

(*a*) Given the angle *BAC*. Use any radius with the vertex *A* as a center, and strike an arc that intersects the sides of the angle at *D* and *E*. With *D* and *E* as centers and a radius larger than one-half of *DE*, draw intersecting arcs. Draw *AF*. Angle *BAF* equals angle *FAC*.

(*b*) Given an angle formed by the lines *KL* and *MN* having an inaccessible point of intersection. Draw *BA* parallel to *KL* and *CA* parallel to *MN* at the same distance from *MN* as *BA* is from *KL*. Bisect angle *BAC* using the method explained in part (*a*). The bisector *FA* of angle *BAC* bisects the angle between the lines *KL* and *MN*.

3.5
To Draw Parallel Curved Lines About a Curved Center Line (Fig. 3.4). Draw a series of arcs having centers located at random along the given center line *AB*. Using the French curve, draw the required curved lines tangent to these arcs.

3.6
To Trisect an Angle (Fig. 3.5). Given the angle *BAC*. Lay off along *AB* any convenient distance *AD*. Draw *DE* perpendicular to *AC* and *DF* parallel to *AC*. Place the scale so that it passes through *A* with a distance equal to twice *AD* intercepted between the lines *DE* and *DF*. Angle *HAC* equals one-third of the angle *BAC*.

3.7
To Divide a Straight Line into a Given Number of Equal Parts (Fig. 3.6). Given the line *LM*, which is to be divided into five equal parts.

(*a*) Step off, with the dividers, five equal divisions along a line making any convenient angle with *LM*. Connect the last point *P* with *M*, and through the remaining points draw lines parallel to *MP* intersecting the given line. These lines divide *LM* into five equal parts.

(*b*) Some commercial draftsmen prefer a modification of this construction known as the scale method. For the first step, draw a vertical *PM* through point *M*. Place the scale so that the first mark of five equal divisions is at *L* and the last mark falls on *PM*. Locate the four intervening division points, and through these draw verticals intersecting the given line. The verticals will divide *LM* into five equal parts.

3.8
To Divide a Line Proportionally (Fig. 3.7). Given the line *AB*. Draw *BC* perpendicular to *AB*. Place the scale across *A* and *BC* so that the number of divisions intercepted is equal to the sum of the numbers representing the proportions. Mark off these proportions and draw lines parallel to *BC* to divide *AB* as required. The proportions in Fig. 3.7 are 1:2:3.

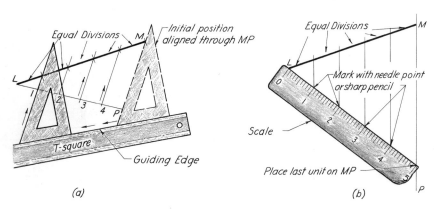

Fig. 3.6 To divide a straight line into a number of equal parts.

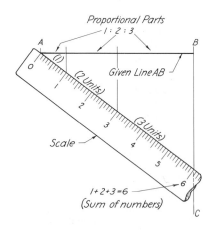

Fig. 3.7 To divide a line proportionally.

3.9

To Construct an Angle Equal to a Given Angle (Fig. 3.8). Given the angle BAC and the line $A'C'$ that forms one side of the transferred angle. Use any convenient radius with the vertex A as a center, and strike the arc that intersects the sides of the angle at D and E. With A' as a center, strike the arc intersecting $A'C'$ at E'. With E' as a center and the chord distance DE as a radius, strike a short intersecting arc to locate D'. $A'B'$ drawn through D' makes angle $B'A'C'$ equal angle BAC.

3.10

To Draw a Line Through a Given Point and the Inaccessible Intersection of Two Given Lines (Fig. 3.9). Given the lines KL and MN, and the point P. Construct any triangle, such as PQR, having its vertices falling on the given lines and the given point. At some convenient location construct triangle STU similar to PQR, by drawing SU parallel to PR, TU parallel to QR, and ST parallel to PQ. PS is the required line.

3.11

To Construct an Angle, Tangent Method (Fig. 3.10). Draftsmen and designers often find it necessary to draw long lines having an angle between them that is not equal to an angle of a triangle. Such an angle may be laid off with a protractor, but it should be remembered that as the lines are extended any error is multiplied. To avoid this situation, the tangent method may be used. The tangent method involves trigonometry but, since it is frequently used, a discussion of it here is pertinent. (See Table 38 of the Appendix.)

In this method, a distance D_1 is laid off along a line that is to form one side of the angle, and a distance D_2, equal to D_1 times the natural tangent of the angle, is marked off along a perpendicular through point P. A line through point X is the required line, and angle A is the required angle. In laying off the distance D_1, unnecessary multiplication will be eliminated if the distance is arbitrarily made 10 full-size units. However, in order to make the construction large enough for accuracy and at the same time keep all the lines on the drawing, it may be necessary to lay off the 10 units to some scale other than full-size. When a decimal-inch scale is being used, the construction may be drawn using either a half-size or quarter-size scale. If a metric scale is used one might decide upon the 1:50 or some other selected scale.

This method is also used for angles formed by short lines whenever a protractor is not available.

3.12

To Construct an Angle, Chord Method (Fig. 3.11). Engineers and draftsmen frequently select the chordal method for constructing an angle accurately. This method, as applied in laying out a given angle, involves the use of an easily determined chord length for a selected length of radius laid off in units. Given the angle (say, 29°), the procedure is as follows: First, lay off any convenient distance, usually 10 units, along a line that is to form one side of the angle and strike an arc of indefinite length using this distance as a radius. Second, obtain the unit chord value for the given angle from Table 39 of the Appendix (29° = .5008) and multiply this value by 10. With the compass or dividers lay off the chord length (.5008 × 10 = 5.008) along the arc from the starting line and complete the angle.

3.13

To Construct a Triangle, Given Its Three Sides (Fig. 3.12). Given the three sides AB, AC, and BC. Draw the side AB in its correct location. Using its end points A and B as centers and radii equal to AC and BC, respectively, strike the two intersecting arcs locating point C. ABC is the required triangle. This construction is particularly useful for developing the surface of a transition piece by triangulation.

3.14

To Construct an Equilateral Triangle (Fig. 3.13). Given the side AB.

(a) Using the end points A and B as centers and a radius equal to the length of AB, strike two intersecting arcs to locate C. Draw lines from A to C and C to B to complete the required equilateral triangle.

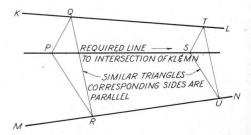

Fig. 3.9 To draw a line through a given point and the inaccessible intersection of two given lines.

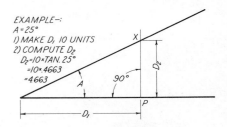

Fig. 3.10 To construct an angle, tangent method.

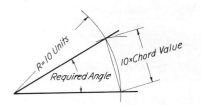

Fig. 3.11 To construct an angle, chord method.

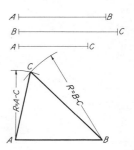

Fig. 3.12 To construct a triangle, given its three sides.

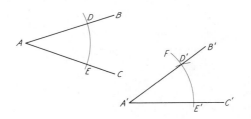

Fig. 3.8 To construct an angle equal to a given angle.

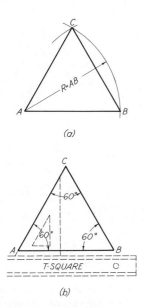

Fig. 3.13 To construct an equilateral triangle.

(*b*) Using a 30°–60° triangle, draw through *A* and *B* lines that make 60° with the given line. If the line *AB* is inclined, the 60° lines should be drawn as shown in Fig. 2.26

3.15
To Transfer a Polygon (Fig. 3.14). Given the polygon *ABCDE*.

(*a*) Enclose the polygon in a rectangle. Draw the "enclosing rectangle" in the new position and locate points *A*, *B*, *C*, *D*, and *E* along the sides by measuring from the corners of the rectangle. A compass may be used for transferring the necessary measurements.

(*b*) To transfer a polygon by the triangle method, divide the polygon into triangles and, using the construction explained in Sec. 3.13, reconstruct each triangle in its transferred position.

3.16
To Construct a Square (Fig. 3.15).

(*a*) Given the side *AB*. Using a T-square and a 45° triangle, draw perpendiculars to line *AB* through points *A* and *B*. Locate point *D* at the intersection of a 45° construction line through *A* and the perpen-

dicular from *B*. Draw *CD* parallel to *AB* through *D* to complete the square. To eliminate unnecessary movements the lines should be drawn in the order indicated.

(*b*) Given the diagonal length *EF*. Using a T-square and a 45° triangle, construct the square by drawing lines through *E* and *F* at an angle of 45° with *EF* in the order indicated.

(*c*) The construction of an inscribed circle is the first step in one method for drawing a square when the location of the center and the length of one side are given.

Using a T-square and a 45° triangle, draw the sides of the square tangent to the circle. This construction is used in drawing square bolt heads and nuts.

3.17
To Construct a Regular Pentagon (Fig. 3.16). Given the circumscribing circle. Draw the perpendicular diameters *AB* and *CD*. Bisect *OB* and, with its midpoint *E* as a center and *EC* as a radius, draw the arc *CF*. Using *C* as a center and *CF* as a radius, draw the arc *FG*. The line *CG* is one of the equal sides of the required pentagon. Locate the remaining vertices by striking off this distance around the circumference.

If the length of one side of a pentagon is given, the construction described in Sec. 3.20 should be used.

3.18
To Construct a Regular Hexagon (Fig. 3.17).

(*a*) Given the distance *AB* across corners. Draw a circle having *AB* as a diameter. Using the same radius and with points *A* and *B* as centers, strike arcs intersecting the circumference. Join these points to complete the construction.

(*b*) Given the distance *AB* across corners. Using a 30°–60° triangle and a

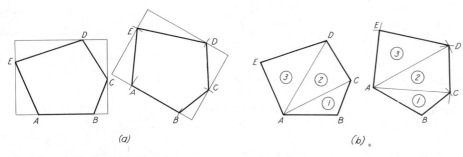

Fig. 3.14 To transfer a polygon.

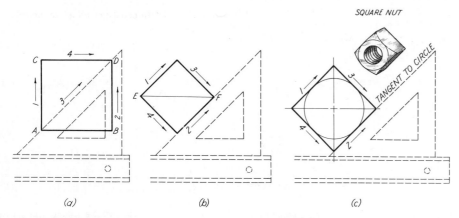

Fig. 3.15 To construct a square.

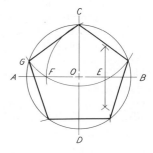

Fig. 3.16 To construct a regular pentagon.

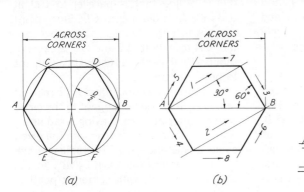

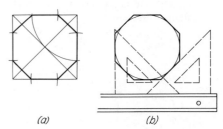

Fig. 3.17 To construct a regular hexagon.

T-square, draw the lines in the order indicated by the numbers on the figure.

(*c*) Given the distance across flats. Draw a circle whose diameter equals the distance across flats. Using a 30°–60° triangle and a T-square, as shown, draw the tangents that establish the sides and vertices of the required hexagon.

This construction is used in drawing hexagonal bolt heads and nuts.

3.19
To Construct a Regular Octagon (Fig. 3.18).

(*a*) Given the distance across flats. Draw the circumscribed square and its diagonals. Using the corners as centers and one-half the diagonal as a radius, strike arcs across the sides of the square. Join these points to complete the required octagon.

(*b*) Given the distance across flats. Draw the inscribed circle; then, using a 45° triangle and T-square, draw the tangents that establish the sides and vertices of the required octagon.

3.20
To Construct Any Regular Polygon, Given One Side (Fig. 3.19). Given the side *LM*.

With *LM* as a radius, draw a semicircle and divide it into the same number of equal parts as the number of sides needed for the polygon. Suppose the polygon is to be seven-sided. Draw radial lines through points 2, 3, and so forth. Point 2 (the second division point) is always one of the vertices of the polygon, and line *L2* is a side. Using point *M* as a center and *LM* as a radius, strike an arc across the radial line *L6* to locate point *N*. Using the same radius with *N* as a center, strike another arc across *L5* to establish *O* on *L5*. Although this procedure may be continued with point *O* as the next center, more

accurate results will be obtained if point *R* is used as a center for the arc to locate *Q*, and *Q* as a center for *P*.

3.21
To Divide the Area of a Triangle or Trapezoid into a Given Number of Equal Parts (Fig. 3.20).

(*a*) Given the triangle *ABC*. Divide the side *AC* into (say, five) equal parts, and draw a semicircle having *AC* the diameter. Through the division points (1, 2, 3, and 4) draw perpendicular lines to points of intersection with the semicircle (5, 6, 7, and 8). Using *C* as a center, strike arcs through these points (5, 6, 7, and 8) that will cut *AC*. To complete the construction, draw lines parallel to *AB* through the points (9, 10, 11, and 12) at which the arcs intersect the side *AC*.

(*b*) Given the trapezoid *DEBA*. Extend the sides of the trapezoid to form the triangle *ABC* and draw a semicircle on *AC* with *AC* as a diameter. Using *C* as a center and *CD* as a radius, strike an arc cutting the semicircle at point *P*. Through *P* draw a perpendicular to *AC* to locate point *Q*. Divide *QA* into the same number of equal parts as the number of equal

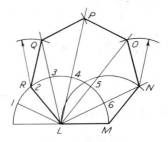

Fig. 3.18 To construct a regular octagon.

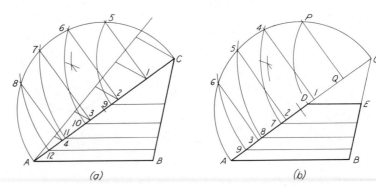

Fig. 3.19 To construct any regular polygon, given one side.

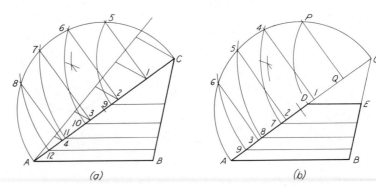

Fig. 3.20 To divide the area of a triangle or trapezoid into a given number of equal parts.

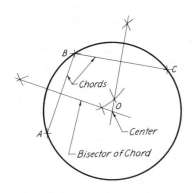

Fig. 3.21 To find the center of a circle through three points.

areas required (in this case, four), and proceed using the construction explained in (*a*) for dividing the area of a triangle into a given number of equal parts.

3.22
To Find the Center of a Circle Through Three Given Points Not in a Straight Line (Fig. 3.21). Given the three points A, B, and C. Join the points with straight lines (which will be chords of the required circle), and draw the perpendicular bisectors. The point of intersection O of the bisectors is the center of the required circle, and OA, OB, or OC is its radius.

3.23
Tangent Circles and Arcs. Figure 3.22 illustrates the geometry of tangent circles. In (*a*) it can be noted that the locus of centers for circles of radius R tangent to AB is a line that is parallel to AB at a distance R from AB. The locus of centers for circles of the same radius tangent to

CD is a line that is parallel to CD at distance R (radius) from CD. Since point O at which these lines intersect is distance R from both AB and CD, a circle of radius R with center at O must be tangent to both AB and CD.

In (*b*) the locus of centers for circles of radius R_3 that will be tangent to the circle with a center at O and having a radius R_1 is a circle that is concentric with the given circle at distance R_3. The radius of the locus of centers will be $R_1 + R_3$. In the case of the circle with center at point P, the radius of the locus of centers will be $R_2 + R_3$. Points Q and Q_1, where these arcs intersect, are points that are distance R_3 from both circles. Therefore, circles of radius R_3 that are centered at Q and Q_1 will be tangent to both circles with centers at O and P.

3.24
To Draw a Circular Arc of Radius R Tangent to Two Lines (Fig. 3.23).

(*a*) Given the two lines AB and CD at right angles to each other, and the radius of the required arc R. Using their point of intersection X as a center and R as a radius, strike an arc cutting the given lines at T_1 and T_2 (tangent points). With T_1 and T_2 as centers and the same radius, strike the intersecting arcs locating the center O of the required arc.

(*b*),(*c*) Given the two lines AB and CD, not at right angles, and the radius R. Draw lines EF and GH parallel to the given lines at a distance R. Since the point of intersection of these lines is distance R from both given lines, it will be the center O of the required arc. Mark the tangent points T_1 and T_2 that lie along perpendiculars to the given lines through O.

These constructions are useful for drawing fillets and rounds on views of machine parts.

3.25
To Draw a Circular Arc of Radius R_1 Tangent of a Given Circular Arc and a Given Straight Line (Fig. 3.24). Given the line AB and the circular arc with center O.

(*a*),(*b*) Draw line CD parallel to AB at a distance R_1. Using the center O of the given arc and a radius equal to its radius plus or minus the radius of the required arc (R_2 plus or minus R_1), swing a parallel arc intersecting CD. Since the line CD and the intersecting arc will be the loci of centers for all circles of radius R_1, tangent respectively to the given line AB and the given arc, their point of intersection P will

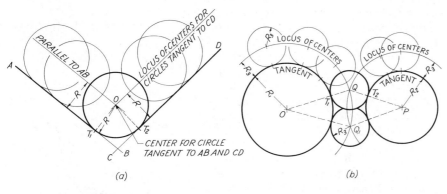

Fig. 3.22 Tangent circles.

(*a*)   (*b*)

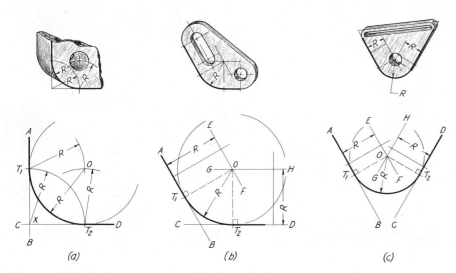

(*a*)   (*b*)   (*c*)

Fig. 3.23 To draw a circular arc tangent to two lines.

be the center of the required arc. Mark the points of tangency T_1 and T_2. T_1 lies along a perpendicular to AB through the center P, and T_2 along a line joining the centers of the two arcs.

This construction is also useful for drawing fillets and rounds on views of machine parts.

3.26

To Draw a Circular Arc of a Given Radius R_1 Tangent to Two Given Circular Arcs (Fig. 3.25). Given the circular arcs AB and CD with centers O and P, and radii R_2 and R_3, respectively.

(a) Using O as a center and R_2 plus R_1 as a radius, strike an arc parallel to AB. Using P as a center and R_3 plus R_1 as a radius, strike an intersecting arc parallel to CD. Since each of these intersecting arcs is the locus of centers for all circular arcs of radius R_1 tangent to the given arc to which it is parallel, their point of intersection S will be the center for the required arc that is tangent to both. Mark the points of tangency T_1 and T_2 that lie on the lines of centers PS and OS.

(b) Using O as a center and R_2 plus R_1 as a radius, strike an arc parallel to AB. Using P as a center and R_3 minus R_1 as a radius, strike an intersecting arc parallel to CD. The point of intersection of these arcs is the center for the required arc.

3.27

To Draw a Reverse (Ogee) Curve (Fig. 3.26).

(a) *Reverse (ogee) curve connecting two parallel lines.* Given the two parallel lines AB and CD. At points B and C, the termini and tangent points of the reverse curve, erect perpendiculars. Join B and C with a straight line and assume a point E as the point at which the curves will be tangent to each other. Draw the perpendicular bisectors of BE and EC. Since an arc tangent to AB at B must have its center on the perpendicular BP, the point of intersection P of the bisector and the perpendicular is the center for the required arc that is to be tangent to the line at B and the other required arc at point E. For the same reason, point Q is the center for the other required arc.

This construction is useful to engineers in laying out center lines for railroad tracks, pipelines, and so forth.

(b) *Reverse (ogee) curve connecting two nonparallel lines.* Given the two nonparallel lines AB and CD. At points B and C, the termini and tangent points,

erect perpendiculars. Along the perpendicular at B lay off the given (or selected) radius R and draw the arc having P as its center. Then draw a construction line through point P perpendicular to CD to establish the location of point X. With the position of X known, join points X and C with a straight line along which will lie the chords of the arcs forming the ogee curve between points X and C. The broken line XY (not a part of the construction) has

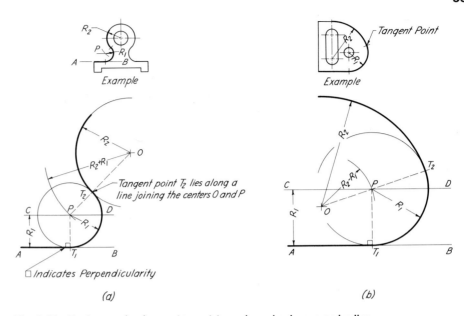

Fig. 3.24 To draw a circular arc tangent to a given circular arc and a line.

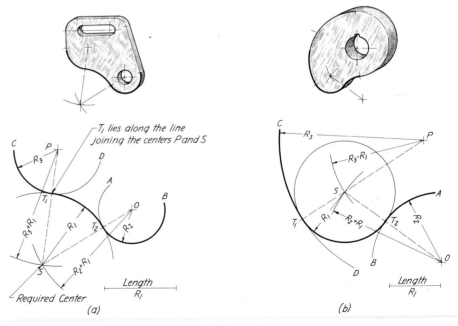

Fig. 3.25 To draw a circular arc tangent to two given arcs.

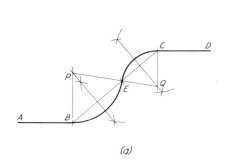

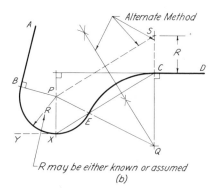

Fig. 3.26 To draw a reverse curve.

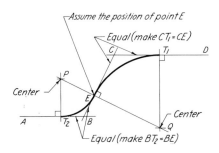

Fig. 3.27 To draw a reverse curve tangent to three lines.

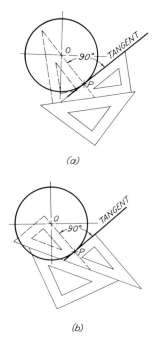

Fig. 3.28 To draw a line tangent to a circle at a point on the circumference.

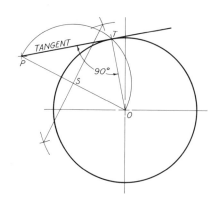

Fig. 3.29 To draw a line tangent to a circle through a given point outside.

been added to show that the procedure to be followed in completing the required curve will be as previously explained for drawing a reverse curve joining two parallel lines. In this case the parallel lines are XY and CD, instead of the lines AB and CD as in (a).

An alternative method for establishing the needed center for the required arc has been added to the illustration in (b). In this method the radius distance R is laid off upward along a perpendicular to CD through C. With point S established by this measurement, the line PS, as drawn, becomes the chord of an arc (not shown) that will have the same center as the required arc EC. The intersection of the perpendicular bisector of PS with the perpendicular erected downward from C will establish the position of point Q, the center of concentric arcs having chords PS and EC.

3.28
To Draw a Reverse Curve Tangent to Three Given Lines (Fig. 3.27). Given the lines AB and CD that are intersected by a third line BC at points B and C. Assume the position of point E (point of tangency) along BC and locate the termini points T_1

and T_2 by making CT_1 equal to CE and BT_2 equal to BE. The intersections of the perpendiculars erected at points T_1, E, and T_2 establish the centers P and Q of the arcs that form the reversed curve.

3.29
To Draw a Line Tangent to a Circle at a Given Point on the Circumference (Fig. 3.28). Given a circle with center O and point P on its circumference. Place a triangle supported by a T-square or another triangle in such a position that one side passes through the center O and point P. When using the method illustrated in (a), align the hypotenuse of one triangle on the center of the circle and the point of tangency; then, with the guiding triangle held in position, revolve the triagnle about the 90° angle and slide into position for drawing the required tangent line.

Another procedure is shown in (b). To draw the tangent by this method, align one leg of a triangle, which is adjacent to the 90° angle, through the center of the circle and the point of tangency; then slide it along the edge of a guiding triangle into position.

This construction satisfies the geometric requirement that a tangent must be perpendicular to a radial line drawn to the point of tangency.

3.30
To Draw a Line Tangent to a Circle Through a Given Point Outside the Circle (Fig. 3.29). Given a circle with center O and an external point P. Join the point P and the center O with a straight line, and bisect it to locate point S. Using S as a center and SO (one-half PO) as a radius, strike an arc intersecting the circle at point T (point of tangency). Line PT is the required tangent.

3.31
To Draw a Tangent Through a Point P on a Circular Arc Having an Inaccessible Center (Fig. 3.30). Draw the chord PB; then erect a perpendicular bisector. With point P as a center swing an arc through point C where the perpendicular bisector cuts the given arc. With C as a center and a radius equal to the chord distance CE, draw an arc to establish the location of point F. A line drawn through points P and F is the required tangent.

3.32
To Draw a Line Tangent to a Circle Through a Given Point Outside the Circle (Fig. 3.31). Place a triangle supported by

a T-square or another triangle in such a
position that one leg passes through point
P tangent to the circle, and draw the tangent. Slide the triangle along the guiding
edge until the other leg coincides with the
center O, and mark the point of tangency.
Although this method is not as accurate
as the geometric one explained in Sec.
3.30, it is frequently employed by commercial draftsmen.

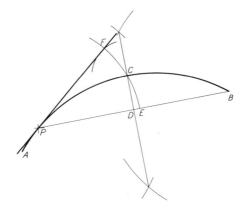

Fig. 3.30 To draw a tangent to a circular arc
having an inaccessible center.

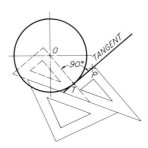

Fig. 3.31 To draw a line tangent to a circle
through a given point outside.

3.33
**To Draw a Line Tangent to Two Given Cir-
cles (Fig. 3.32).** Given two circles with
centers O and P and radii R_1 and R.

(a) *Open belt*. Using P as a center and
a radius equal to R minus R_1, draw an arc.
Through O draw a tangent to this arc
using the method explained in Sec. 3.30.
With the location of tangent point T established, draw line PT and extend it to
locate T_1. Draw OT_2 parallel to PT_1. The
line from T_2 to T_1 is the required tangent
to the given circles.

(b) *Crossed belt*. Using P as a center
and a radius equal to R plus R_1, draw an
arc. With the location of tangent point T
determined through use of the method
shown in Fig. 3.29, locate tangent point T_1
on line TP and draw OT_2 parallel to PT.
The line $T_1 T_2$, drawn parallel to OT, is the
required tangent.

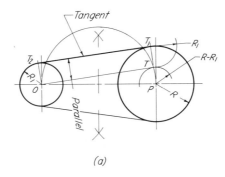

(a)   (b)

Fig. 3.32 To draw a line tangent to two given circles.

3.34
**To Approximate a Curve with Tangent Cir-
cular Arcs (Fig. 3.33).** Draftsmen often
find it desirable to approximate a noncircular curve with a series of tangent
arcs. If the curve consists of a number of
points, a pleasing curve should be
sketched lightly through these points before starting to draw the arcs. The centers
and radii are selected by trial, but it must
be remembered after the first arc has
been drawn (as far as it coincides with the
sketched curve) that when arcs are tangent, the centers are on a common normal through their point of tangency.
Sometimes draftsmen use this method to
draw curves in ink instead of using a
French curve.

3.35
**To Lay Off the Approximate Length of the
Circumference of a Circle (Fig. 3.34).**
Draw a line through point A tangent to the
circle and lay off along it a distance AB
equal to three times the diameter ($3D$).
Using point E on the circumference as a
center and a radius equal to the radius of
the circle, strike an arc to establish the

location of point C. Draw CD perpendicular to the vertical center line through point
A. DB is the rectified length of the circumference; however, it is slightly longer
than the true circumference by a negligible amount (approximate error $1/21{,}800$).

3.36
**To Lay Off the Approximate Length of a
Circular Arc on Its Tangent (Fig. 3.35).**
Given the arc AB.

(a) Draw the tangent through A, and
extend the chord BA. Locate point C by
laying off AC equal to one-half the length
of the chord AB. With C as a center and a

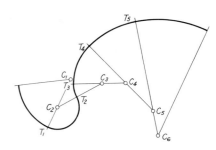

Fig. 3.33 To approximate a curve with
tangent circular arcs.

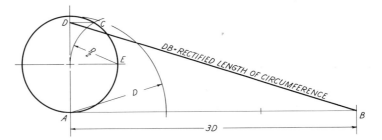

Fig. 3.34 To lay off the approximate length of the circumference of a
circle.

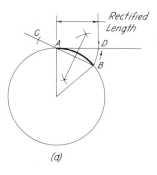

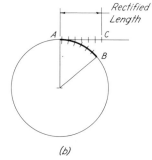

(a)

(b)

Fig. 3.35 To lay off the approximate length of a circular arc on its tangent.

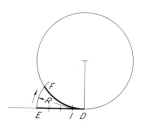

Fig. 3.36 To lay off a specified length along an arc.

radius equal to *CB*, strike an arc intersecting the tangent at *D*. The length *AD* along the tangent is slightly shorter than the true length of the arc *AB* by an amount that may be disregarded, for when the angle is less than 60°, the length of *AD* differs from the true length of the arc *AB* by less than 6 ft in 1 mile; when 30°, the error is 4½ in. in 1 mile.

(*b*) Draw the tangent through *A*. Using the dividers, start at *B* and step off equal chord distances around the arc until the point nearest *A* is reached. From this point (without raising the dividers) step off along the tangent an equal number of distances to locate point *C*. If the point nearest *A* is indented into the tangent instead of the arc, the almost negligible error in the length of *AC* will be still less.

Since the small distances stepped off are in reality the chords of small arcs, the length *AC* will be slightly less than the true length of the arc. For most practical purposes the difference may be disregarded.

When the central angle (θ) and the radius of an arc are known, the length of the arc may be computed by the formula $L = 2\pi R(\theta/360°) = 0.01745R\theta$.

3.37
To Lay Off a Specified Length Along a Given Circle Arc (Fig. 3.36). On the tangent to the arc, lay off the distance *DE* representing the specified length of arc. Divide *DE* into four equal parts. Then, using point 1 as a center and with the length 1–*E* as the radius *R*, strike an arc intersecting the given arc at *F*. The arc *DF* is approximately equal in length to the line *DE*. For large angles, it is advisable to make the construction for one-half of *DE*.

3.38
Conic Sections (Fig. 3.37). When a right circular cone of revolution is cut by planes at different angles, four curves of intersection and obtained that are called *conic sections*.

When the intersecting plane is perpendicular to the axis, the resulting curve of intersection is a *circle*.

If the plane makes a greater angle with the axis than do the elements, the intersection is an *ellipse*.

If the plane makes the same angle with the axis as the elements, the resulting curve is a *parabola*.

Finally, if the plane makes a smaller angle with the axis than do the elements or is parallel to the axis, the curve of intersection is a *hyperbola*.

The geometric methods for constructing the ellipse, parabola, and hyperbola are discussed in succeeding sections.

3.39
Ellipse (Fig. 3.37). Mathematically the ellipse is a curve generated by a point moving so that at any position the sum of its distances from two fixed points (foci) is a constant (equal to the major diameter). It is encountered very frequently in orthographic drawing when holes and circular forms are viewed obliquely. Ordinarily, the major and minor diameters are known.

3.40
To Construct an Ellipse, Foci Method (Fig. 3.38). Draw the major and minor axes (*AB*, *CD*) and locate the foci F_1 and F_2 by striking arcs, centered at *C* and having a radius equal to *OA* (one-half of the major diameter). The construction is as follows: Determine the number of points needed along the circumference of each quadrant of the ellipse for a relatively accurate layout (say, four) and mark off this number of division points (*P*, *Q*, *R*, and *S*) between *O* and F_1 on the major axis. In many cases it may be desirable to use additional points spaced closer together nearer F_1 in order to form accurately the sharp curvature at the end of the ellipse. Next, with F_1

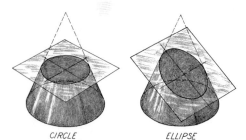

CIRCLE   ELLIPSE   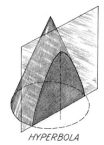 PARABOLA   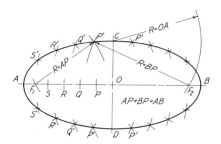 HYPERBOLA

Fig. 3.37 Conic sections.

Fig. 3.38 To construct an ellipse, foci (definition) method.

and F_2 as centers and the distances AP and BP as radii, respectively, strike intersecting arcs to locate P' on the circumference of the ellipse. Distances AQ and BQ are radii for locating points Q'. Locate points R' and S' in a similar manner and complete the ellipse using a French curve.

This method is sometimes known as the definition method, since it is based on the mathematical definition of the ellipse as given in Sec. 3.39.

3.41
To Construct an Ellipse, Trammel Method (Fig. 3.39). Given the major axis AB and the minor axis CD. Along the straight edge of a strip of paper or cardboard, locate the points O, C, and A so that the distance OA is equal to one-half the length of the major axis, and the distance OC is equal to one-half the length of the minor axis. Place the marked edge across the axes so that point A is on the minor axis and point C is on the major axis. *Point O will fall on the circumference of the ellipse.* Move the strip, keeping A on the minor axis and C on the major axis, and mark at least five other positions of O on the ellipse in each quadrant. Using a French curve, complete the ellipse by drawing a smooth curve through the points. The ellipsograph, which draws ellipses mechanically, is based on this same principle. The trammel method is an accurate method.

An alternative method for marking off the location of points A, O, and C is given in Fig. 3.39.

3.42
To Construct an Ellipse, Concentric Circle Method (Fig. 3.40). Given the major axis AB and the minor axis CD. Using the center of the ellipse (point O) as a center, describe circles having the major and minor axes as diameters. Divide the circles into equal central angles and draw diametrical lines such as P_1P_2. From point P_1 on the circumference of the larger circle, draw a line parallel to CD, the minor axis, and from point P_1' at which the diameter P_1P_2 intersects the inner circle, draw a line parallel to AB, the major axis. The point of intersection of these lines, point E, is on the required ellipse. At points P_2 and P_2' repeat the same procedure and locate point F. Thus, two points are established by the line P_1P_2. Locate at least five points in each of the four quadrants. The ellipse is completed by drawing a smooth curve through the points.

This is one of the most accurate methods used to form ellipses.

3.43
To Construct an Ellipse, Four-Center Method (Fig. 3.41). Given the major axis AB and the minor axis CD. Draw the line AC. Using the center of the ellipse O as a center and OC as a radius, strike an arc intersecting OA at point E. Using C as a center and EA as a radius, strike an arc intersecting the line AC at F. Draw the perpendicular bisector of the line AF. The points G and H, at which the perpendicular bisector intersects the axes AB and CD (extended), are the centers of two of the arcs forming the ellipse. Locate the other two centers, J and K, by laying off OJ equal to OH and OK equal to OG. To determine the junction points (tangent points) T, T_1, T_2, and T_3, for the arcs, draw lines through the centers of the tangent arcs. The figure thus formed by the four circle arcs approximates a true ellipse. When an accurate ellipse is required, this method should not be used.

3.44
To Construct an Ellipse, Parallelogram Method (Fig. 3.42). Given the major axis AB and the minor axis CD. Construct the circumscribing parallelogram. Divide AO and AE into the same number of equal parts (say, four) and number the division points from A. From C draw a line through point 3 on line AE, and from D draw a line through point 3 on line AO. The point of intersection of these lines is on the required ellipse. Similarly, the intersections of lines from C and D through points numbered 1 and 2 are on the ellipse. A similar construction will locate points in the other three quadrants of the ellipse. Use of a French curve will permit a

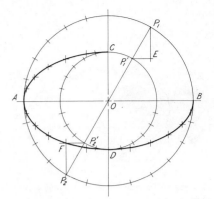

Fig. 3.40 To construct an ellipse, concentric circle method.

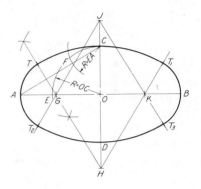

Fig. 3.41 To construct an ellipse, center method.

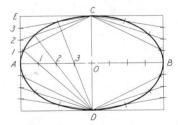

Fig. 3.42 To construct an ellipse, parallelogram method.

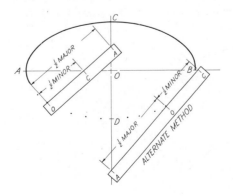

Fig. 3.39 To construct an ellipse, trammel method.

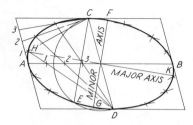

Fig. 3.43 To draw the major and minor axes of an ellipse, given the conjugate diameters.

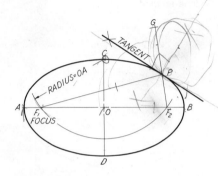

Fig. 3.44 To draw a tangent to an ellipse.

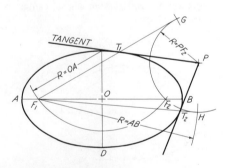

Fig. 3.45 To draw a tangent to an ellipse through a point outside the ellipse.

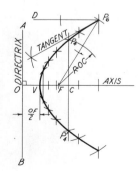

Fig. 3.46 To construct a parabola.

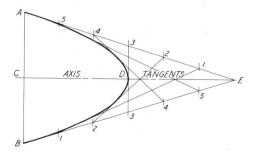

Fig. 3.47 To construct a parabola, tangent method.

smooth curve to be drawn through the points.

Had the circumscribing parallelogram not been a rectangle as in Fig. 3.42, the completed construction would appear as in Fig. 3.43, and AB and CD would be conjugate axes. To establish the major and minor axes, draw a semicircle on CD as a diameter, intersecting the ellipse at E. FG, running parallel to CE through the center of the ellipse, will be the required minor axis. HK, running through the center of the ellipse parallel to DE and perpendicular to FG, will be the major axis.

3.45

To Draw a Tangent to an Ellipse at Any Given Point (Fig. 3.44). Given any point, such as P, on the perimeter of the ellipse $ABCD$. Using C as a center and a radius equal to OA (one-half the major diameter), strike arcs across the major axis at F_1 and F_2. From these points, which are foci of the ellipse, draw F_1P and F_2G. The bisector of the angle GPF_1 is the required tangent to the ellipse.

3.46

To Draw a Tangent to an Ellipse From a Given Point P Outside the Ellipse (Fig. 3.45). With the end of the minor axis as a center and a radius R equal to one-half the length of the major axis, strike an arc to find the foci F_1 and F_2. With point P as a center and the distance PF_2 as a radius, draw an arc. Using F_1 as a center and the length AB as a radius, strike arcs cutting the arc with center P at points G and H. Draw lines GF_1 and HF_1 to establish the location of the tangent points T_1 and T_2. Draw the required tangent.

3.47

Parabola (Fig. 3.37). Mathematically the parabola is a curve generated by a point moving so that at any position its distance from a fixed point (the focus) is always

exactly equal to its distance to a fixed line (the directrix). The construction shown in Fig. 3.46 is based on this definition.

In engineering design, the parabola is used for parabolic sound and light reflectors, for vertical curves on highways, and for bridge arches.

3.48

To Construct a Parabola (Fig. 3.46). Given the focus F and the directrix AB. Draw the axis of the parabola perpendicular to the directrix. Through any point on the axis, for example, point C, draw a line parallel to the directrix AB. Using F as a center and the distance OC as a radius, strike arcs intersecting the line at points P_4 and P_4'. Repeat this procedure until a sufficient number of additional points have been located to determine a smooth curve. The vertex V is located at a point midway between O and F.

To construct a tangent to a parabola, say, at point P_6, draw the line P_6D parallel to the axis; then bisect the angle DP_6F. The bisector of the angle is the required tangent.

3.49

To Construct a Parabola, Tangent Method (Fig. 3.47). Given the points A and B and the distance CD from AB to the vertex. Extend the axis CD, and set off DE equal to CD. EA and EB are tangents to the parabola at A and B, respectively.

Divide EA and EB into the same number of equal parts (say, six) and number the division points as shown. Connect the corresponding points 1 and 1, 2 and 2, 3 and 3, and so forth. These lines, as tangents of the required parabola, form its envelope. Draw the tangent curve.

3.50

To Construct a Parabola, Offset Method (Fig. 3.48). Given the enclosing rectangle $A'ABB'$. Divide VA' into any number of equal parts (say, ten) and draw from the division points the perpendiculars parallel to VC, along which offset distances are to be laid off. The offsets vary as the square of their distances from V. For example, since V to 2 is two-tenths of the distance from V to A', 2–$2'$ will be $(.2)^2$ or $.04$ of $A'A$. Similarly, 6–$6'$ will be $(.6)^2$ or $.36$ of $A'A$; and 8–$8'$ will be $.64$ of $A'A$. To complete the parabola, lay off the computed offset values along the perpendiculars and form the figure with a French curve.

The entire construction can be done graphically (as illustrated) by first calculating the values for the squared distances and then dividing the depth distance (along the axis) proportionally using these values. The graphical method shown in Fig. 3.7 was used.

The offset method is preferred by civil engineers for laying out parabolic arches and computing vertical curves for highways. The parabola shown in Fig. 3.48 could represent a parabolic reflector.

3.51

To Construct a Curve of Parabolic Form Through Two Given Points (Fig. 3.49). Given the points A and B. Assume a point C. Draw the tangents CA and CB, and construct the parabolic curve using the tangent method shown in Fig. 3.47. This method is frequently used in machine design to draw curves that are more pleasing than circular arcs.

3.52

To Construct a Parabola, Parallelogram Method. Since the dimensions for a parallelogram that will enclose a given parabola are generally known, a parabola may be constructed by the parallelogram method illustrated in Fig. 3.50. With the enclosing rectangle drawn to the given width and depth dimensions, divide VA and AB into the same number of equal divisions (say, five) and number the division points as shown in (a). Draw light construction lines from point V to each of the division points along AB. Then draw lines parallel to the axis from points 1, 2, 3, and 4 on VA. The intersection of the construction lines from points numbered 1 is on the parabola.

Likewise, the intersection of the lines from points numbered 2 is on the parabola. The complete parabolic outline passes through the additional points at the intersection of the lines from points numbered 3 and the lines from points numbered 4. The method a explained for a parabola enclosed in a rectangle may be applied to a nonrectangular shape, as shown in (b).

3.53

To Locate the Directrix and Focus of a Given Parabolic Curve (Fig. 3.51). With the location of the axis known [see (a)] draw the tangent PY by locating point Y on the axis extended at a distance D_1 from the vertex V. Draw PX parallel to the axis and construct angle FPY equal to YPX. The point at which the line forming the newly constructed angle cuts the given axis is the focus F. The directrix through O

is perpendicular to the axis at the same distance from the vertex V as the focus F.

When the position of the axis is not known, the procedure illustrated in (b) will establish the location of the axis, focus, and directrix. As the initial step, draw two parallel chords at random and locate the midpoint of each, points R and S. Draw line RS to establish the direction of the axis that will be parallel to RS. Next, draw a chord perpendicular to RS at any location and through the midpoint M draw the axis of the parabola as required. The position of the vertex is now known and one might follow the procedure given in (a) to locate the focus and directrix. However, the position of the directrix may be found easily and quickly merely by drawing a tangent to the parabola at 45° with the axis. The directrix passes through O at the intersection of the tangent and the axis extended.

3.54

Hyperbola (Fig. 3.37). Mathematically the hyperbola can be described as a curve generated by a point moving so that at any position the difference of its distances

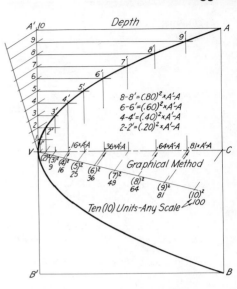

Fig. 3.48 To construct a parabola, offset method.

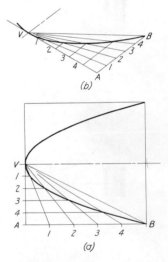

Fig. 3.50 To construct a parabola, parallelogram method.

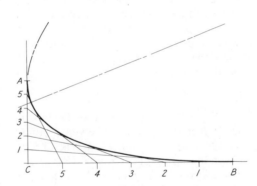

Fig. 3.49 To construct a curve of parabolic form.

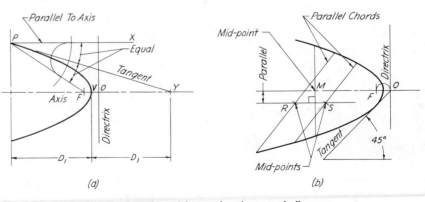

Fig. 3.51 To locate the directrix and focus of a given parabolic curve.

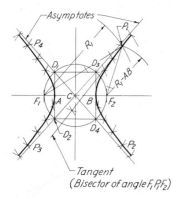

Fig. 3.52 To construct a hyperbola.

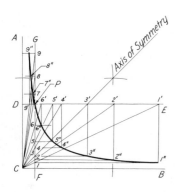

Fig. 3.53 To construct an equilateral hyperbola.

from two fixed points (foci) is a constant (equal to the transverse axis of the hyperbola). This definition is the basis for the construction shown in Fig. 3.52.

3.55

To Construct a Hyperbola (Fig. 3.52). Given the foci F_1 and F_2, and the transverse axis AB. Using F_1 and F_2 as centers, and any radius R_1 greater than F_1B, strike arcs. With these same centers and a radius equal to R_1–AB, strike arcs intersecting the first arcs. These intersecting arcs establish the positions of four symmetrically located points (P_1, P_2, P_3, and P_4) using a single pair of radii. Additional sets of four points are obtained by assuming a different initial radius each time. Repeat this procedure, as outlined, until a sufficient number of points have been located to determine a smooth curve.

The tangent to the hyperbola at any point, such as P_1, is the bisector of the angle between the focal radii F_1P_1 and F_2P_1.

As hyperbolic curves extend toward infinity they gradually approach two straight lines known as *asymptotes*. These may be located by drawing a circle having the distance F_1–F_2 as a diameter and erecting perpendiculars to the transverse axis through points A and B. The points at which these perpendicular lines intersect the circle are points (D_1, D_2, D_3, and D_4) on the asymptotes.

3.56

To Construct an Equilateral, or Rectangular, Hyperbola Through a Given Point P —Asymptotic to Given Axes (Fig. 3.53). Given the asymptotes CA and CB and any point P on the curve. Bisect the angle ACB to establish the position of the axis of symmetry of the hyperbola. Then, through point P, draw the lines PD and PF perpendicular to CA and CB, respectively. Along PF extended, mark off points 1, 2,

3, 4, and so forth, as needed. Through these points draw rays converging at C and intersecting DE (DP extended) at points 1', 2', 3', 4', 5', and so forth. From each of the numbered points along DE draw perpendiculars to DP and through the numbered points along FG draw perpendiculars to PF. The intersections of corresponding lines (perpendiculars) establish the location of the points 1'', 2'', 3'', 4'', and so forth, along the required curve.

3.57

Involute. The spiral curve traced by a point on a cord as it unwinds from around a circle or a polygon is an *involute curve*. Figure 3.54 (*a*) shows an involute of a circle, while (*b*) shows that of a square. The involute of a polygon is obtained by extending the sides and drawing arcs using the corners, in order, as centers. The circle in (*a*) may be considered to be a polygon having an infinite number of sides.

3.58

To Draw an Involute of a Circle [Fig. 3.54(*a*)]. Divide the circumference into a number of equal parts. Draw tangents through the division points. Then, along each tangent, lay off the rectified length of the corresponding circular arc, from the starting point to the point of tangency. The involute curve is a smooth curve through these points. The involute of a circle is used in the development of tooth profiles in gearing.

3.59

To Draw the Involute of a Polygon [Fig. 3.54(*b*)]. Extend the sides of the polygon as shown in (*b*). With the corners as centers, in order around the polygon, draw arcs terminating on the extended sides. The first radius is equal to the length of one side of the polygon. The radius of each successive arc is the distance from the center to the terminating point of the previous arc.

3.60

Cycloid. A cycloid is the curve generated by a point on the circumference of a moving circle when the circle rolls in a plane along a straight line, as shown in Fig. 3.55.

3.61

To Draw a Cycloid (Fig. 3.55). Draw the generating circle and the line AB tangent

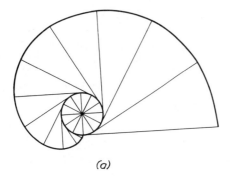

(a)   *(b)*

Fig. 3.54 Involute.

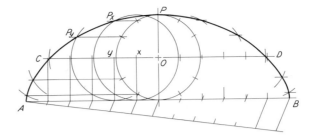

Fig. 3.55 Cycloid.

to it. The length AB should be made equal to the circumference of the circle. Divide the circle and the line AB into the same number of equal parts. With this much of the construction completed, the next step is to draw the line of centers CD through point O and project the division points along AB to CD by drawing perpendiculars. Using these points as centers for the various positions of the moving circle, draw circle arcs. For the purpose of illustration, assume the circle is moving to the left. When the circle has moved along CD to x, point P will have moved to point P_x. Similarly, when the center is at y, P will be at P_y. To locate positions of P along the cycloidal curve, project the division points of the divided circle in their proper order, across to the position circles. A smooth curve through these points will be the required cycloid.

3.62

Epicycloid (Fig. 3.56). An epicycloid is the curve generated by a point on the circumference of a circle that rolls in a plane on the outside of another circle. The method used in drawing an epicycloid is similar to the one used in drawing the cycloid.

3.63

Hypocycloid (Fig. 3.57). A hypocycloid is the curve generated by a point on the circumference of a circle that rolls in a plane on the inside of another circle. The method used to draw a hypocycloid is similar to the method used to draw the cycloid.

Additional information on the use of cycloidal curves to form the outlines of cycloidal gear teeth may be found in Chapter 21.

3.64

Spiral of Archimedes. Archimedes' spiral is a plane curve generated by a point moving uniformly around and away from a fixed point. In order to define this curve more specifically, it can be said that it is generated by a point moving uniformly along a straight line while the line revolves with uniform angular velocity about a fixed point.

The definition of the spiral of Archimedes is applied in drawing this curve as illustrated in Fig. 3.58. To find a sufficient number of points to allow the use of an irregular curve for drawing the spiral it is the practice to divide the given circle into a number of equal parts (say, 12) and draw radial lines to the division points. Next,

divide a radial line into the same number of equal parts as the circle and number the division points on the circumference of the circle beginning with the radial line adjacent to the divided one. With the center of the circle as a center, draw concentric arcs that in each case will start at a numbered division point on the divided radial line and will end at an intersection with the radial line that is numbered correspondingly. The arc starting at point 1 gives a point on the curve at its intersection with radial line 1, the arc starting at 2 gives an intersection point on radial line 2, etc. The spiral is a smooth curve drawn through these intersection points.

3.65

Helix (Fig. 3.59). The cylindrical helix is a space curve that is generated by a point moving uniformly on the surface of a cylinder. The point must travel parallel to the axis with uniform linear velocity while at the same time it is moving with uniform angular velocity around the axis. The curve can be thought of as being generated by a point moving uniformly along a straight line while the line is revolving with uniform angular velocity around the axis of the given cylinder. Study the pictorial drawing, Fig. 3.59.

The first step in drawing a cylindrical helix is to lay out the two views of the cylinder. Next, the lead should be measured along a contour element and divided into a number of equal parts (say, 12). Divide the circular view of the cylinder into the same number of parts and number the division points.

The division lines of the lead represent the various positions of the moving point as it travels in a direction parallel to the axis of the cylinder along the moving line. The division points on the circular view are the related position of the moving line. For example, when the line has moved from the 0 to the 1 position, the point has traveled along the line a distance equal to one-twelfth of the lead; when the line is in

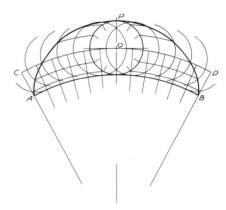

Fig. 3.56 Epicycloid.

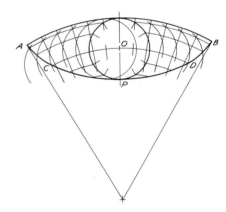

Fig. 3.57 Hypocycloid.

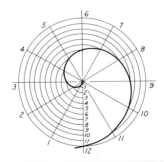

Fig. 3.58 Spiral of Archimedes.

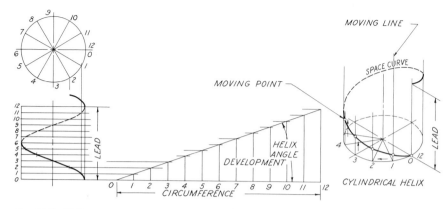

Fig. 3.59 Helix.

the 2 position, the point has traveled one-sixth of the lead. (See pictorial drawing, Fig. 3.59.) In constructing the curve the necessary points are found by projecting from a numbered point on the circular view to the division line of the lead that is numbered similarly.

A helix may be either right-hand or left-hand. The one shown in Fig. 3.59 is a left-hand helix.

When the cylinder is developed, the helix becomes a straight line on the development, as shown. It is inclined to the base line at an angle known as the "helix angle." A screw thread is an example of a practical application of the cylindrical helix.

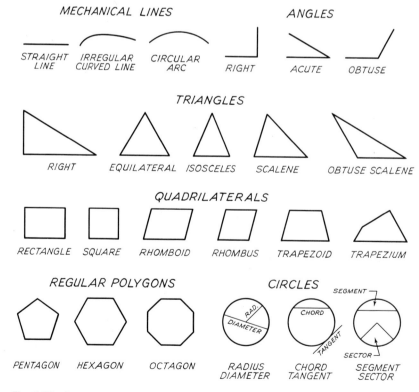

Fig. 3.60 Geometric shapes.

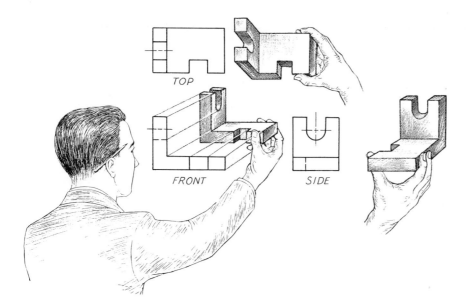

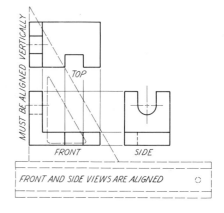

Fig. 5.2 **Obtaining three views of an object.**

Fig. 5.3 **Position of views.**

the top and side views must be arranged in their natural positions relative to the front view.

Figure 5.3 illustrates the natural relationship of views. Note that the top view is *vertically above* the front view, and the side view is *horizontally in line with* the front view. In both of these views *the front of the block is toward the front view*.

5.5
"Glass Box" Method. An imaginary glass box is used widely by instructors to explain the arrangement of orthographic views. An explanation of this scheme can best be made by reviewing the use of planes of projection (Chapter 4). It may be considered that planes of projection placed parallel to the six faces of an object form an enclosing glass box (Fig. 5.4). The observer views the enclosed object from the outside. The views are obtained by running projectors from points on the object to the planes. This procedure is in accordance with the theory of orthographic projection explained in Sec. 4.3, as well as the

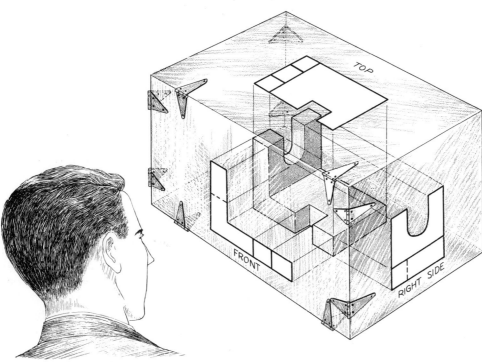

Fig. 5.4 **"Glass box."**

definition in Sec. 5.2. The top, front, and right side of the box represent the H (horizontal), F (frontal), and P (profile) projection planes.

Since the projections on the sides of the three-dimensional transparent box are to appear on a sheet of drawing paper, it

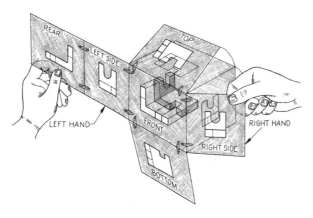

Fig. 5.5 Opening the glass box.

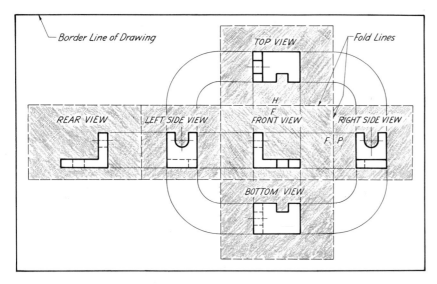

Fig. 5.6 Six views of an object on a sheet of drawing paper.

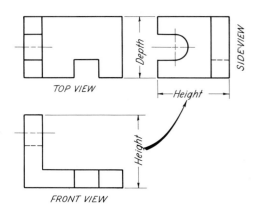

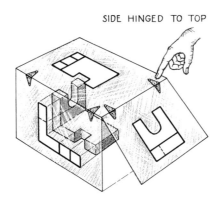

Fig. 5.7 "Second position" for the side view.

must be assumed that the box is hinged (Fig. 5.5) so that, when it is opened outward into the plane of the paper, the planes assume the positions illustrated in Figs. 5.5 and 5.6. Note that all of the planes, except the back one, are hinged to the frontal plane. In accordance with this universally recognized assumption, the top projection must take a position directly above the front projection, and the right-side projection must lie horizontally to the right of the front projection. To identify the separate projections, engineers call the one on the frontal plane the *front view* or *front elevation*, the one on the horizontal plane the *top view* or *plan*, and the one on the side or profile plane the *side view*, *side elevation*, or *end view*. Figure 5.6 shows the six views of the same object as they would appear on a sheet of drawing paper. Ordinarily, only three of these views are necessary (front, top, and right side). A bottom or rear view will be required in comparatively few cases.

5.6
"Second Position." Sometimes, especially in the case of a broad, flat object, it is desirable to hinge the sides of the box to the horizontal plane so that the side view will fall to the right of the top view, as illustrated in Fig. 5.7. This arrangement conserves space on the paper and gives the views better balance.

5.7
Principles of Multiview Drawing. The following principles should be studied carefully and understood thoroughly before any attempt is made to prepare an orthographic drawing:

1. The front and top views are *always* in line vertically (Fig. 5.3).

2. The front and side views are in line horizontally, except when the second position is used (Fig. 5.3).

3. The front of the object in the top view faces the front view (Fig. 5.5).

4. The front of the object in the side view faces the front view (Fig. 5.5).

5. The depth of the top view is the same as the depth of the side view (or views) (see Fig. 5.8).

6. The width of the top view is the same as the width of the front view (Fig. 5.8).

7. The height of the side view is the same as the height of the front view (Fig. 5.8).

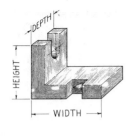

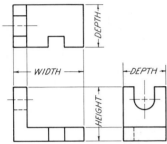

Fig. 5.8 View terminology.

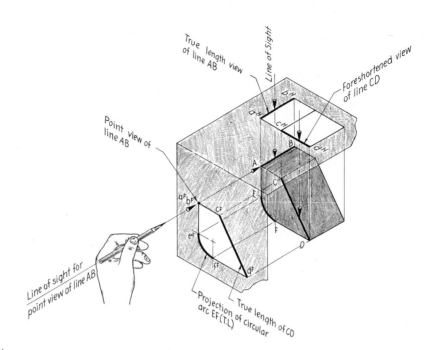

Fig. 5.9 Projected views of lines.

8. A view taken from above is a top view and *must* be drawn above the front view (Fig. 5.6).

9. A view taken from the right, in relation to the selected front, is a right-side view and *must* be drawn to the right of the front view (Fig. 5.6).

10. A view taken from the left is a left-side view and *must* be drawn to the left of the front view (Fig. 5.6).

11. A view taken from below is a bottom view and *must* appear below the front view (Fig. 5.6).

5.8
Projection of Lines. A line may project either in true length, foreshortened, or as a point in a view depending on its relationship to the projection plane on which the view is projected (see Fig. 5.9). In the top view, the line projection $a^H b^H$ shows the true length of the edge AB (see pictorial) because AB is parallel to the horizontal plane of projection. Looking directly at the frontal plane, along the line, AB projects as a point ($a^F b^F$). Lines, such as CD, that are inclined to one of the planes of projection, will show a foreshortened projection in the view on the projection plane to which the line is inclined and true length in the view on the plane of projection to which the line is parallel. The curved line projection $e^F f^F$ shows the true length of the curved edge.

The student should study Fig. 5.10 and attempt to visualize the space position of each of the given lines. It is very necessary both in preparing and reading graphical representations to recognize the position of a point, line, or plane and to know whether the projection of a line is true length or foreshortened, and whether the projection of a plane shows the true size and shape. The indicated reference lines may be thought of as representing the edges of the glass boxes shown. The projections of a line are identified as being on either a frontal, horizontal, or profile plane by the use of the letters F, H, or P with the lowercase letters that identify the end points of the line. For example, in Fig.

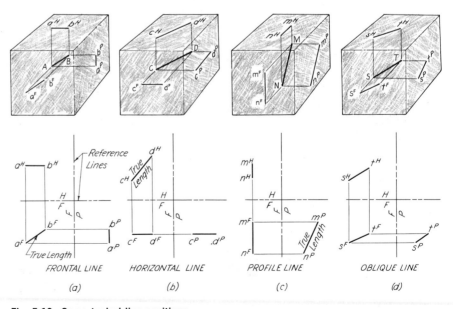

FRONTAL LINE *(a)*   HORIZONTAL LINE *(b)*   PROFILE LINE *(c)*   OBLIQUE LINE *(d)*

Fig. 5.10 Some typical line positions.

5.10(*a*), $a^H b^H$ is the horizontal projection of line AB, $a^F b^F$ is the frontal projection, and $a^P b^P$ is the profile projection.

It is suggested that the student hold a pencil before him and move it into the following typical line positions to observe the conditions under which the pencil representing a line, appears in true length.

1. *Vertical line.* The vertical line is perpendicular to the horizontal and will therefore appear as a point in the H (top) view. It will appear in true length in the F (frontal) view and in the P (profile) view.

2. *Horizontal line* [Fig. 5.10(*b*)]. The horizontal line will appear in true length when viewed from above because it is parallel to the H-plane of projection and its end points are theoretically equidistant from an observer looking downward.

3. *Inclined line* [Fig. 5.10(*c*)]. The inclined line is any line not vertical or horizontal that is parallel to either the frontal plane or the profile plane of projection. An inclined line will show true length in the F (frontal) view or P (profile) view.

4. *Oblique line* [Fig. 5.10(*d*)]. The oblique line will not appear in true length in any of the principal views because it is inclined to all of the principal planes of projection. It should be apparent in viewing the pencil alternately from the directions used to obtain the principal views, namely, from the front, above, and side, that one end of the pencil is always farther away from the observer than the other. Only when looking directly at the pencil from such

a position that the end points are equidistant from the observer can the true length be seen. On a drawing, the true length projection of an oblique line will appear in a supplementary A (auxiliary) view projected on a plane that is parallel to the line (Sec. 9.7).

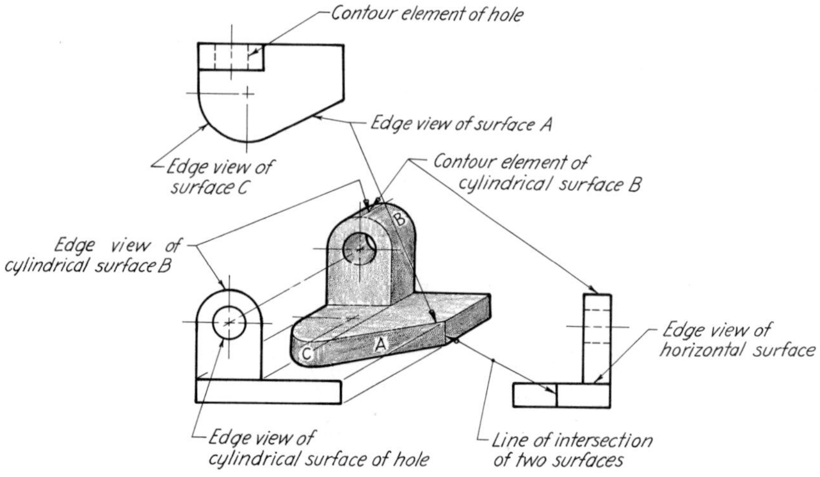

5.9
Meaning of Lines. On a multiview drawing a visible or invisible line may represent either the intersection of two surfaces, the edge view of a surface, or it may be the limiting element of a surface. These three different meanings of a line are illustrated in Fig. 5.11. In the top view, the curved line is an edge view of surface C, while a straight line is the edge view of surface A. The full circle in the front view may be considered as the edge view of the cylindrical surface of the hole. In the side view, the top line, representing the contour element of the cylindrical surface, indicates the limits for the surface and therefore can be thought of as being a surface limit line. The short vertical line in this same view represents the intersection of two surfaces. In reading a drawing, one can be sure of the meaning of a line on a view only after an analysis of the related view or views. All views must be studied carefully.

5.10
Projection of Surfaces. The components of most machine parts are bounded by either plane or single-curved surfaces. Plane surfaces bound cubes, prisms, and pyramids, while single-curved surfaces, ruled by a moving straight line, bound cylinders and cones. The projected representations (lines or areas) of both plane and single-curved surfaces are shown in Fig. 5.12. From this illustration the student should note that (1) when a surface is parallel to a plane of projection, it will appear in true size in the view on the plane of projection to which it is parallel; (2) when it is perpendicular to the plane of projection, it will project as a line in the view; and (3) when it is positioned at an angle, it will appear foreshortened. A surface will always project either as a line or an area on a view. The area representing the surface may be either a full-size or foreshortened representation.

In Fig. 5.12 the cylindrical surface A appears as a line in the side (profile) view and as an area in the top and front views. Surface B shows true size in the top view and as a line in both the front and side

Fig. 5.11 Meaning of lines.

views. Surface C, a vertical surface, will appear as a line when observed from above.

5.11
Analysis of Surfaces, Lines, and Points in Three Principal Views. An analysis of the representation of the surfaces of a mutilated block is given pictorially in Fig. 5.13. It can be noted that each of the surfaces A, B, and C appears in true size and shape in one view and as a line in each of the other two related views. Surface D, which is inclined, appears with foreshortened length in the top and side views and as an inclined line in the front view.

Three views of each of the visible points are shown on the multiview drawing. At the very beginning of an elementary course in drawing, a student will often find it helpful to number the corners of an object in all views.

5.12
Selection of Views. Careful study should be given to the outline of an object before the views are selected; otherwise there is no assurance that the object will be described completely from the reader's viewpoint (Fig. 5.14). Only those views that are necessary for a clear and complete description should be selected. Since the repetition of information only tends to confuse the reader, superfluous views should be avoided.

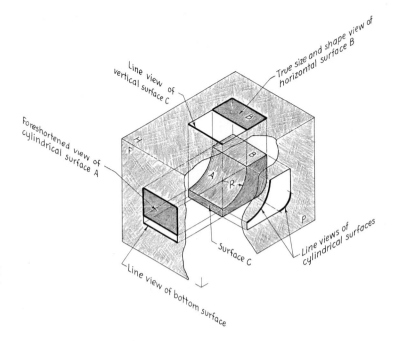

Fig. 5.12 Projected views of surfaces.

Although some objects, such as cylinders, bushings, bolts, and so forth, require only two views (front and side), more complicated pieces may require an auxiliary or sectional view in addition to the ordinary three views.

The space available for arranging the views often governs the choice between the use of a top or side view. The difference between the descriptive values of the

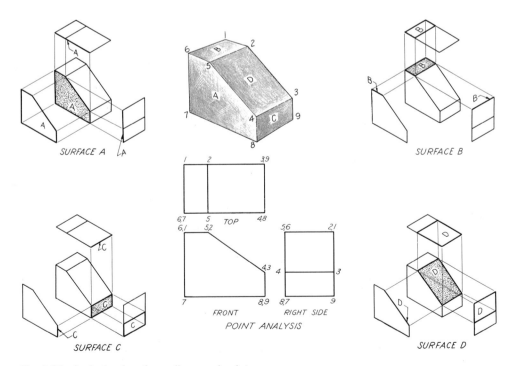

Fig. 5.13 Analysis of surfaces, lines, and points.

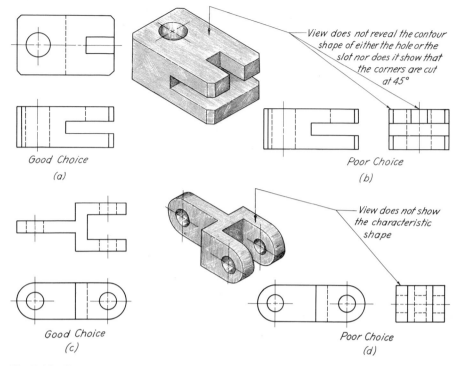

Good Choice
(a)

*View does not reveal the contour
shape of either the hole or the
slot nor does it show that
the corners are cut
at 45°*

Poor Choice
(b)

Good Choice
(c)

*View does not show
the characteristic
shape*

Poor Choice
(d)

Fig. 5.14 Choice of views.

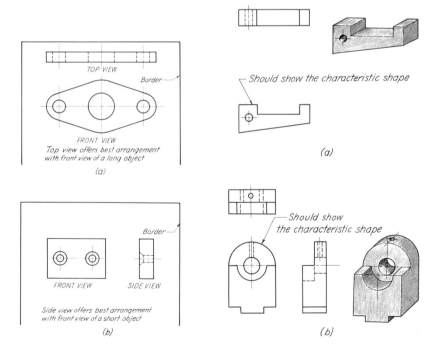

TOP VIEW

Border

FRONT VIEW

*Top view offers best arrangement
with front view of a long object*

(a)

Border

FRONT VIEW   SIDE VIEW

*Side view offers best arrangement
with front view of a short object*

(b)

Fig. 5.15 Choice of views.

Should show the characteristic shape

(a)

*Should show
the characteristic shape*

(b)

Fig. 5.16 Principal view of an object.

two frequently is not great. For example, a draftsman often finds that the views of a long object will have better balance if a top view is used [see Fig. 5.15(a)]; while in the case of a short object [see (b)], the use of a side view may make possible a more pleasing arrangement. It should be remembered that the choice of views for many objects is definitely fixed by the contour alone, and no choice is offered as far as spacing is concerned. It is more important to have a set of views that describes an object clearly than one that is artistically balanced.

Often there is a choice between two equally important views, such as between a right-side and left-side view or between a top and bottom view (Fig. 5.6). In such cases, one should adhere to the following rule: *A right-side view should be used in preference to a left-side view and a top view in preference to a bottom view.* When this rule is applied to irregular objects, the front (contour) view should be drawn so that the most irregular outline is toward the top and right side.

Another rule, one that must be considered in selecting the front view, is as follows: *Place the object to obtain the smallest number of hidden lines.*

5.13

Principal (Front) View. The principal view is the one that shows the characteristic contour of the object [see Fig. 5.16(a) and (b)]. Good practice dictates that this be used as the front view on a drawing. It should be clearly understood that the view of the natural front of an object is not always the principal view, because frequently it fails to show the object's characteristic shape. Therefore, another rule to be followed is: *Ordinarily, select the view showing the characteristic contour shape as the front view, regardless of the normal or natural front of the object.*

When an object does have a definite normal position, however, the front view should be in agreement with it. In the case of most machine parts, the front view can assume any convenient position that is consistent with good balance.

5.14

Invisible Lines. Dotted lines are used on an external view of an object to represent surfaces and intersections invisible at the point from which the view is taken. In Fig. 5.17(a), one invisible line represents a line of intersection or edge line, while the other invisible line may be considered to represent either the surface or lines of intersection. On the side view in (b) there are invisible lines, which represent the contour elements of the cylindrical holes.

5.15
Treatment of Invisible Lines. The short dashes that form an invisible line should be drawn carefully in accordance with the recommendations in Sec. 5.24. An invisible line always starts with a dash in contact with the object line from which it starts, unless it forms a continuation of a visible line. In the latter case, it should start with a space, in order to establish at a glance the exact location of the endpoint of the visible line (see Fig. 5.18C). Note that the effect of definite corners is secured at points A, B, E, and F, where, in each case, the end dash touches the intersecting line. When the point of intersection of an invisible line and another object line does not represent an actual intersection on the object, the intersection should be open as at points C and D. An open intersection tends to make the lines appear to be at different distances from the observer.

Parallel invisible lines should have the breaks staggered.

The correct and incorrect treatment for starting invisible arcs is illustrated at G and G'. Note that an arc should start with a dash at the point of tangency. This treatment enables the reader to determine the exact end points of the curvature.

5.16
Omission of Invisible Lines. Although it is common practice for commercial draftsmen to omit hidden lines when their use tends to confuse further an already overburdened view or when the shape description of a feature is sufficiently clear in another view, it is not advisable for a beginning student to do so. The beginner, until he has developed the discrimination that comes with experience, will be wise to show all hidden lines.

5.17
Precedence of Lines. When one discovers in making a multiview drawing that two lines coincide, the question arises as to which line should be shown or, in other words, which line must have precedence if the drawing is to be read intelligently. For example, as revealed in Fig. 5.19, a solid line may have the same position as an invisible line representing the contour element of a hole, or an invisible line may

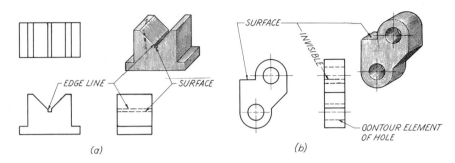

Fig. 5.17 Invisible lines.

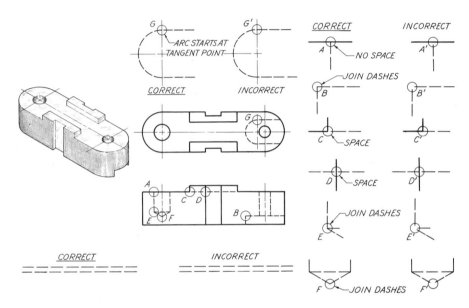

Fig. 5.18 Correct and incorrect junctures of invisible outlines.

occur at the same place as a center line for a hole. In these cases the decision rests on the relative importance of each of the two lines that can be shown. The precedence of lines is as follows:

Solid lines (visible object lines) take precedence over all other lines.

Dashed lines (invisible object lines) take precedence over center lines, although evidence of center lines may be indicated as shown in both the top and side views of Fig. 5.19.

A *cutting-plane line* takes precedence over a center line where it is necessary to indicate the position of a cutting plane.

5.18
Projection of Angles. When an angle lies in a plane that is parallel to one of the

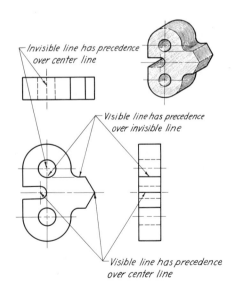

Fig. 5.19 Precedence of lines.

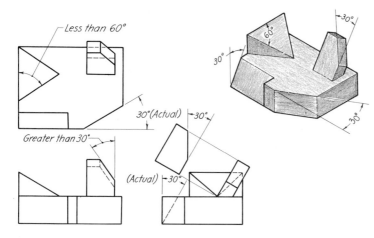

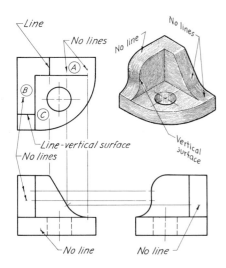

Fig. 5.20 Projection of angles.

Fig. 5.21 Treatment of tangent surfaces.

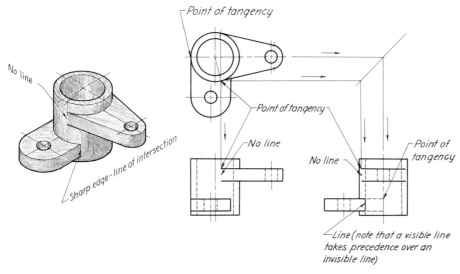

Fig. 5.22 Treatment of tangent surfaces.

planes of projection, the angle will show in true size in the view on that particular plane of projection to which the angle is parallel. In Fig. 5.20 those angles indicated as actual show in their true size. The 60° angle, which lies in a surface that is not parallel to the H-plane, appears at less than 60° in the top view. The 30° angle for the sloping line on the component portion that is inclined backward projects at greater than 30° in the front view. It may be said that, except for a 90° angle having one leg as a normal line, angles lying on inclined planes will project either larger or smaller than true size, depending on the position of the plane in which the angle lies. A 90° angle always projects in true size, even on an inclined plane, if the line forming one side of the angle is parallel to the plane of projection and a normal view of the line results. The normal view of a line is any view of a line that is obtained

with the direction of sight perpendicular to the line.

What has been stated concerning angles on inclined surfaces can be easily verified by the student if he will observe what happens to the angles of a 30° × 60° triangle resting on the long leg, as it is revolved from a vertical position downward onto the surface of his desk top.

5.19

Treatment of Tangent Surfaces. When a curved surface is tangent to a plane surface, as illustrated in several ways on the pictorial drawing in Fig. 5.21, no line should be shown as indicated at A and B in the top view and as noted for the front and side views. At C in the top view the line represents a small vertical surface that must be shown even though the upper and lower lines for this surface may be omitted in the front view, depending on the decision of the draftsman. In the top view a line has been drawn to represent the intersection of the inclined and horizontal surfaces at the rear, even though they meet in a small round instead of a sharp edge. The presence of this line emphasizes the fact that there are two surfaces meeting here that are at a definite angle, one to the other. Several typical examples of tangencies and intersections have been illustrated in Fig. 5.22.

5.20

Parallel Lines. When parallel surfaces are cut by a plane, the resulting lines of intersection will be parallel, as shown by the pictorial drawing in Fig. 5.23(b), where the near corner of the object has been removed by the oblique plane ABC. It can be observed from the multiview drawing in (c) that "when two lines are parallel in space, their projections will be parallel in all of the views," even though at times both lines may appear as points on one view.

In Fig. 5.23, three views are to be drawn that show the block after the near front corner has been removed [see (b)]. Several of the required lines of intersection can be readily established through the given points A, B, and C that define the oblique plane. For example, $c^F b^F$ can be drawn in the front view and the line through a^F can be drawn parallel to it. In the top view, $a^H b^H$ should be drawn first and the intersection line through c^H should then be drawn parallel to this H-view of AB. The drawing can now be com-

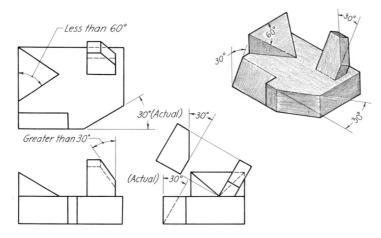

Fig. 5.20 Projection of angles.

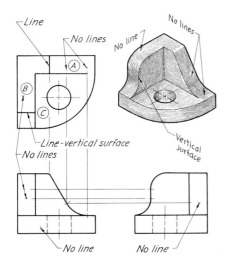

Fig. 5.21 Treatment of tangent surfaces.

planes of projection, the angle will show in true size in the view on that particular plane of projection to which the angle is parallel. In Fig. 5.20 those angles indicated as actual show in their true size. The 60° angle, which lies in a surface that is not parallel to the H-plane, appears at less than 60° in the top view. The 30° angle for the sloping line on the component portion that is inclined backward projects at greater than 30° in the front view. It may be said that, except for a 90° angle having one leg as a normal line, angles lying on inclined planes will project either larger or smaller than true size, depending on the position of the plane in which the angle lies. A 90° angle always projects in true size, even on an inclined plane, if the line forming one side of the angle is parallel to the plane of projection and a normal view of the line results. The normal view of a line is any view of a line that is obtained

with the direction of sight perpendicular to the line.

What has been stated concerning angles on inclined surfaces can be easily verified by the student if he will observe what happens to the angles of a 30° × 60° triangle resting on the long leg, as it is revolved from a vertical position downward onto the surface of his desk top.

5.19

Treatment of Tangent Surfaces. When a curved surface is tangent to a plane surface, as illustrated in several ways on the pictorial drawing in Fig. 5.21, no line should be shown as indicated at A and B in the top view and as noted for the front and side views. At C in the top view the line represents a small vertical surface that must be shown even though the upper and lower lines for this surface may be omitted in the front view, depending on the decision of the draftsman. In the top view a line has been drawn to represent the intersection of the inclined and horizontal surfaces at the rear, even though they meet in a small round instead of a sharp edge. The presence of this line emphasizes the fact that there are two surfaces meeting here that are at a definite angle, one to the other. Several typical examples of tangencies and intersections have been illustrated in Fig. 5.22.

5.20

Parallel Lines. When parallel surfaces are cut by a plane, the resulting lines of intersection will be parallel, as shown by the pictorial drawing in Fig. 5.23(b), where the near corner of the object has been removed by the oblique plane ABC. It can be observed from the multiview drawing in (c) that "when two lines are parallel in space, their projections will be parallel in all of the views," even though at times both lines may appear as points on one view.

In Fig. 5.23, three views are to be drawn that show the block after the near front corner has been removed [see (b)]. Several of the required lines of intersection can be readily established through the given points A, B, and C that define the oblique plane. For example, $c^F b^F$ can be drawn in the front view and the line through a^F can be drawn parallel to it. In the top view, $a^H b^H$ should be drawn first and the intersection line through c^H should then be drawn parallel to this H-view of AB. The drawing can now be com-

Fig. 5.22 Treatment of tangent surfaces.

Point of tangency

No line

Sharp edge-line of intersection

Point of tangency

No line

No line

Point of tangency

Line (note that a visible line takes precedence over an invisible line)

sent either the surface or lines of intersection. On the side view in (b) there are invisible lines, which represent the contour elements of the cylindrical holes.

5.15

Treatment of Invisible Lines. The short dashes that form an invisible line should be drawn carefully in accordance with the recommendations in Sec. 5.24. An invisible line always starts with a dash in contact with the object line from which it starts, unless it forms a continuation of a visible line. In the latter case, it should start with a space, in order to establish at a glance the exact location of the endpoint of the visible line (see Fig. 5.18C). Note that the effect of definite corners is secured at points A, B, E, and F, where, in each case, the end dash touches the intersecting line. When the point of intersection of an invisible line and another object line does not represent an actual intersection on the object, the intersection should be open as at points C and D. An open intersection tends to make the lines appear to be at different distances from the observer.

Parallel invisible lines should have the breaks staggered.

The correct and incorrect treatment for starting invisible arcs is illustrated at G and G'. Note that an arc should start with a dash at the point of tangency. This treatment enables the reader to determine the exact end points of the curvature.

5.16

Omission of Invisible Lines. Although it is common practice for commercial draftsmen to omit hidden lines when their use tends to confuse further an already overburdened view or when the shape description of a feature is sufficiently clear in another view, it is not advisable for a beginning student to do so. The beginner, until he has developed the discrimination that comes with experience, will be wise to show all hidden lines.

5.17

Precedence of Lines. When one discovers in making a multiview drawing that two lines coincide, the question arises as to which line should be shown or, in other words, which line must have precedence if the drawing is to be read intelligently. For example, as revealed in Fig. 5.19, a solid line may have the same position as an invisible line representing the contour element of a hole, or an invisible line may

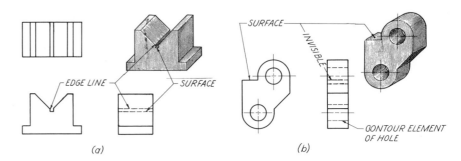

Fig. 5.17 Invisible lines.

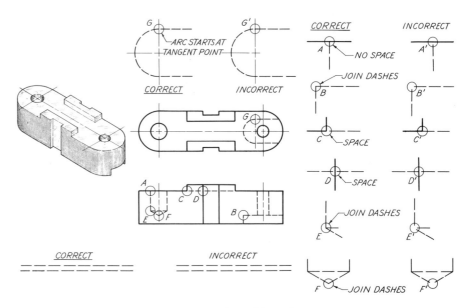

Fig. 5.18 Correct and incorrect junctures of invisible outlines.

occur at the same place as a center line for a hole. In these cases the decision rests on the relative importance of each of the two lines that can be shown. The precedence of lines is as follows:

> *Solid lines* (visible object lines) take precedence over all other lines.

> *Dashed lines* (invisible object lines) take precedence over center lines, although evidence of center lines may be indicated as shown in both the top and side views of Fig. 5.19.

> A *cutting-plane line* takes precedence over a center line where it is necessary to indicate the position of a cutting plane.

5.18

Projection of Angles. When an angle lies in a plane that is parallel to one of the

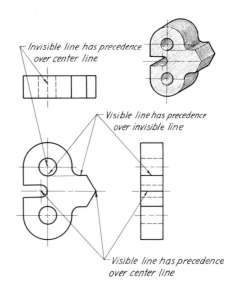

Fig. 5.19 Precedence of lines.

pleted by working back and forth from view to view while applying the rule that a plane intersects parallel planes along lines of intersection that are parallel. The remaining lines are thus drawn parallel to either AB or CB [see the pictorial in (b)].

5.21

Plotting an Elliptical Boundary. The actual intersection of a circular cylinder or cylindrical hole with a slanting surface (inclined plane) is an ellipse (Fig. 5.24). The elliptical boundary in (a) appears as an ellipse in the top view, as a line in the front view, and as a semicircle in the side view. The ellipse was plotted in the top view by projecting selected points (such as points A and B) from the circle arc in the side view, as shown. For example, point A was projected first to the inclined line in the front view and then to the top view. The mitre line shown was used to project the depth distance for A in the top view for illustrative purposes only. Ordinarily, dividers should be used to transfer measurements to secure great accuracy.

In (b) the intersection of the hole with the sloping surface is represented by an ellipse in the side view. Points selected around the circle in the top view (such as points C and D) projected to the side view as shown permit the draftsman to form the elliptical outline. It is recommended that a smooth curve be sketched freehand through the projected points before the French curve is applied to draw the finished ellipse, because it is easier to fit a curved ruling edge to a line than to scattered points.

5.22

Projecting a Curved Outline (Space Curve). When a boundary curve lies in an inclined plane, the projection of the curve may be found in another view by projecting points along the curve, as illustrated in Fig. 5.25. In the example, points selected along the arcs forming the curve in the top view were first located in the side view, using distances taken from the top view, as shown by the X and Y measurements. Then the front view positions of these points, through which the front view of the curve must pass, were established by projecting horizontally from the side view and downward from the top view.

5.23

Treatment of Intersecting Finished and Unfinished Surfaces. Figure 5.26 illustrates the removal of material when ma-

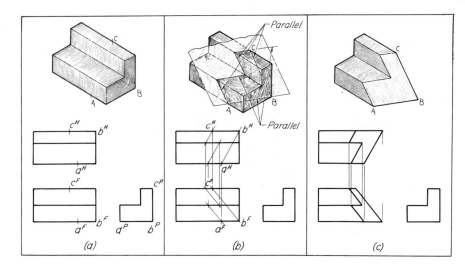

Fig. 5.23 Parallel lines.

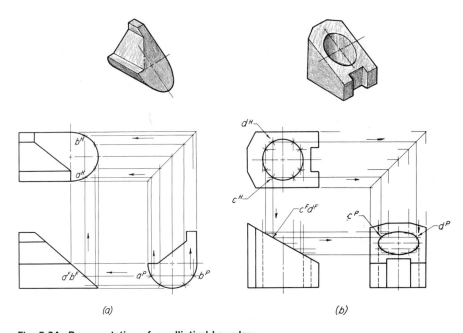

Fig. 5.24 Representation of an elliptical boundary.

chining surfaces, cutting a slot, and drilling a hole in a small part. The italic f on a surface of a pictorial drawing in this text indicates that the surface has been machined. The location of sharp and rounded corners, as illustrated in (b) and (c), are noted on the multiview drawing. A discussion covering rounded internal and external corners is given in Sec. 5.44. At this time, the f is not recommended for use on a detail drawing in any ANSI standard. See Sec. 13.7.

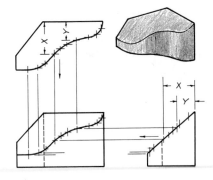

Fig. 5.25 Projecting a space curve.

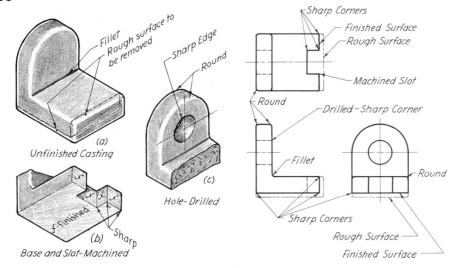

Fig. 5.26 Rough and finished surfaces on a casting.

5.24

To Make an Orthographic Drawing. The location of all views should be determined before a drawing is begun. This will ensure balance in the appearance of the finished drawing. The contour view is usually started first. After the initial start, the draftsman should construct his views simultaneously by projecting back and forth from one to the other. It is poor practice to complete one view before starting the others, as much more time will be required to complete the drawing. Figure 5.27 shows the procedure for laying out a three-view

drawing. The general outline of the views first should be drawn lightly with a hard pencil and then heavied with a medium-grade pencil. Although experienced persons sometimes deviate from this procedure by drawing in the lines of known length and location in finished weight while constructing the views, it is not recommended that beginners do so (see Fig. 5.27, step III).

Although a 45° mitre line is sometimes used for transferring depth dimensions from the top view to the side view, or vice versa, as shown in Fig. 5.28(b), it is better practice to use dividers, as in (a). Continuous lines need not be drawn between the views and the mitre line, as in the illustration, for one may project from short dashes across the mitre line. The location of the mitre line may be obtained by extending the construction lines representing the front edge of the top view and the front edge of the side view to an intersection.

When making an orthographic drawing in pencil, the beginner should endeavor to use the line weights recommended in Sec. 2.31. The object lines should be made very dark and bright, to give snap to the drawing as well as to create the contrast necessary to cause the shape of the object to stand out. Special care should be taken to gauge the dashes and spaces in invisible object lines. On ordinary drawings, .12-in. dashes and .03-in. spaces are recommended (Fig. 5.29).

Center lines consist of alternate long and short dashes. The long dashes are from .80 to 1.60 in. long, the short dashes .12 in., and the spaces .03 in. (Fig. 5.29). The following technique is recommended in drawing center lines:

1. Where center lines cross, the short dashes should intersect symmetrically (Fig. 5.29). (In the case of very small circles the breaks may be omitted.)

2. The breaks should be so located that they will stand out and allow the center line to be recognized as such.

3. Center lines should extend approximately .12 in. beyond the outline of the part whose symmetry they indicate (Fig. 5.29).

4. Center lines should not end at object lines.

5. Center lines that are aligned with object lines should have not less than a .06-in. space between the end of the center line and the object line.

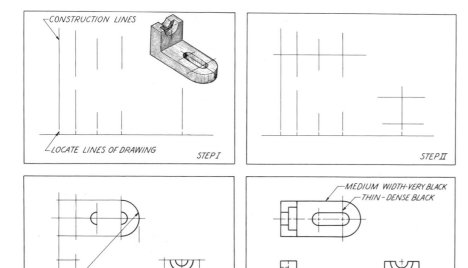

Fig. 5.27 Steps in making a three-view drawing of an object.

For a finished drawing to be pleasing in appearance, all lines of the same type must be uniform, and each type must have proper contrast with other symbolic types. The contrast between the types of pencil lines is similar to that of ink lines (Fig. 2.58), except that pencil lines are never as wide as ink lines (read Sec. 2.31). On commercial drawings, the usual practice is to "burn in" the object lines by applying heavy pressure.

If reasonable care is taken not to soil a drawing, it will not be necessary to clean any part of it with an eraser. Since the practice in most commercial drawing rooms is not to erase construction lines if they have been drawn lightly, the student, at the very beginning of his first course, should try to acquire habits that ensure cleanliness.

When constructing a two-view drawing of a circular object, the pencil work must start with the drawing of the center lines, as shown in Fig. 5.30. This is necessarily the first step, because the construction of the circular (contour) view is based on a horizontal and a vertical center line. The horizontal object lines of the rectangular view are projected from the circles.

5.25

Visualizing an Object from Given Views. Most students in elementary graphics courses find it difficult to visualize an object from two or more views. This trouble is largely due to the lack of systematic procedure for analyzing complex shapes.

The simplest method of determining shape is illustrated pictorially in Fig. 5.31. This method of "breaking down" may be applied to any object, since all objects may be thought of as consisting of elemental geometric forms, such as prisms, cylinders, cones, and so on. These imaginary component parts may be additions in the form of projections or subtractions in the form of cavities. Following such a detailed geometric analysis, a clear picture of an entire object can be obtained by mentally assembling a few easily visualized forms.

It should be realized, when analyzing component parts, that it is impossible ordinarily to determine whether a form is an addition or a subtraction by looking at one view. For example, the small circles in the top view in Fig. 5.31 indicate a cylindrical form, but they do not reveal whether the form is a hole or a projection. By consulting the front view, however, the

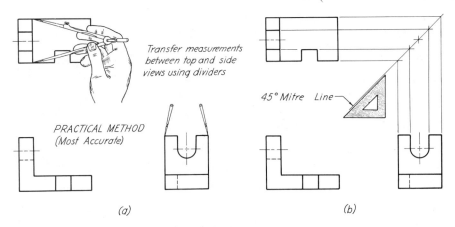

Fig. 5.28 Methods for transferring depth dimensions.

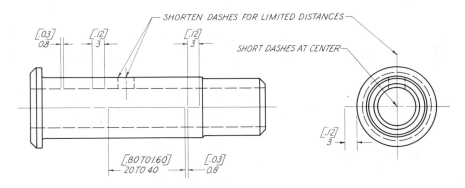

Fig. 5.29 Invisible lines and center lines.

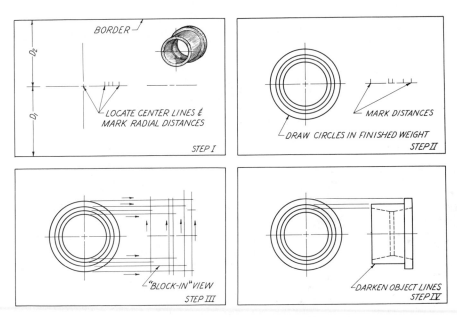

Fig. 5.30 Steps in making a two-view drawing of a circular object.

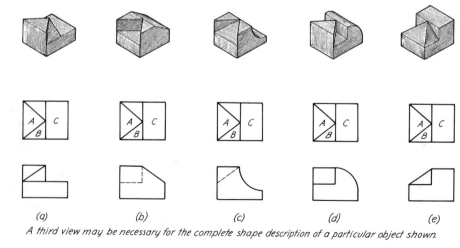

Fig. 5.31 "Breaking down" method.

form is shown to be a hole (subtracted cylinder).

The graphic language is similar to the written language in that neither can be read at a glance. A drawing must be read patiently by referring systematically back and forth from one view to another. At the same time the reader must imagine a three-dimensional object and not a two-dimensional flat projection.

A student usually will find that a pictorial sketch will clarify the shape of a part that is difficult to visualize. The method for preparing quick sketches in isometric is explained in Secs. 6.14 and 6.15.

5.26

Interpretation of Adjacent Areas of a View. To obtain a full understanding of the true geometric shape of a part, all the areas on a given view must be carefully analyzed, because each area represents a surface on the part. For example, in reading a drawing it must be determined whether a particular area in a top view

represents a surface that is inclined or horizontal and whether the surface is higher or lower than adjacent ones. Five distinctly different objects are shown in Fig. 5.32, all having the same top view. In determining the actual shape of these objects, memory and previous experience can be a help, but one can easily be misled if he does not approach the analysis with an open mind, for it is only by trial-and-error effort and by referring back and forth from view to view that a drawing can be read. In considering area A in (a) it might be thought that the triangular surface could be high and horizontal, which would be correct because of the arrangement of lines in the front view. However, in considering the top view alone, A could be either sloping, as in (c) and (e), or low and horizontal, as in (b). An analysis of the five parts reveals that the surface represented by area B can also be either sloping, high and horizontal, or low and horizontal. Area C offers even a wider variety of possibilities in that it may be either low and horizontal (a), sloping (b), cylindrical [(c) and (d)], or high and horizontal (e).

The student must realize at this point that since there are infinite possibilities for the shape, position, and arrangement of surfaces that form objects, he must learn to study tediously the views of any object with which he is not familiar until he is sure of the exact shape. Multiview drawings cannot be read with ease of our written language, which lists all the components in a dictionary.

5.27

True-length Lines. Students who, lacking a thorough understanding of the principles of projection (Sec. 5.8), find it difficult to determine whether or not a projection of a line in one of the principal views

| (a) | (b) | (c) | (d) | (e) |

A third view may be necessary for the complete shape description of a particular object shown.

Fig. 5.32 Meaning of areas.

shows the true length of the line, should study carefully the following facts:

1. If the projection of a line shows the true length of the line, one of the other projections must appear as a horizontal line, a vertical line, or a point, on one of the other views of the drawing.

2. If the top and front views of a line are horizontal, then both views show the true length.

3. If the top view of a line is a point, the front and side views show the true length.

4. If the front view of a line is a point, the top and side views show the true length.

5. If the top and front views of a line are vertical, the side view shows the true length.

6. If the side projection of a line is a point, the top and front views show the true length.

7. If the front view of a line is horizontal and the top view is inclined, the top inclined view shows the true length.

8. If the top view of a line is horizontal and the fron view is inclined, the front inclined view shows the true length.

5.28

Representation of Holes. In preparing drawings of parts of mechanisms, a draftsman finds it necessary to represent machined holes, which most often are either drilled, drilled and reamed, drilled and countersunk, drilled and counterbored, or drilled and spotfaced. Graphically, a hole is represented to conform with the finished form. The form may be completely specified by a note attached to the view showing the circular contour (Fig. 5.33). The shop note, as prepared by the draftsman, usually specifies the several shop operations in the order that they are to be performed in the shop. For example, in (*d*) the hole, as specified, is drilled before it is counterbored. When depth has not been given in the note for a hole, it is understood to be a through hole; that is, the holes goes entirely through the piece [(*a*), (*c*), (*d*), and (*e*)]. A hole that does not go through is known as a *blind hole* (*b*). For such holes, depth is the length of the cylindrical portion. Drilled, bored, reamed, cored, or punched holes are always specified by giving their diameters, never their radii. Drill diameters for number- and letter-size drills may be found in Table 35 in

the Appendix. Metric drills have been listed in Table 36.

In drawing the hole shown in (*a*), which must be drilled before it is reamed, the limits are ignored and the diameter is scaled to the nearest regular inch or millimeter size. In (*b*) the 30° × 60° triangle is used to draw the approximate representation of the conical hole formed by the drill point. In (*c*) a 45° triangle has been used to draw an approximate representation of the outline of the conical enlargement. The actual angle of 82° is ignored in order to save time in drawing. The spotface in (*e*) is most often cut to a depth of .06 in. (1.5 mm); however, the depth is usually not specified.

The beginner should now scan the several sections in the chapter on shop processes to obtain some general information on the production of holes. Complete information on the preparation of shop notes for holes may be found in Chapters 13, 14, and 15.

B □ Revolution

5.29

To Find the True Length of a Line by Revolution. In engineering drafting, it is frequently necessary to determine the true length of a line when constructing the development of a surface (Chapter 10). The true lengths must be found of those lines that are not parallel to any coordinate plane and therefore appear foreshortened in all the principal views. (see Sec. 5.8). The practical as well as theoretical procedure is to revolve any such

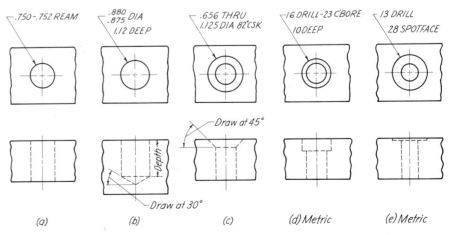

(*For representations of threaded holes see chapter covering screw threads and fasteners*)

Fig. 5.33 Representation of holes.

oblique line into a position parallel to a coordinate plane such that its projection on that particular plane will be the same length as the line. In Fig. 5.34(a), this is illustrated by the edge AB on the pyramid. AB is oblique to the coordinate planes, and its projections are foreshortened. If this edge line is imagined to be revolved until it becomes parallel to the frontal plane, then the projection ab_r in the front view will be the same length as the true length of AB.

A practical application of this method is shown in Fig. 5.34(c). The true length of the edge AB in Fig. 5.34(a) would be found by revolving its top projection into the position ab_r, representing AB revolved parallel to the frontal plane, and then projecting the end point b_r down into its new position along a horizontal line-through b. The horizontal line represents the horizontal plane of the base, in which the point B travels as the line AB is revolved.

Commercial draftsmen who are unfamiliar with the theory of coordinate planes find the true-length projection of a line by visualizing the line's revolution. They think of an edge as being revolved until it is in a plane perpendicular to the line of sight of an observer stationed an infinite distance away (Figs. 10.10, 10.11, and 10.12). The process corresponds to that used in drawing regular orthographic views (Sec. 5.4). Usually this method is more easily understood by a student.

Note in Fig. 5.34(a) and (b) that the true length of a line is equal to the hypotenuse of a right triangle whose altitude is equal to the difference in the elevation of the end points and whose base is equal to the

top projection of the line. With this fact in mind, many draftsmen determine the true length of a line by constructing a true-length triangle similar to the one illustrated in Fig. 5.34(d).

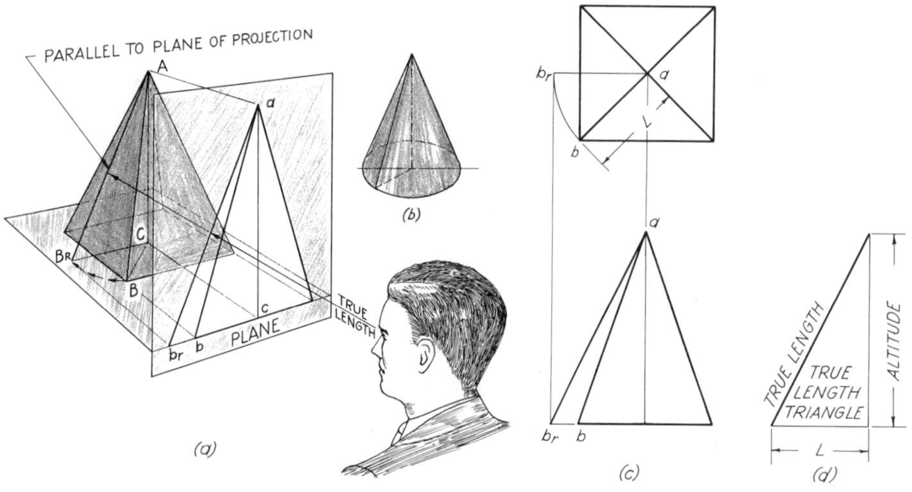

5.30

Revolution of an Object. Although in general the views on a working drawing represent a machine part satisfactorily when shown in a natural position, it is sometimes desirable to revolve an element until it is parallel to a coordinate plane in order to improve the respresentation or to reveal the true size and shape of a principal surface, or true length of a line.

The distinguishing difference between this method and the method of auxiliary projection (Chapter 8) is that, in the procedure of revolution, the observer turns (revolves) the object with respect to the customary planes of projection, instead of shifting his viewing position with respect to either an inclined or an oblique surface of the object.

Despite the fact that the revolution of an entire object, as illustrated in Fig. 5.35, rarely has a practical application in industry, the making of such a drawing provides excellent drill in projection. Therefore, since the several articles that follow are intended primarily for training students, the practical applications have been omitted, while the procedures for revolving simple objects are explained in detail.

5.31

Simple (Single) Revolution. When the regular views are given, an object may be shown in another position, as may be required, by imagining it to be revolved about an axis perpendicular to one of the principal (horizontal, frontal, or profile) planes. A single revolution about such an axis is known as a "simple revolution." The three general cases are shown in Fig. 5.35, spaces II, III, and IV.

5.32

To Determine the True-shape Projection (Normal View) of a Plane Surface—Revolution Method. If a surface is parallel to one of the principal planes of projection, its projection on that plane will show its true shape, as has been explained in Sec. 5.10. In Fig. 5.35, space I, the inclined surface $ABCD$ appears as a line ($a^Fb^Fc^Fd^F$) in the front view and is shown foreshortened in the top and side views. Line A–B was taken as the axis of rotation and the surface was revolved into a posi-

Fig. 5.34 True length of a line, revolution method.

tion parallel to the *H*-plane, as shown by the pictorial drawing (ABC_RD_R). First, the edge view of the surface, as represented by the line $a^Fb^Fc^Fd^F$, was revolved about a^Fb^F into a horizontal position, as shown by the line $a^Fb^Fc_r^Fd_r^F$. The complete *H*-view ($a^Hb^Hc_r^Hd_r^H$) of the surface as revolved shows the true shape of *ABCD*.

5.33
Revolution About a Vertical Axis Perpendicular to the Horizontal Plane. A simple revolution about an axis perpendicular to the horizontal plane is illustrated in Fig. 5.35, space II. The object is first revolved about the assumed imaginary axis until it is in the desired position (see pictorial). The views of the part in its revolved position then are obtained by orthographic projection, as in the case of any ordinary multiview drawing. Since the top view will not be changed in shape by the revolution, it must be drawn first in its revolved position (at 30° in this case) and the front and side views should be projected from it. Since the heights to all of the points on the object also remain unchanged by the revolution, height distances could be conveniently projected from the initial views in I.

The top view may be drawn directly in revolved position, without first drawing the usual orthographic views. If this procedure is followed, the height distances for the front and side views may be set off to known dimensions.

5.34
Revolution about a Horizontal Axis Perpendicular to the Frontal Plane. If an object is revolved about an imaginary axis perpendicular to the frontal plane, as shown in Fig. 5.35, space III, the front view changes in position but not in shape. The front view, therefore, should be drawn first in its revolved position, and the top and side views should be projected from it. The depth of the top view and the side view remain unchanged since the depth distance is parallel to the axis. If the usual unrevolved views are not drawn first, the front view may be drawn directly in its revolved position and depth distances can then be laid off to known dimensions.

5.35
Revolution about a Horizontal Axis Perpendicular to the Profile Plane. A single revolution of an object about an axis perpendicular to the profile plane is illustrated in Fig. 5.35, space IV. Since in this

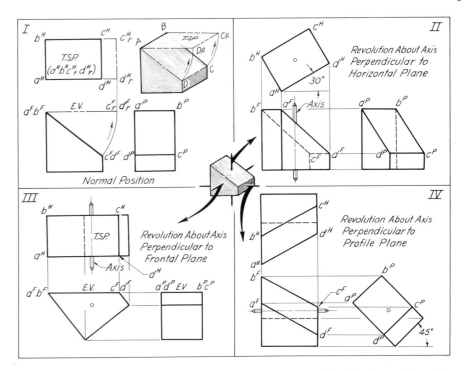

Fig. 5.35 Single revolution about an axis perpendicular to a principal (horizontal, frontal, or profile) plane.

case it is the side view that is perpendicular to the axis and revolves parallel to the coordinate plane of projection, it is the side view that remains unchanged in its shape. The width of the top and front views is not affected by the revolution. Therefore, horizontal dimensions for these views may be set off by using known measurements.

From these general cases of simple revolution, two principles have emerged that can be stated as follows:

1. The view that is perpendicular to the axis of revolution changes only in position.

2. The lengths of the lines parallel to the axis do not change during the revolution and, therefore, may be either laid off to known measurements or projected from the usual orthographic views of the object.

5.36
Clockwise and Counterclockwise Revolution. An object may be revolved either clockwise or counterclockwise about an axis of revolution. The direction is indicated by the view to which the axis is perpendicular. For example, front views, when revolved as in Fig. 5.36(*b*), show a

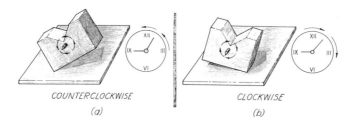

COUNTERCLOCKWISE
(a)

CLOCKWISE
(b)

Fig. 5.36 Direction of revolution.

clockwise revolution. When revolved as in (*a*), their revolution is counterclockwise. Top views show a clockwise revolution when revolved to the right. Right-side views indicate a clockwise direction of revolution when they have been revolved to the right and a counterclockwise direction when revolved to the left.

5.37

Successive (Multiple) Revolution. Since it is possible to show an object in any position relative to the coordinate planes of projection, it can be drawn as may be required by making a series of successive simple revolutions. Usually such a series is limited to three or four stages. Figure 5.37 shows an object revolved successively about two separate axes. The usual orthographic views of the given object are shown at the left in space I. In space II, the object has been revolved (counterclockwise) about a vertical axis to obtain an edge view (EV) of the oblique surface *ABC* (see pictorial). Note that the line *AC* projects as a point ($a^F c^F$) in the front view. The identification of each corner may not be necessary in the case of a simple object. However, if the object is in the least complex, possible confusion is avoided if each corner point is either numbered or identified by a letter, as was done for the oblique surface in the illustration. In space III at the right, the object is represented after it has been revolved

from its previous position in space II. In this case, since the edge view (EV) of surface *ABC* is now horizontal, a true-shape projection (TSP) of the surface will appear in the top view.

A convenient method of copying a view in a new revolved position is first to trace it on a small piece of tracing paper and then to place the traced view in correct position and draw the new view. An alternative method is to place the traced view in correct position and prick the corner points, using the dividers opened out. With the traced view removed, the revolved view on the final drawing may be completed by joining the prick-points.

C □ Conventional practices

5.38

To reduce the high cost of preparing engineering drawings and at the same time to convey specific and concise information without a great expenditure of effort, some generally recognized systems of symbolic representation and conventional practices have been adopted by American industry.

A standard symbol or conventional representation can express information that might not be understood from a true-line representation unless accompanied by a lettered statement. In many cases, even though a true-line representation would convey exact information, very little more would be gained from the standpoint of better interpretation. Some conventional practices have been adopted for added clearness. For instance, they can eliminate awkward conditions that arise from strict adherence to the rules of projection.

These idioms of drawing have slowly developed with the graphic language until at the present time they are universally recognized and observed and appear in the various standards of the American National Standards Institute.

Professional men and skilled workmen have learned to accept and respect the use of the symbols and conventional practices, for they can interpret these representations accurately and realize that their use saves valuable time in both the drawing room and the shop.

5.39

Half Views and Partial Views. When the available space is insufficient to allow a

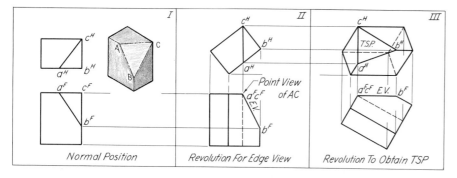

Normal Position   *Revolution For Edge View*   *Revolution To Obtain TSP*

Fig. 5.37 Successive revolution to show the true shape of an oblique surface.

satisfactory scale to be used for the representation of a symmetrical piece, it is considered good practice to make one view either a half view or a partial view, as shown in Fig. 5.38. The half view, however, must be the top or side view and not the front view, which shows the characteristic contour. The half view should be the front half of the top or side view. In the case of the partial view shown in (b), a break line is used to limit the view.

5.40

Accepted Violations of True Projection in the Representation of Boltheads, Slots, and Holes for Pins. A departure from true projection is encountered in representing a bolthead. For example, on a working drawing, it is considered the best practice to show the head across corners in both views, regardless of the fact that in true projection one view would show "across flats." This method of treatment eliminates the possibility of a reader's interpreting a hexagonal head to be a square head (Fig. 5.39). Furthermore, the showing of a head the "long way" in both views clearly reveals the space needed for proper clearance.

In the case of the slotted head fasteners, the slots are shown at 45° in the end views in order to avoid placing a slot on a center line, where it is usually difficult to draw so that the center line passes accurately through the center (Fig. 5.40). This practice does not affect the descriptive value of the drawing, because the true size and shape of the slot is shown in the front view. The hole for a pin is shown at 45° for the same reason. In such a position it may be more quickly observed.

5.41

Treatment of Unimportant Intersections. The conventional methods of treating various unimportant intersections are shown in Fig. 5.41. To show the true line of intersection in each case would add little to the value of the drawing. Therefore, in the views designated as preferred, true projection has been ignored in the interest of simplicity. On the side views, in (a) and (b), for example, there is so little difference between the descriptive values of the true and approximate representations of the holes that the extra labor necessary to draw the true representation is unwarranted.

5.42

Aligned Views. Pieces that have arms, ribs, lugs, or other features at an angle

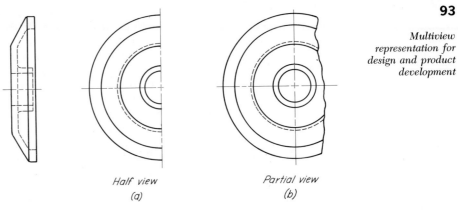

Half view
(a)

Partial view
(b)

Fig. 5.38. Half views and partial views.

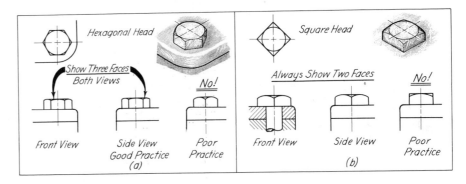

Fig. 5.39 Treatment of bolt heads.

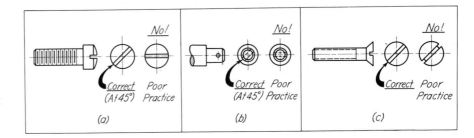

Fig. 5.40 Treatment of slots and holes in fasteners and pins.

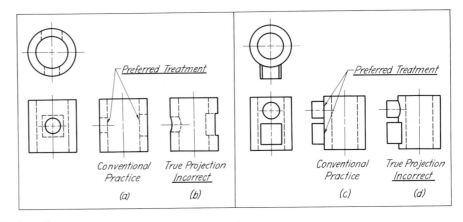

Fig. 5.41 Treatment of unimportant intersections.

are shown aligned or "straightened out" in one view, as illustrated in Fig. 5.42. By this method, it is possible to show the true shape as well as the true position of such features. In Fig. 5.43, the front view has been drawn as though the slotted arm had been revolved into alignment with the element projecting outward to the left. This practice is followed to avoid drawing an element—that is at an angle—in a foreshortened position.

5.43
Conventional Treatment of Radially Arranged Features. Many objects that have radially arranged features may be shown more clearly if true projection is violated, as in Fig. 5.42(*b*). Violation of true projection in such cases consists of intentionally showing such features swung out of position in one view to present the idea of symmetry and show the true relationship of the features at the same time. For example, while the radially arranged holes in a flange (Fig. 5.44) should always be shown in their true position in the circular view, they should be shown in a revolved position in the other view in order to show their true relationship with the rim.

Radial ribs and radial spokes are similarly treated [Fig. 5.42(*a*)]. The true projection of such features may create representations that are unsymmetrical and misleading. The preferred conventional method of treatment, by preserving symmetry, produces representations that are more easily understood and that at the same time are much simpler to draw. Figure 5.45 illustrates the preferred treatment for radial ribs and holes.

5.44
Representations of Fillets and Rounds. Interior corners, which are formed on a casting by unfinished surfaces, are always filled in (filleted) at the intersection in order to avoid possible fracture at that point. Sharp corners are also difficult to obtain and are avoided for this reason as well. Exterior corners are rounded for appearance and for the comfort of persons who must handle the part when assembling or repairing the machine on which the part is used. A rounded internal corner is known as a *fillet*; a rounded external corner is known as a *round* (Fig. 5.46).

When two intersecting surfaces are ma-

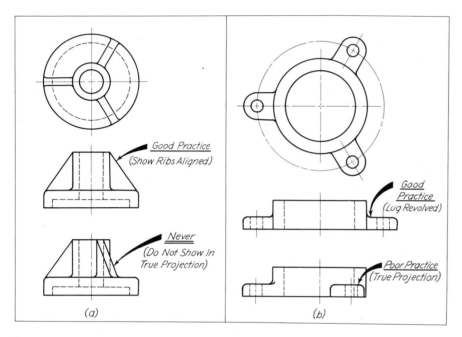

Fig. 5.42 Conventional practice of representing ribs and lugs.

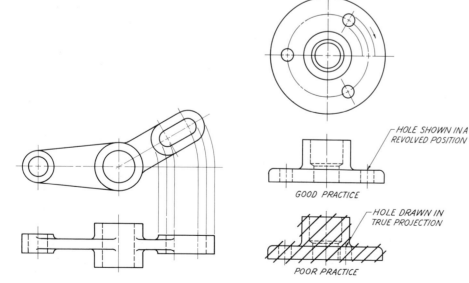

Fig. 5.43 Aligned views.

Fig. 5.44 Radially arranged holes.

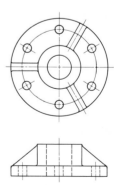

Fig. 5.45 Conventional treatment of radially arranged ribs.

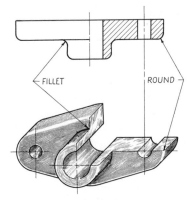

Fig. 5.46 Fillets and rounds.

chined, however, their intersection will become a sharp corner. For this reason, all corners formed by unfinished surfaces should be shown "broken" by small rounds, and all corners formed by two finished surfaces, or one finished surface and one unfinished surface, should be shown "sharp." Although in the past it has been the practice to allow pattern-makers to use their judgment about the size of fillets and rounds, many present-day companies require their designers and draftsmen to specify their size even though their exact size may not be important.

Since fillets and rounds eliminate the intersection lines of intersecting surfaces, they create a special problem in orthographic representation. To treat them in the same manner as they would be treated if they had large radii results in views that are misleading. For example, the true-projection view in Fig. 5.47(c) confuses the reader, because at first glance it does not convey the idea that there are abrupt changes in direction. To prevent such a probable first impression and to improve the descriptive value of the view, it is necessary to represent these theoretically nonexisting lines. These characteristic lines are projected from the approximate intersections of the surfaces, with the fillets disregarded.

Figure 5.48 illustrates the accepted conventional method of representing the "run-out" intersection of a fillet in cases where a plane surface is tangent to a cylindrical surface. Although run-out arcs such as these are usually drawn freehand, a French curve or a bow instrument may be used. If they are drawn with the bow instrument, a radius should be used that is equal to the radius of the fillet, and the completed arc should form approximately one-eighth of a circle.

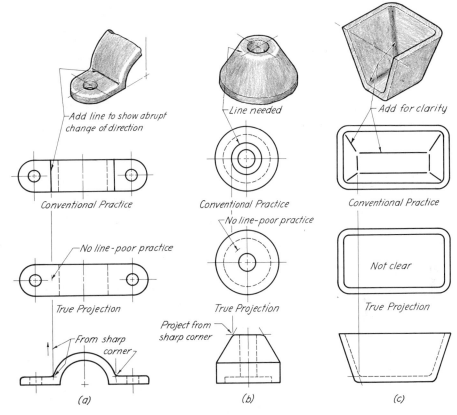

Fig. 5.47 Conventional practice of representing nonexisting lines of intersection.

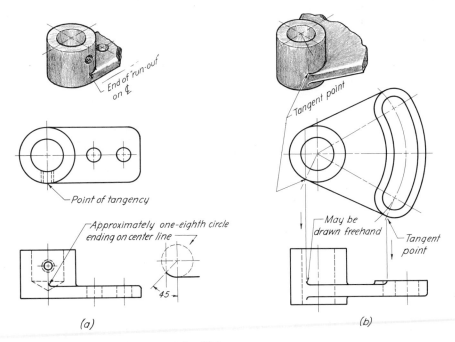

Fig. 5.48 Conventional treatment for fillets.

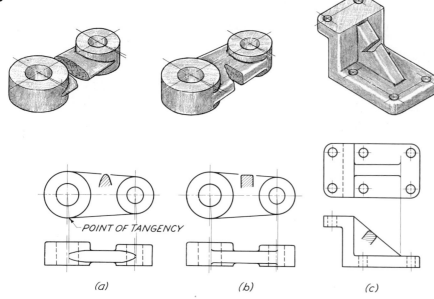

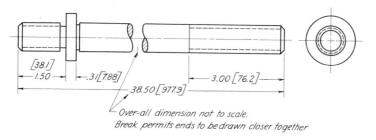

Fig. 5.49 Approximate methods of representing run-outs for intersecting fillets and rounds.

POINT OF TANGENCY

(a)   (b)   (c)

[38.1]
1.50   .31 [7.88]   3.00 [76.2]

38.50 [977.9]

Over-all dimension not to scale.
Break permits ends to be drawn closer together

Fig. 5.50 Broken-out view.

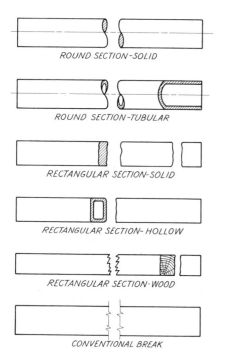

ROUND SECTION–SOLID

ROUND SECTION–TUBULAR

RECTANGULAR SECTION–SOLID

RECTANGULAR SECTION– HOLLOW

RECTANGULAR SECTION–WOOD

CONVENTIONAL BREAK

Fig. 5.51 Conventional breaks.

detail or assembly drawings. The break representations for indicating the broken ends of rods, shafts, tubes, and so forth, are designed to reveal the characteristic shape of the cross section in each case. Although break lines for round sections may be drawn freehand, particularly on small views, it is better to draw them with either an irregular curve or a bow instrument. The breaks for wood sections, however, always should be drawn freehand.

5.46
Ditto Lines. When it is desirable to minimize labor in order to save time, ditto lines may be used to indicate a series of identical features. For example, the threads on the shaft shown in Fig. 5.52 are just as effectively indicated by ditto lines as by a completed profile representation. When ditto lines are used, a long shaft of this type may be shortened without actually showing a conventional break.

5.47
A Conventional Method for Showing a Part in Alternative Positions. A method frequently used for indicating an alternative position of a part or a limiting position of a moving part is shown in Fig. 5.53. The dashes forming the object lines of the view showing the alternative position should be of medium weight. The phantom line shown in Fig. 2.58 is recommended for representing an alternative position.

5.48
Conventional Representation. Symbols are used on topographic drawings, architectural drawings, electrical drawings, and machine drawings. No engineer serving in a professional capacity can very well escape their use.

Most of the illustrations that are shown in Fig. 5.54 should be easily understood. However, the crossed-lines (diagonals) symbol has two distinct and different meanings. First, this symbol may be used on a drawing of a shaft to indicate the position of a surface for a bearing or, second, it may indicate that a surface perpendicular to the line of sight is flat. These usages are illustrated with separate examples.

The generally accepted methods of representing intersecting fillets and rounds are illustrated in Fig. 5.49. The treatment, in each of the cases shown, is determined by the relationship existing between the sizes of the intersecting fillets and rounds.

5.45
Conventional Breaks. A relatively long piece of uniform section may be shown to a larger scale, if a portion is broken out so that the ends can be drawn closer together (Fig. 5.50). When such a scheme is employed, a conventional break is used to indicate that the length of the representation is not to scale. The American National Standard conventional breaks, shown in Fig. 5.51, are used on either

Fig. 5.52 Ditto lines.

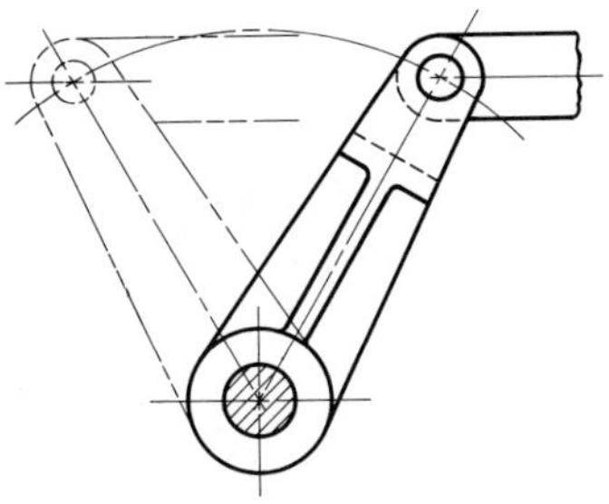

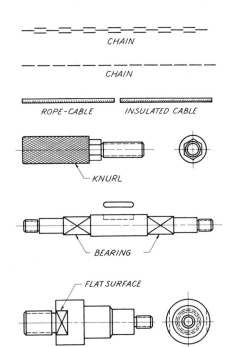

CHAIN

CHAIN

ROPE-CABLE   INSULATED CABLE

KNURL

BEARING

FLAT SURFACE

Fig. 5.53 Alternative positions.

Fig. 5.54 Conventional symbols.

Problems

The problems that follow are intended primarily to furnish study in multiview projection through the preparation of either sketches or instrumental drawings. Many of the problems in this chapter, however, may be prepared in more complete form. Their views may be dimensioned as are the views of working drawings, if the student will study carefully the beginning of the chapter covering dimensioning before attempting to record size description (Chapter 13). All

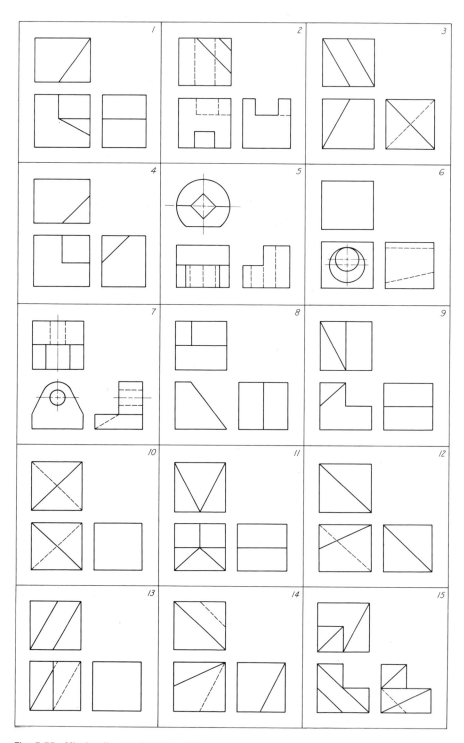

Fig. 5.55 Missing-line (or lines) exercises.

dimensions should be placed in accordance with the general rules of dimensioning. The problems given at the end of Chapter 6 offer further study in multiview representation.

The views shown in a sketch or drawing should be spaced on the paper with aim for balance within the borderlines. Ample room should be allowed between the views for the necessary dimensions. If the views are not to be dimensioned, the distance between them may be made somewhat less than would be necessary otherwise.

Before starting to draw, the student should reread Sec. 5.24 and study Fig. 5.27, which shows the steps in making a multiview drawing. The preparation of a preliminary sketch always proves helpful to the beginner.

All construction work should be done in light lines with a sharp hard pendil. A drawing should be checked by an instructor before the lines are "heavied in," unless the preliminary sketch was checked beforehand.

Metric: When a problem has been dual-dimensioned and the metric values represent inch values converted, use the nearest full millimeter value for a dimension except in the case of a critical dimension or a machined hole. For drilled holes see Tables 35 and 36 in the Appendix. For a dimension that will be divided by two, the dimension may be taken to the nearest even number of millimeters.

1. (Fig. 5.55). Add the missing line or lines in one view of each of the three-view drawings. When the missing line or lines have been determined, the three views of each object will be consistent with one another.

2. (Fig. 5.56). Draw or sketch the third view for each of the given objects.

3–6. (Figs. 5.57–5.60). Reproduce the given views and draw the required view. Show all hidden lines.

7. (Fig. 5.61). Make an orthographic drawing or sketch of the bench stop. The views may be dimensioned. The shaft portion that fits into the hole in the bench top is 19 mm. in diameter and 50 mm. long.

8–39. (Figs. 5.62 and 5.93). Make multiview drawings of the given objects. The views of a drawing may or may not be dimensioned.

40. (Fig. 5.94). Make a complete orthographic drawing of the anchor bracket.

41. (Fig. 5.95). Make a complete orthographic drawing of the tube holder.

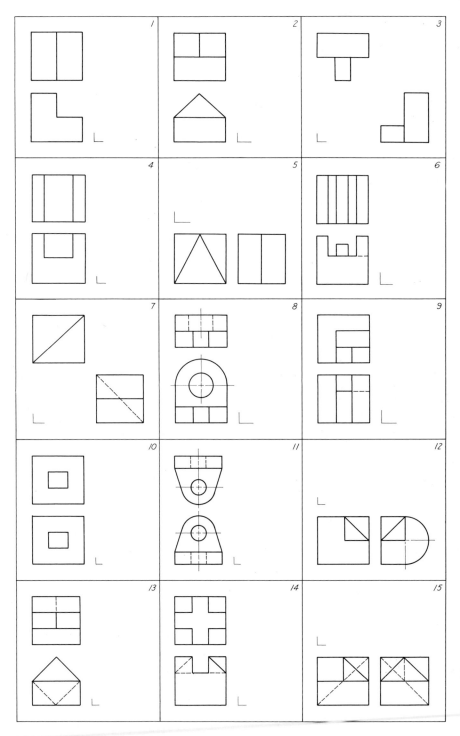

Fig. 5.56 Third-view problems.

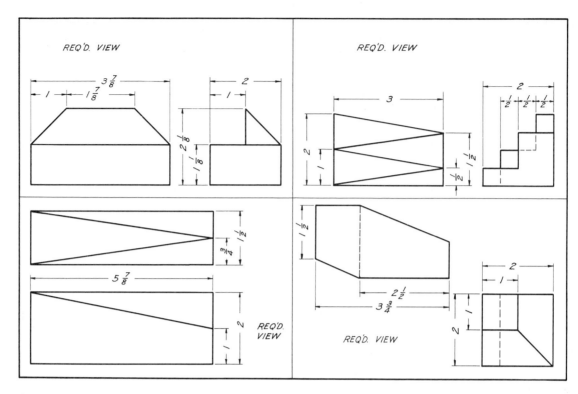

Fig. 5.57

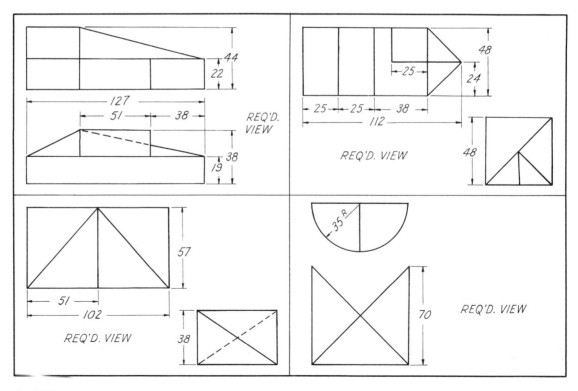

Fig. 5.58

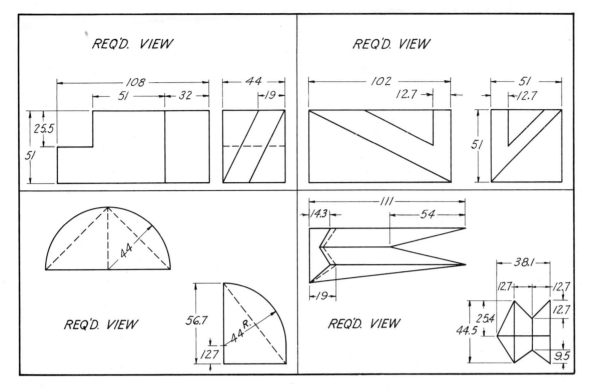

Fig. 5.59

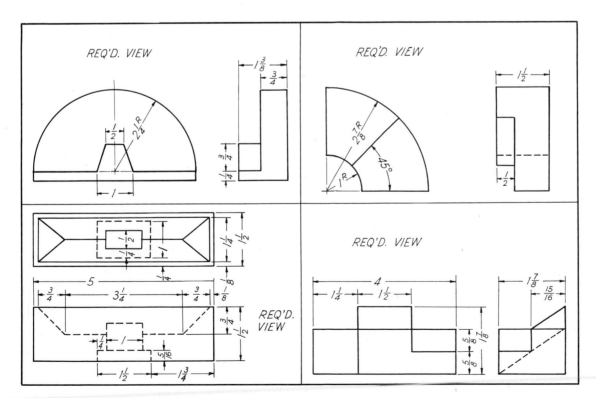

Fig. 5.60

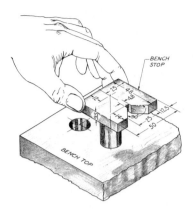

Fig. 5.61   Bench stop.

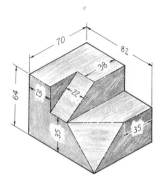

Fig. 5.62   Corner block.

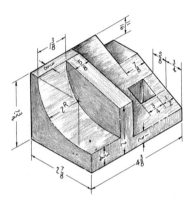

Fig. 5.63   Stop block.

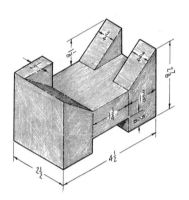

Fig. 5.64   Rest block.

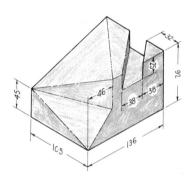

Fig. 5.65   Adjustment block.

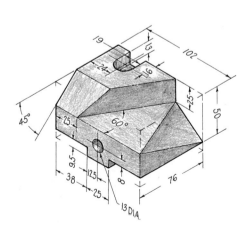

Fig. 5.66   Locating block.

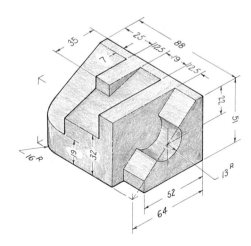

Fig. 5.67   Safety block.

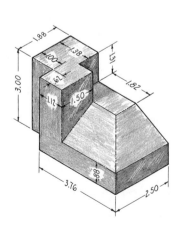

Fig. 5.68   Angle block.

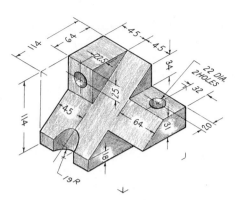

Fig. 5.69   Cross stop.

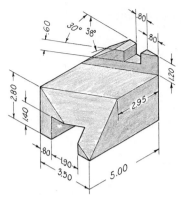

Fig. 5.70   End block.

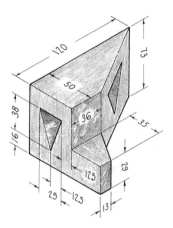

Fig. 5.71   Bevel block.

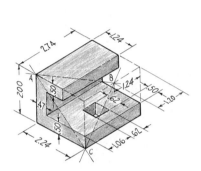

Fig. 5.72   Stabilizer block.

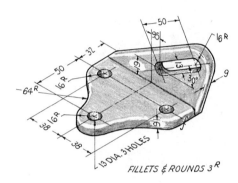

FILLETS & ROUNDS 3ᴿ

Fig. 5.73   Mounting bracket.

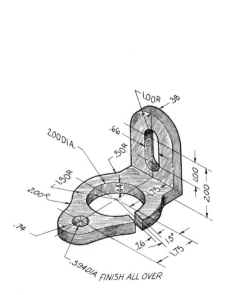

.594 DIA FINISH ALL OVER

Fig. 5.74   Index guide.

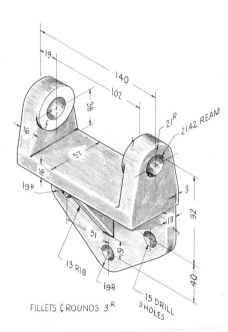

FILLETS & ROUNDS 3ᴿ

Fig. 5.75   Guide bracket.

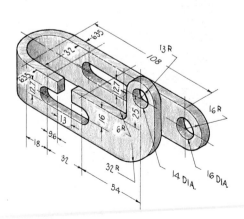

Fig. 5.76   Control guide.

*Multiview
representation for
design and product
development*

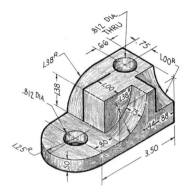

Fig. 5.77   Shoe block.

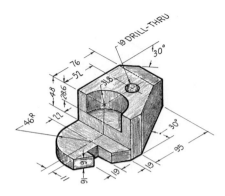

Fig. 5.78   Holder block.

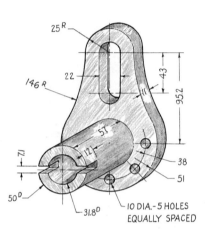

Fig. 5.79   Slotted guide.

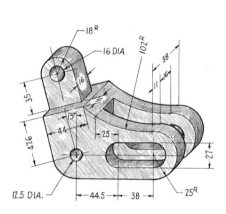

Fig. 5.80   Pivot guide.

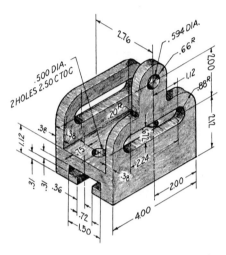

Fig. 5.81   Control guide.

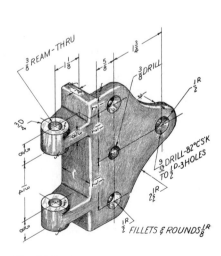

Fig. 5.82   Corner bracket.

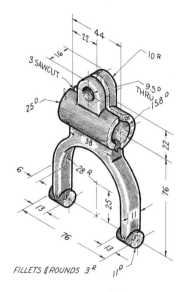

Fig. 5.83   Auxiliary fork.

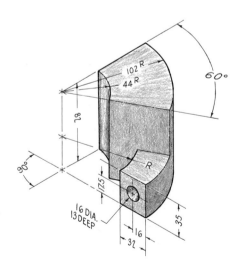

Fig. 5.84   Jaw block.

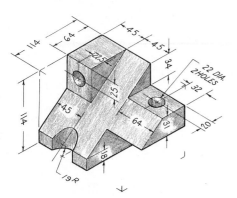

Fig. 5.69 Cross stop.

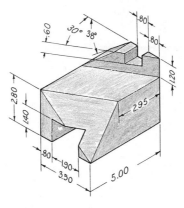

Fig. 5.70 End block.

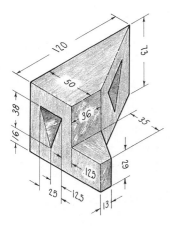

Fig. 5.71 Bevel block.

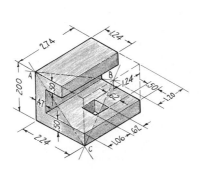

Fig. 5.72 Stabilizer block.

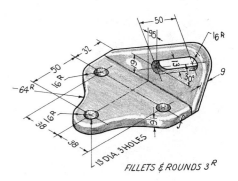

Fig. 5.73 Mounting bracket.

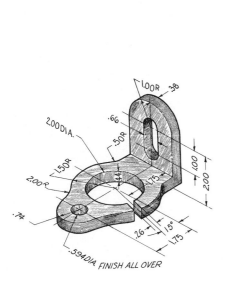

Fig. 5.74 Index guide.

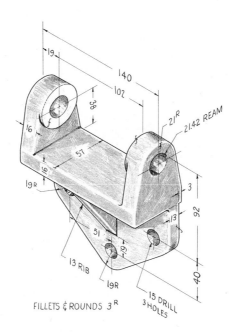

Fig. 5.75 Guide bracket.

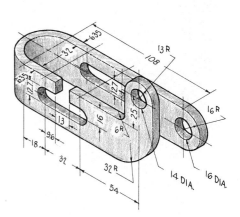

Fig. 5.76 Control guide.

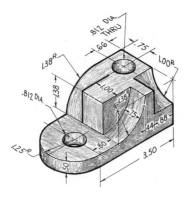

Fig. 5.77   Shoe block.

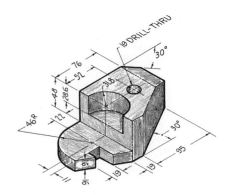

Fig. 5.78   Holder block.

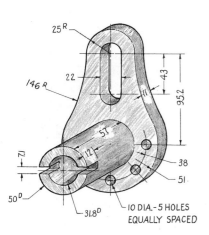

Fig. 5.79   Slotted guide.

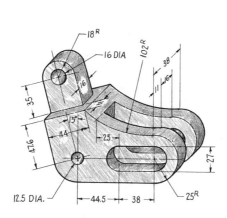

Fig. 5.80   Pivot guide.

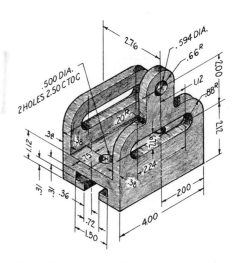

Fig. 5.81   Control guide.

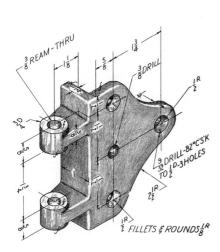

Fig. 5.82   Corner bracket.

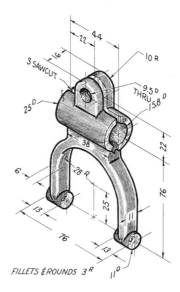

Fig. 5.83   Auxiliary fork.

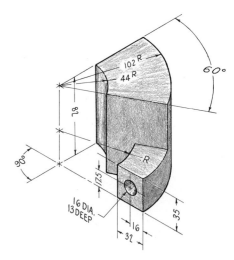

Fig. 5.84   Jaw block.

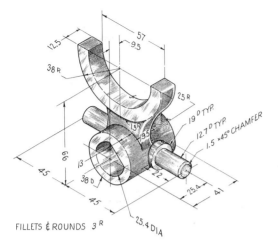

Fig. 5.85 Shifter.

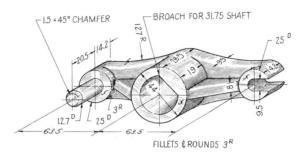

Fig. 5.86 Stud guide.

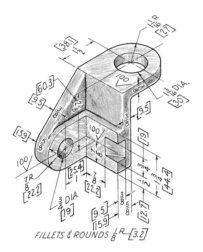

Fig. 5.87 Ejector bracket. (Dimension values shown in [ ] are in millimeters.)

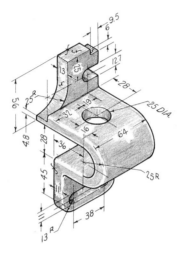

Fig. 5.88 Guide clip.

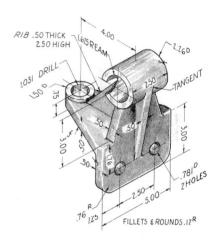

Fig. 5.89 Control rod guide.

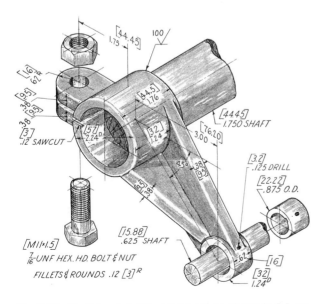

Fig. 5.90 Shaft bracket. (Dimension values shown in [ ] are in millimeters.)

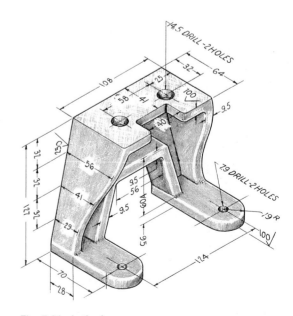

Fig. 5.91 Lathe leg.

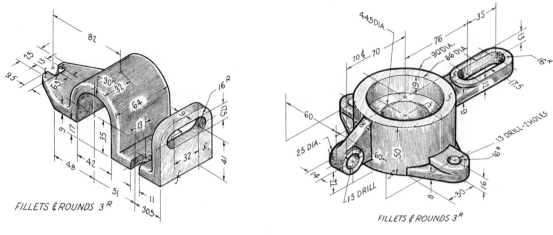

FILLETS & ROUNDS 3^R

Fig. 5.92 Feed guide.

FILLETS & ROUNDS 3^R

Fig. 5.93 Control bracket.

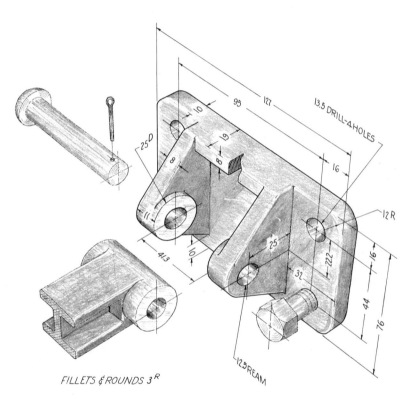

FILLETS & ROUNDS 3^R

Fig. 5.94 Anchor bracket.

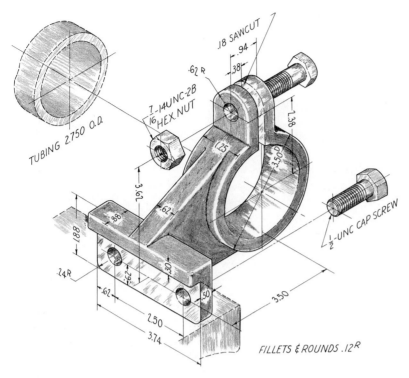

Fig. 5.95 Tube holder.

TUBING 2.750 O.D.

.18 SAWCUT
.94
.62 R
.38
7_-14UNC-2B
16 HEX. NUT
3.62
.25
3.50 D.
1.38
.88
.88
.24 R
.62
2.94
.30
.50
.62
2.50
3.74
3.50

½-UNC CAP SCREW

FILLETS & ROUNDS .12ᴿ

At a time when fossil fuels are becoming scarce and costly, designers and technologists employed in the transportation field are directing their attention to the development of new and different mass transit systems that will meet the needs of our expanding population. One answer to this problem has been the development of the BART **ba** train that is now in operation at San Francisco.

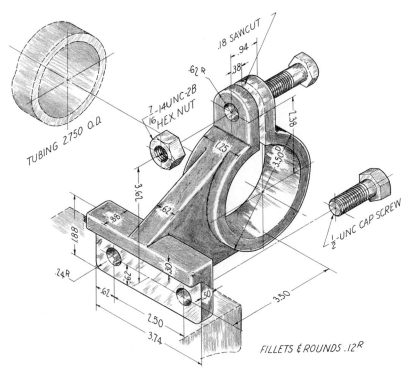

TUBING 2.750 O.D.

.18 SAWCUT

.94

.38

.62 R

7 -14UNC-2B
16 HEX. NUT

2.38

1.25

3.50 D

3.62

.62

.88

1.88

.24 R

.62

.29

.30

.50

2.50

3.74

3.50

1/2 -UNC CAP SCREW

FILLETS & ROUNDS .12 R

Fig. 5.95 Tube holder.

At a time when fossil fuels are becoming scarce and costly, designers and technologists employed in the transportation field are directing their attention to the development of new and different mass transit systems that will meet the needs of our expanding population. One answer to this problem has been the development of the BART **ba** train that is now in operation at San Francisco.

6

Freehand sketching for visualization and communication

A □ Sketching and design

6.1

Introduction. From the earliest times pictorial representations have been the means of conveying the ideas of one person to another person and of one group to another group (Fig. 6.2). There is little doubt that our ancestors traced out in the dust on the cave floor many crude pictures to supplement their guttural utterances. On their cave walls these same primitive men and women drew pictures, which today convey to others the stories of their lives. They used the only permanent means they were aware of at that time.

At present, we have at our command the spoken languages, which no doubt developed from limited semiintelligent throat sounds, and the written languages, graphical and symbolic in form. The descriptive powers of the various forms of presentation may be compared in Fig. 6.1. The use of sign language representation, (*a*), is rather easy to learn and may be quickly executed but interpretation is restricted to persons who understand the particular language in which it is presented. The multiview representation, shown in (*b*), may be understood universally by persons who have been trained in its use. However, the given views will prove to be almost meaningless to the many who have not had the advantage of needed training. What then is the one form of representation that can be understood and used by all? It is the pictorial form shown in (*c*).

It is necessary that a designer or engineering technologist be capable of executing well-proportioned and understand-

110

控馭火箭

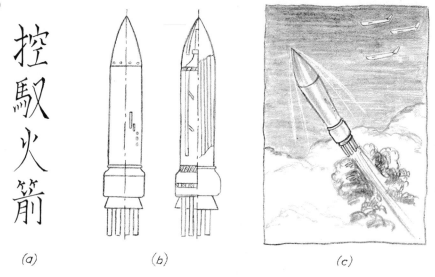

(a)   (b)   (c)

Fig. 6.1 Graphic methods for presenting ideas (symbolic, multiview, and pictorial).

Fig. 6.2 **Designer's idea sketch.** In the future heli-lifted classrooms may be transported to every part of the earth. The lessons would be projected in 3-D. Transmittal would be carried by laser beams, relayed by satellite. (*Courtesy Raymond Loewy/William Smith, Inc., and Charles Bruning Company*)

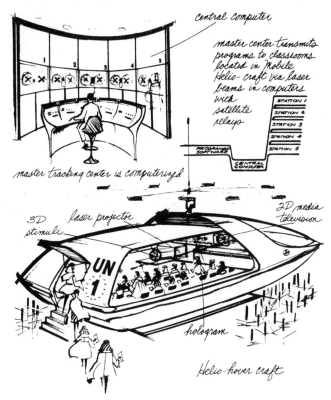

able freehand pictorial sketches, for it is one of the most important modes of expression that he has available for use. Like other means for conveying ideas, when depended on alone, it will usually prove to be inadequate. However, when used in combination with the written or spoken language and related graphical representations, it makes a full understanding by others become sure and not just possible (Fig. 6.3). Each method of expression is at

hand to supplement another to convey the intended idea.

The designer should be capable of speaking his own language and possibly one foreign language fluently; he should be able to write to present his ideas clearly and accurately; he should be familiar with the graphical method of presenting shape through the use of multiviews; and, finally, he should be competent to execute well-proportioned and understandable pictorial sketches, which are needed to clarify and ensure complete transfer of his ideas to others.

The designer is a creative person living in a world where all that he creates must exist in space. He must visualize space conditions, space distances, and movement in space. In addition he must be able to retain as well as alter the image of his idea, which will at the very start exist only in his mind.

As his idea forms the designer nearly always resorts to sketching to organize his thoughts quickly and to more clearly visualize the problems that appear. The first ideas may be sketched in pictorial form as they are visualized. Later, in making a preliminary study, a combination of orthographic design sketches and pictorial sketches may quickly pile up on his desk as problems are recognized and possible solutions are recorded for reference and for conferences with others.

The designer's use of sketches, both pictorial and orthographic (Fig. 6.4), continues throughout the preliminary design stages and into the development and detailing stages. This comes about because the designer is usually called on to serve as both planner and director. Throughout all stages in the development of a product he must solve problems and clarify instructions. Very often a pictorial sketch of some detail of construction will prove to be more intelligible and will convey the idea much better than an orthographic sketch, even when dealing with an experienced draftsman or detailer (Fig. 6.22).

Design sketches may be done in the quiet of the designer's office or amid the confusion of the conference table. To meet the requirement of speed of preparation, one must resist all temptation to use instruments of any type and rely on the pencil alone, for the true measure of the quality of a finished sketch is neatness and good proportion rather than the straightness of the lines. A pictorial sketch need not be an artistic masterpiece to be useful.

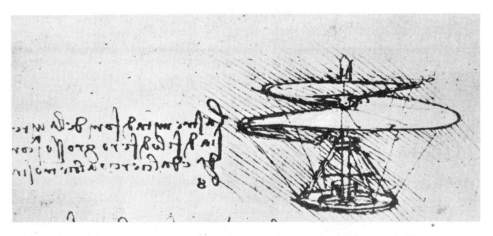

Fig. 6.3 Idea sketch of a helicopter prepared by Leonardo da Vinci (1452–1519). (*From Collections of Fine Arts Department, International Business Machines Corporation*)

Training for making pictorial sketches must include the presentation of basic fundamentals, as is done with other how-to-do-it subjects. As learning the mechanics of English does not make one a creative writer, so training in sketching will not make one a creative engineer. However, sketching is the means of recording creative thoughts.

Some design sketches drawn by an electrical engineer are shown in Fig. 12.21. These sketches were prepared in making a study of the wiring to the electronic control panel for an automatic machine.

6.2

Thinking with a Pencil. As an attempt is made to bring actuality to a plan, sketches undergo constant change as different ideas develop. An eraser may be in constant use or new starts may be made repeatedly, even though one should think much and sketch only when it would appear to be worthwhile. Sketching should be done as easily and freely as writing, so that the mind is always centered on the idea and not on the technique of sketching. To reach the point where one can "think with the pencil" is not easy. Continued practice is necessary until one can sketch with as little thought about how it is done as he gives to how he uses a knife and fork at the dinner table.

6.3

Value of Freehand Drawing. Freehand technical drawing is primarily the language of those in charge of the development of technical designs and plans. Chief engineers, chief draftsmen, designers, and squad bosses have found that the best way to present their ideas for either a simple or complex design is through the medium of sketches. Sketches may be schematic, as are those that are original expressions of new ideas (Fig. 6.3), or they may be instructional, their purpose being to convey ideas to draftsmen or shopmen. Some sketches, especially those prepared for the manufacture of parts that are to replace worn or broken parts on existing machines, may resemble complete working drawings (Fig. 6.5).

6.4

Projections. Although freehand drawing lacks the refinement given by mechanical instruments, it is based on the same principles of projection and conventional practices that apply to multiview, pictorial, and the other divisions of mechanical drawing. For this reason, one must be thoroughly familiar with projection, in all its many forms, before he is adequately trained to prepare sketches.

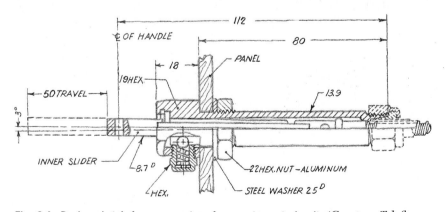

Fig. 6.4 Design sketch for a connector of a remote control unit. (*Courtesy Teleflex, Inc.*)

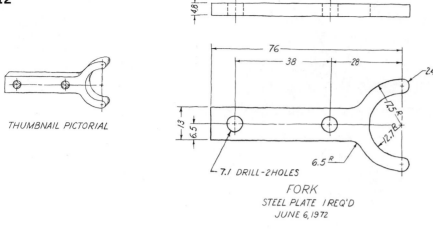

THUMBNAIL PICTORIAL

76
38 — 28
24ᴿ
13
6.5
7.1 DRILL-2HOLES
15 R
12.7 R
6.5 ᴿ
4.8

FORK
STEEL PLATE 1 REQ'D
JUNE 6, 1972

Fig. 6.5 Freehand sketch for the manufacture of a part.

cheap pocket compasses that they could well dispense with if they would adopt the correct technique. Preparing sketches with instruments consumes much unnecessary time.

For the person who cannot produce a satisfactory sketch without guide lines, cross-section paper is helpful (Fig. 6.6).

6.6
Technique of Lines. Freehand lines quite naturally will differ in their appearance from mechanical ones. A well-executed freehand line will never be perfectly straight and absolutely uniform in weight, but an effort should be made to approach *exacting uniformity*. As in the case of mechanical lines, they should be black and clear, not broad and fuzzy (Fig. 6.7).

6.7
Sharpening the Sketching Pencil. A sketching pencil should be sharpened on a file or piece of sandpaper to a conical point. The point then should be rounded slightly, on the back of the sketch pad or on another sheet of paper, to the correct degree of dullness. When rounding the point, rotate the pencil to prevent the formation of sharp edges.

B □ Sketching techniques

6.5
Sketching Materials. For the type of sketching discussed here, the required materials are an F pencil, a soft eraser, and some paper. In the industrial field, men who have been improperly trained in sketching often use straightedges and

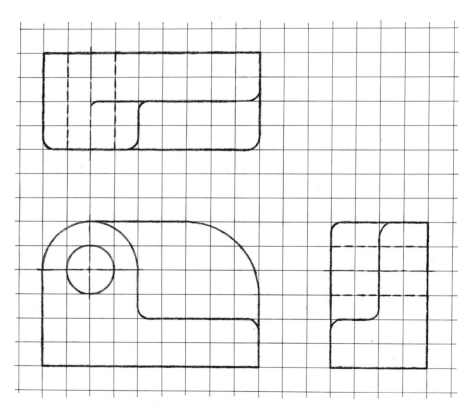

Fig. 6.6 Sketch on cross-section paper.

Straight Lines. The pencil should rest on the second finger and be held loosely by the thumb and index finger about 1–1½ in. (25–40 mm) above the point.

Horizontal lines are sketched from left to right with an easy arm motion that is pivoted about the muscle of the forearm. The straight line thus becomes an arc of infinite radius. When sketching a straight line, it is advisable first to mark the end points with light dots or small crosses (Fig. 6.8).

The complete procedure for sketching a straight line is as follows:

1. Mark the end points.

2. Make a few trial motions between the marked points to adjust the eye and hand to the contemplated line.

3. Sketch a *very* light line between the points by moving the pencil in two or three sweeps. When sketching the trial line, the eye should be on the point toward which the movement is directed. With each stroke, an attempt should be made to correct the most obvious defects of the stroke preceding, so that the finished trial line will be relatively straight.

4. Darken the finished line, keeping the eye on the pencil point on the trial line. The final line, replacing the trial line, should be distinct, black, uniform, and straight.

It is helpful to turn the paper through a *convenient angle* so that the horizontal and vertical lines assume a slight inclination (Fig. 6.9). A horizontal line, when the paper is in this position, is sketched to the right and upward, thus allowing the arm to be held slightly away from the body and making possible a free arm motion.

Short vertical lines may be sketched either downward or upward, without changing the position of the paper. When sketching downward, the arm is held slightly away from the body and the movement is toward the sketcher (Fig. 6.10). To sketch vertical lines upward, the arm is held well away from the body. By turning the paper, a long vertical line may be made to assume the position of a horizontal line and can be sketched with the same general movements used for the latter.

Inclined lines running upward from lower left to upper right may be sketched upward with the same movements used for horizontal lines, but those running

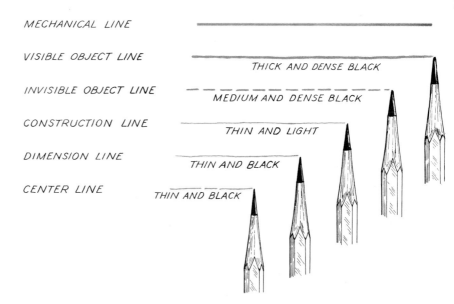

Fig. 6.7 Pencil points and sketch lines.

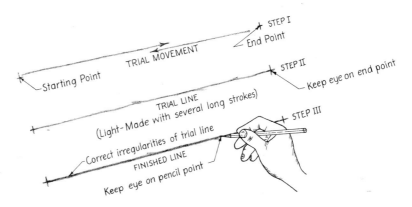

Fig. 6.8 Steps in sketching a straight line.

Fig. 6.9 Sketching horizontal lines.

downward from upper left to lower right are sketched with the general movements used for either horizontal or vertical lines, depending on their inclination (Fig. 6.11). Inclined lines may be more easily sketched by turning the paper to make them conform to the direction of horizontal lines.

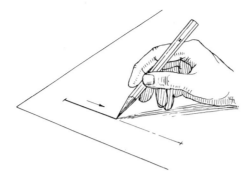

Fig. 6.10 Sketching vertical lines.

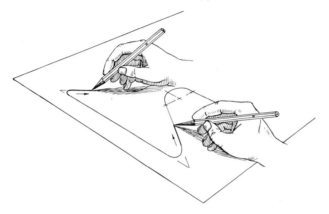

Fig. 6.11 Sketching inclined lines.

6.9

Circles. Small circles may be sketched by marking radial distances on perpendicular center lines. When additional points are needed, the distances can be marked off either by eye or by measuring with a marked strip of paper (Fig. 6.12). Larger circles may be constructed more accurately by sketching two or more diagonals, in addition to the center lines, and by sketching short construction lines perpendicular to each, equidistant from the center. Tangent to these lines, short arcs are drawn perpendicular to the radii. The circle is completed with a light construction line, and all defects are corrected before darkening (Fig. 6.13).

C □ Multiview sketches

6.10

Making a Multiview Sketch (Fig. 6.14). When making orthographic working sketches a systematic order should be followed, and all the rules and conventional practices used in making working drawings should be applied. The following procedure is recommended:

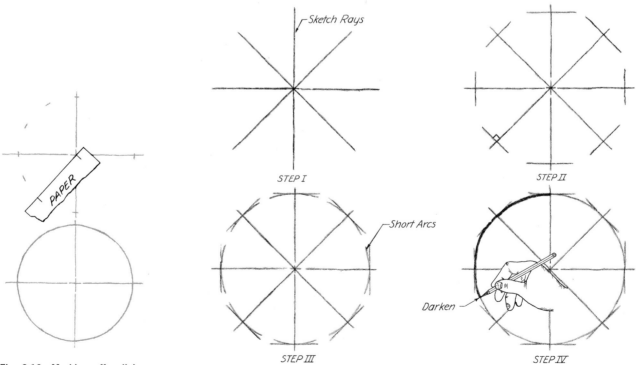

Fig. 6.12 Marking off radial distances.

Fig. 6.13 Sketching large circles.

1. Examine the object, giving particular attention to detail.

2. Determine which views are necessary.

3. "Block-in" the views, using light construction lines.

4. Complete the detail and darken the object lines.

5. Sketch extension lines and dimension lines, including arrowheads.

6. Complete the sketch by adding dimensions, notes, title, date, sketcher's name or initials, and so on.

7. Check the entire sketch carefully to see that no dimensions have been omitted.

The beginning student should read Part A of Chapter 5 before he attempts to make a multiview sketch.

6.11

Proportions. The beginner must recognize the importance of being able to estimate comparative relationships between the width, height, and depth of an object being sketched. The complete problem of proportioning a sketch also involves relating the estimated dimensions for any component parts, such as slots, holes, and projections, to the over-all dimensions of the object. It is not the practice to attempt to estimate actual dimensions, for sketches are not usually made to scale. Rather one must decide, for example, that the width of the object is twice its height, that the width of a given slot is equal to one-half the width of the object, and that its depth is approximately one-fourth the overall height.

To become proficient at sketching one must learn to recognize proportions and be able to compare dimensions "by eye." Until one is able to do so, he can not really "think with his pencil." Some people can develop a keen eye for proportion with only a limited amount of practice and can maintain these estimated proportions when making the views of a sketch. Others have alternately discouraging and encouraging experiences. Discouragement comes when one's knowledge of sketching is ahead of his ability and he has not had as much practice as he needs. The many who find it difficult to make the proportions of the completed sketch agree with the estimated proportions of the object may begin by using the graphical method shown in Figs. 6.15(*a*), (*b*), and (*c*). This

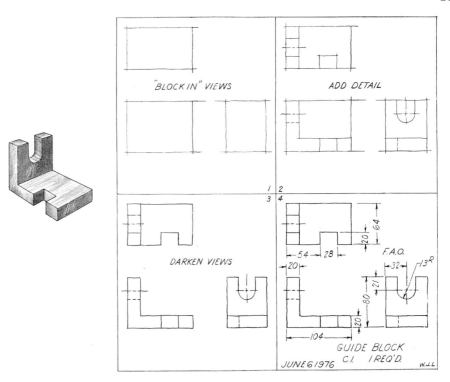

Fig. 6.14 Steps in sketching.

method is based on the fact that a rectangle (enclosing a view) may be divided to obtain intermediate distances along any side that are in such proportions to the total length as one-half, one-fourth, one-third, and so on. Those who start with this rectangle method as an aid in proportioning should abandon its use when they have developed their eye and sketching skills so that it is no longer needed.

Sketching must be done rapidly, and the addition of unnecessary lines consumes much valuable time. Furthermore, the addition of construction lines distracts the reader, and it is certain that they do not contribute to the neatness of the sketch.

The midpoint of a rectangle is the point of intersection of the diagonals, as shown in Fig. 6.15(*a*). A line sketched through

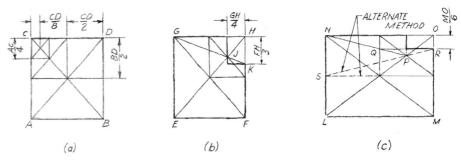

Fig. 6.15 Methods of proportioning a rectangle representing the outline of a view.

this point that is perpendicular to any side will establish the midpoint of that side. Should it be necessary to determine a distance that is equal to one-fourth the length of a side, (say, *AC*), the quarter-point may be located by repeating this procedure for the small rectangle representing the upper left-hand quarter of *ABCD*.

With the midpoint *J* located by the intersecting diagonals of the small rectangle [representing one-fourth of the larger rectangle *EFGH*, as in (*b*)], the one-third point along *FH* may be located by sketching a line from point *G* through *J* and extending it to line *FH*. The point *K* at the intersection of these lines establishes the needed one-third distance.

To determine one-sixth the length of a side of a rectangle, as in (*c*), sketch a line from *N* through point *P*, as was done in (*b*), to determine a one-third distance. Point *Q* at which the line *NP* crosses the center line of the rectangle establishes the one-sixth distance along the center line.

Figure 6.16 shows how this method for dividing the sides of a rectangle might be used to proportion an orthographic sketch.

The square may be used to proportion a view after one dimension for the view has been assumed. In this method, additional squares are added to the initial one having the assumed length as one side (Fig. 6.17). As an example, suppose that it has been estimated that the front view of an object should be three times as long as it is high. In Fig. 6.17 the height of the view has been represented by the line *AB* sketched to an assumed length. The first step in making the construction is to

sketch the initial square *ABCD* and extend *AC* and *BD* to indefinite length, being certain that the overall length from *A* and *B* will be slightly greater than three times the length of *AB*. Then the center line must be sketched through the intersection of *AD* and *BC*. Now *BX* extended to *E* locates *EF* to form the second square, and *DY* extended to point *G* locates the line *GH*. Line *AG* will be three times the length of *AB*.

D □ Sketching in isometric oblique and perspective

6.12

Pictorial Sketching. Students may employ pictorial sketches to advantage as an aid in visualizing and organizing problems. Sales engineers may frequently include pictorial sketches with orthographic sketches when preparing field reports on the needs and suggestions of the firm's customers.

With some training anyone can prepare pictorial sketches that will be satisfactory for all practical purposes. Artistic ability is not needed. This fact is important, for many persons lack only the necessary confidence to start making pictorial sketches.

6.13

Mechanical Methods of Sketching. Many engineers have found that they can produce satisfactory pictorial sketches by using one of the so-called mechanical methods. They rely on these methods because of their familiarity with the procedures used in making pictorial drawings with instruments.

The practices presented in Chapter 11 for the mechanical methods, axonometric, oblique, and perspective, are followed generally in pictorial sketching, except that angles are assumed and lengths are estimated. For this reason, one must develop an eye for good proportion before he

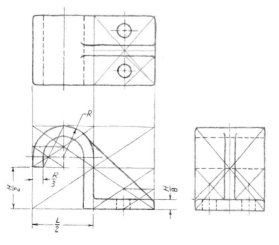

Fig. 6.16 Rectangle method applied in making an orthographic sketch.

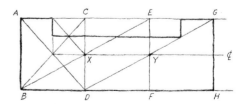

Fig. 6.17 Build-up method.

will be able to create a satisfactory pictorial sketch that will be in no way misleading.

A student having difficulty in interpreting a multiview drawing usually will find that a pictorial sketch, prepared as illustrated in Fig. 6.18, will clarify the form that he is trying to visualize, even before the last lines of the sketch have been drawn.

6.14

Isometric Sketching. Isometric sketching starts with three isometric lines, called axes, which represent three mutually perpendicular lines. One of these axes is sketched vertically, the other two at 30° with the horizontal. In Fig. 6.18 (step I), the near front corner of the enclosing box lies along the vertical axis, while the two visible receding edges of the base lie along the axes receding to the left and to the right.

If the object is of simple rectangular form, as in Fig. 6.18, it may be sketched by drawing an enclosing isometric box (step I) on the surfaces of which the orthographic views may be sketched (step II). Care must be taken in assuming lengths and distances so that the finished view (step III) will have relatively correct proportions. In constructing the enclosing box (step I), the vertical edges are parallel to the vertical axis, and edges receding to the right and to the left are parallel to the right and left axes, respectively.

Objects of more complicated construction may be "blocked in," as shown in Fig. 6.19. Note that the projecting cylindrical features are enclosed in "isometric" prisms and that the circles are sketched within isometric squares. The procedure in Fig. 6.19 is the same as in Fig. 6.18, except that three enclosing isometric boxes are needed in the formation of the final representation instead of one.

In sketching an ellipse to represent a circle pictorially, an enclosing "isometric square" (rhombus) is drawn having sides equal approximately to the diameter of the true circle (step I, Fig. 6.20). The ellipse is formed by first drawing arcs tangent to the midpoints of the sides of the isometric square in light sketchy pencil lines (step II). In finishing the ellipse (step III) with a dark heavy line, care must be taken to obtain a nearly elliptical shape.

Figure 6.21 shows the three positions for an isometric circle. Note that the major axis is horizontal for an ellipse on a horizontal plane (I).

An idea sketch prepared in isometric is shown in Fig. 6.22.

6.15

Proportioning. As stated in Sec. 6.11, one should eventually be able to judge lengths and recognize proportions. Until this ultimate goal has been reached, the graphical method presented in Fig. 6.15 may be used with pictorial sketching (Fig. 6.23).

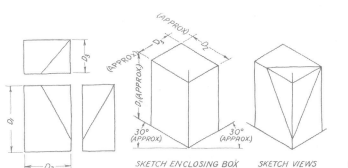
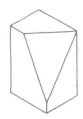
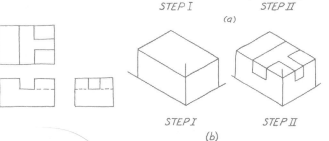

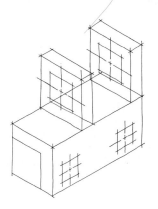
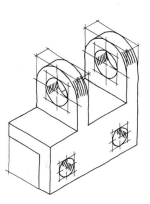

Fig. 6.18 Steps in isometric sketching.

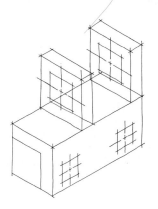

Fig. 6.19 Blocking in an isometric sketch.

*SKETCH "ISOMETRIC SQUARE"
STEP I*   *SKETCH SHORT ARCS
STEP II*   *COMPLETE ELLIPSE
STEP III*

Fig. 6.20 Isometric circles.

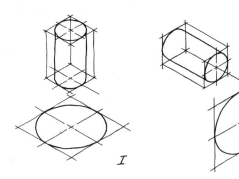

Fig. 6.21 Isometric circles.

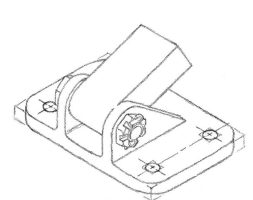

Fig. 6.22 Idea sketch in isometric.

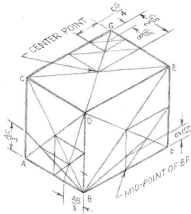

Fig. 6.23 Method for proportioning a rhomboid.

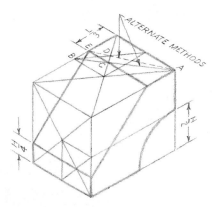

Fig. 6.24 Proportioning method applied.

6.16

Sketches in Oblique. A sketch in oblique shows the front face without distortion, in its true shape. It has this one advantage over a representation prepared in isometric, even though the final result usually will not present so pleasing an appearance. It is not recommended for objects having circular or irregularly curved features on any but the front plane or in a plane parallel to it.

The beginner who is familiar with axonometric sketching will have very little difficulty in preparing a sketch in oblique, for, in general, the methods of preparation presented in the previous sections apply to both. The principal difference between these two forms of sketching is in the position of the axes, oblique sketching being unlike the isometric in that two of the axes are at right angles to each other. The third axis may be at any convenient angle, as indicated in Fig. 6.25.

Figure 6.26 shows the steps in making an oblique sketch using the proportioning methods previously explained for dividing a rectangle and a rhomboid. The receding lines are made parallel when a sketch is made in oblique projection.

The distortion and illusion of extreme elongation in the direction of the receding axis may be minimized by foreshortening to obtain proportions that are more realistic to the eye and by making the receding lines converge slightly. The resulting sketch will then be in a form of pseudo-perspective, which resembles parallel perspective to some extent.

6.17

Perspective Sketching. A sketch that has been prepared in accordance with the concepts of perspective will present a somewhat more pleasing and realistic effect than one in oblique or isometric. A perspective sketch actually presents an object as it would appear when observed from a particular point. The recognition of

The procedures as used are identical, the only recognizable difference being that the rectangle in the first case now becomes a rhomboid. Figure 6.24 illustrates how the method might be applied in making a sketch of a simple object. The enclosing box was sketched first with light lines, and then the graphical method was applied as shown to locate the points at one-quarter and one-half of the height. To establish the line of the top surface that is at a distance equal to one-third of the length from the end, a construction line was sketched from A to the midpoint B to locate C at the point of intersection of AB with the diagonal. Point C will fall on the required line.

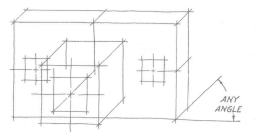

Fig. 6.25 Blocking in an oblique sketch.

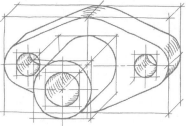

this fact, along with an understanding of the concepts that an object will appear smaller at a distance than when it is close and that horizontal lines converge as they recede until they meet at a vanishing point, should enable one to produce sketches having a perspective appearance. In sketching an actual object, a position should be selected that will show it to the best advantage. When the object exists only in one's mind or on paper in orthographic form, then the object must be visualized and the viewing position assumed.

At the start, the principal lines should be sketched in lightly, each line extending for some length toward its vanishing point. After this has been accomplished, the enclosing perspective squares for circles should be blocked in and the outline for minor details added. When the object lines have been darkened, the construction lines extending beyond the figure may be erased.

Figure 6.27 shows a parallel or one-point perspective that bears some resemblance to an oblique sketch. All faces in planes parallel to the front show their true shape. All receding lines should meet at a single vanishing point. Figure 6.28 is an angular or two-point perspective.

As stated previously, a two-point perspective sketch shows an object as it would appear to the human eye at a fixed point in space and not as it actually exists. All parallel receding lines converge (Fig. 6.29). Should these receding lines be horizontal, they will converge at a vanishing point on the eyeline. Those lines extending toward the right converge to a vanishing point to the right (VP_R), and those to the left converge to the left (VP_L). These vanishing points are at the level of the observer's eye (Fig. 6.29). A system of lines that is neither perpendicular nor horizontal will converge to a VP for inclined lines.

In one-point perspective one of the principal faces is parallel to the picture plane. All of the vertical lines will appear as vertical, and the receding horizontal lines will converge to a single vanishing point (Fig. 6.30).

Those interested in a complete discussion of the geometry of perspective drawing should read Secs. 11.27–11.40. The beginner should make two or three mechanically drawn perspectives at the start to fix the fundamentals of the methods of perspective projection in his mind, even though there is some difference between sketching what one sees or imagines and true geometrical perspective.

In making a sketch in artist's perspective, several fundamental concepts must be recognized.

First, a circle sketched in perspective will resemble an ellipse (Fig. 6.29). The long diameter of the repesentation of a circle on a horizontal plane is always in a horizontal direction (Fig. 6.31).

Second, if an object or a component part of an object is above the eyeline, it will be seen from below. Should the object be below the eyeline, it will be seen from above (Fig. 6.29). The farther the object is removed below or above the eyeline, the more one can see of the top or the bottom surface.

Third, the nearest vertical edge of an object will be the longest vertical line of the view, as shown in Fig. 6.29. When two or more objects of the same actual height appear in a perspective sketch, their represented heights will decrease in the view as they near the vanishing point.

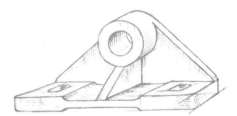

Fig. 6.27 Sketch in parallel perspective.

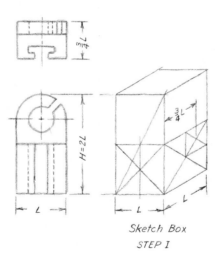

Sketch Box
STEP I

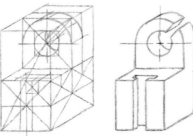

Block-in Outline   *Complete Pictorial*
STEP II   *STEP III*

Fig. 6.26 Steps in oblique sketching.

Fig. 6.28 Sketch in angular perspective.

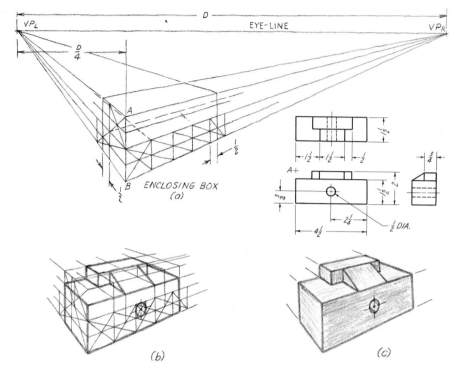

Fig. 6.29 Preparing a perspective sketch.

6.18

Preparing a Perspective Sketch. The application of proportioning methods to the construction and division of an enclosing box is shown in Fig. 6.29. Read Secs. 6.11 and 6.15. The construction of a required perspective by steps is as follows:

> **Step I.** Sketch the eyeline. This line should be well toward the top of the sheet of sketch paper.
>
> **Step II.** Locate VP_L and VP_R on the eyeline. These vanishing points should be placed as far apart as possible.

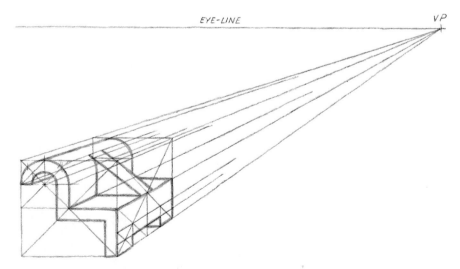

Fig. 6.30 Sketch made in one-point perspective.

Step III. Assume the position and length for the near front edge AB. The length of this line, along with the spacing of the vanishing points, establishes the size of the finished sketch. The position of AB determines how the visible surfaces are to appear. For instance, if the line AB had been moved downward from the position shown in (a), much more of the top surface would be seen in (b) and (c). Should AB have been moved to the right from its position shown in (a), the left side would have become more prominent. If AB were placed midway between the two vanishing points, then both the front and left-side surfaces would be at 45° with the picture plane for the perspective. As it has been placed in (a), the side face is at 60° to the picture plane, while the front face is at 30°. Before establishing the position for the near front edge, one must decide which surfaces are the most important surfaces and how they may best be displayed in the sketch.

Step IV. Sketch light construction lines from points A and B to each vanishing point.

Step V. Determine the proportions for the enclosing box, in this case $4\frac{1}{2}$, 2, and $1\frac{1}{2}$, and mark off two equal units along AB.

Step VI. Sketch perspective squares, representing the faces of 1-unit cubes, starting at AB and working toward each vanishing point. In cases where an overall length must be completed with a partial unit, a full-perspective square must be sketched at the end.

Step VII. Subdivide any of the end squares if necessary and sketch in the enclosing box (a).

Step VIII. Locate and block in the details, subdividing the perspective squares as required to establish the location of any detail. When circles are to be sketched in perspective by a beginner, it is advisable to sketch the enclosing box first using light lines (b).

Step IX. Darken the object lines of the sketch. Construction lines may be removed and some shading added to the surfaces as shown in (c), if desired.

A sketch in one-point perspective might be made as shown in Fig. 6.30. For this particular sketch the enclosing box was made to the assumed overall proportions for the part. Then the location of the details was established by subdividing the regular rectangle of the front face of the enclosing box and the perspective rectangle of the right side.

Pencil Shading. The addition of some shading to the surfaces of a part will force its form to stand out against the white surface of the sketching paper and will increase the effect of depth in a view that might otherwise appear to be somewhat flat.

Seldom are technologists and engineers able to do creditable work in artistic shading, with cast shadows included, as they could many years ago when training in art was part of an engineer's education. It is unfortunate that they lack this training at the present time, for art and design go hand in hand. This is especially true today, for a pleasing and appealing styling sells more products than a good mechanical design.

Within the scope of this chapter, written for beginning students, it will only be possible to present a few simple rules as a guide for those making a first attempt at surface shading. However, continued practice and some thought should lead one to the point where he can do a creditable job of shading and definitely improve a pictorial sketch.

When shading an imaginary part, a designer may consider the source of the light to be located in a position to the left, above, and in front of the object. Of course, if the part actually exists and is being sketched by viewing it, then the sketcher should attempt to duplicate the degrees of shade and shadows as they are observed.

With the light source considered to be to the left, above, and in front of the object, a rectangular part would be shaded as shown in Fig. 6.31(a). The use of gradation of tone on the surfaces gives additional emphasis to the depth. To secure this added effect by shading, the darkest tone on the surface that is away from the light must be closest to the eye. As the surface recedes, the tone must be made lighter in value with a faint trace of reflected light showing along the entire length of the back edge. On the lighted side, the lightest area must be closest to the eye, as indicated by the letter L_1 in (a). To make this lighted face appear to go into the distance, it is made darker as it recedes, but it should never be made as dark as the lightest of the dark tones on the dark surface.

Shading a cylindrical part is not as easy to do as shading a rectangular part but, if it is realized that practically half of the cylinder is in the light and half in the dark

and that the lightest light and the darkest dark fall along the elements at the quarter-points of each half, then one should not find the task too difficult (b). The two extremes are separated by lighter values of shade. The first quarter on the lighted side must be made lighter than the last quarter on the dark side. In starting at the left and going counterclockwise there is a dark shade of light blending into the full light at the first quarter-point. From this point and passing the center to the dark line, the tone should become gradually darker. If vertical lines are used for shading, they should be spaced closer and closer together as they approach the dark line. The extreme-right-hand quarter should show the tones of reflected light.

There are two ways that pencil shading may be applied. If the paper has a medium-rough surface, solid tone shading may be used with one shade blending into the other. For the best results, the light tones are put on first over all areas to be shaded. The darker tones are then added by building up lighter tones to the desired intensity for a particular area. For this form of shading, a pencil with flattened point is used.

The other form of shading, and the one that is best suited for quick sketches, is produced with lines of varied spacing and weight. Light lines with wide spacing are used on the light areas and heavy lines that are closely spaced give the tone for the darkest areas. No lines are needed for the lightest of the light areas.

6.20

Conventional Treatment of Fillets, Rounds, and Screw Threads. Sketches that are not given full pencil shading may be given a more or less realistic appearance by representing the fillets and rounds of the unfinished surfaces as shown in Fig. 11.41. The conventional treatment for screw threads is shown in (b) and (c) of the same illustration.

L_1 – LIGHTEST LIGHT
L_2 – HALFTONE (HALF LIGHT)
D_1 – DARKEST DARK
D_2 – DARK

(a) (b)

Fig. 6.31 Shading rectangular and cylindrical parts.

6.21

Use of an Overlay Sheet. In preparing a design sketch, an overlay sheet may be used to advantage in making a sketch that is complicated by many details (Fig. 6.32). In this case, a quick sketch showing the general outline of the principal parts is made first in a rather rough form. Then an overlay sheet is placed over this outline sketch and the lines are retraced. In doing so, slight corrections can be made for any errors existing in the proportions of the parts or in the position of any of the lines of the original rough sketch. When this has been done, the representation of the

related minor parts are added. If at any time one becomes discouraged with a sketch (multiview or pictorial) that he is making and feels that he should make a new start, he should use an overlay sheet, for there are usually many features on his existing sketch that may be retraced with a great saving of time.

6.22

Illustration Sketches Showing Mechanisms Exploded. A sketch of a mechanism showing the parts in exploded positions along the principal axes is shown in Fig. 6.33. Through the use of such sketches those who have not been trained to read multiview drawings may readily understand how a mechanism should be assembled, for both the shapes of the parts and their order of assembly, as denoted by their space relationship, is shown in pictorial form.

Illustration sketches may be made for discussions dealing with ideas for a design, but more frequently they are prepared for explanatory purposes to clarify instructions for preparing illustration drawings in a more finished form as assembly illustrations, advertising illustrations, catalogue illustrations, and illustrations for service and repair charts.

Many persons find that it is desirable, when preparing sketches of exploded mechanisms, to first block in the complete mechanism with all of the parts in position. At this initial stage of construction, the parts are sketched in perspective in rough outline and the principal axes and object lines are extended partway toward their vanishing points.

When the rough layout has been completed to the satisfaction of the person preparing the sketch, an overlay sheet is placed over the original sketch and the parts are traced directly from the sketch beneath in exploded positions along the principal axes and along the axes of holes. Frequently some beginners make a traced sketch of each individual part and place the sketches in exploded positions before preparing the finished sketch. Others with more experience accomplish the same results by first tracing the major part along with the principal axes and then moving the overlay sheet as required to trace off the remaining parts in their correct positions along the axes.

It should be recognized that in preparing sketches of exploded mechanisms in this manner the parts are not shown perspectively reduced although they are

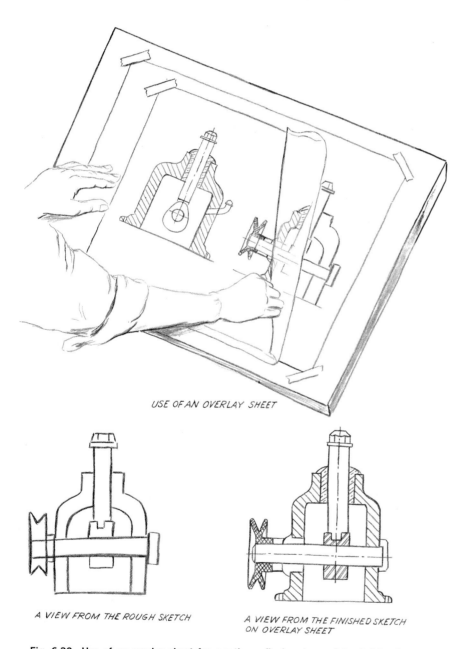

USE OF AN OVERLAY SHEET

A VIEW FROM THE ROUGH SKETCH

A VIEW FROM THE FINISHED SKETCH ON OVERLAY SHEET

Fig. 6.32 Use of an overlay sheet for creating a final and complete sketch of a mechanism.

removed from their original place in the pictorial assembly outward and toward the vanishing points. To prepare a sketch of this type with all of the parts shown in true geometrical perspective would result in a general picture that would be misleading and one that would be apt to confuse the nontechnical person because some parts would appear much too large or too small to be mating parts.

6.23
Pictorial Sketching on Ruled Paper. Although one should become proficient in sketching on plain white bond paper, a specially ruled paper, shown in Fig. 6.34, can be used by those who need the help of guidelines.

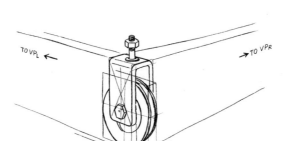

Block-in assembly in outline

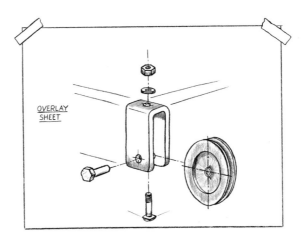

Retrace parts on overlay sheet in exploded positions from assembly sketch

Fig. 6.33 Sketch showing the parts of a mechanism in exploded positions.

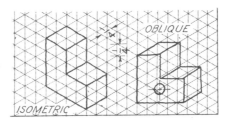

Fig. 6.34 Sketches on isometric paper.

Problems

The problems presented with this chapter have been selected to furnish practice in freehand drawing. The individual pieces that appear in pictorial form have been taken from a wide variety of mechanisms used in different fields of engineering. The student may be required to prepare complete working sketches of these parts if his instructor desires. Such an assignment, however, would presuppose an understanding of the fundamentals of dimensioning as they are presented in the beginning sections of Chapter 13.

The problems presented in Fig. 6.53–6.60 were selected to give practice in preparing pictorial sketches—isometric, oblique, and perspective. In addition to developing proficiency in sketching, these problems offer the student further opportunity to gain experience in reading drawings. Additional problems that are suitable for pictorial sketching may be found in Chapter 11.

1. (Fig. 6.35). Reproduce the one-view sketch on a sheet of sketching paper.

2–4. (Figs. 6.36–6.38). Sketch, freehand, the necessary views of the given objects as assigned. The selected length for the unit will determine the size of the views. Assume any needed dimensions that are not given in units.

5–14. (Figs. 6.39–6.48). These problems are designed to give the student further study in multiview representation and, at the same time, offer him the opportunity to apply good line technique to the preparation of sketches.

Only the necessary views on which all of the hidden lines are to be shown should be drawn.

If dimensions are to be given, ample space must be allowed between the views for their placement. The beginning sections of Chapter 13 present the basic principles of size description.

15. (Fig. 6.49). Make a complete three-view sketch of the motor base. The ribs are $\frac{3}{8}$ in. thick. At points A, four holes are to be drilled for $\frac{1}{2}$-in. bolts that are to be $2\frac{1}{2}$ in. center to center in one direction and $3\frac{1}{8}$ in. in the other. At points B four holes are to be drilled for $\frac{1}{2}$-in. bolts that fasten the motor base to a steel column. Fillets and rounds are $\frac{1}{8}$ R.

16. (Fig. 6.50). Make a three-view orthographic sketch of the motor bracket.

17. (Fig. 6.51). Make a complete three-view sketch of the tool rest and/or the tool rest bracket. The rectangular top surface of the tool rest is to be $1\frac{1}{8}$ in. above the center line of the hole for the $\frac{7}{16}$ in. bolt. The overall dimensions

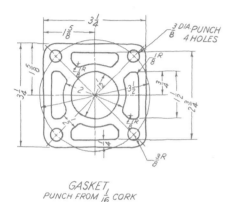

Fig. 6.35

Fig. 6.36 Wedge block.

Fig. 6.37 End block.

Fig. 6.38 Corner block.

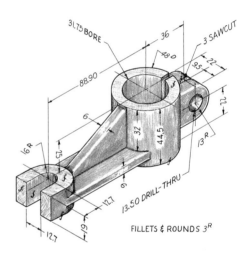

Fig. 6.39 Shifter.

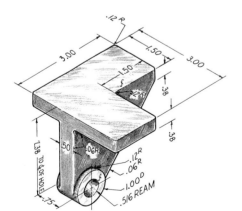

Fig. 6.40 Tool rest.

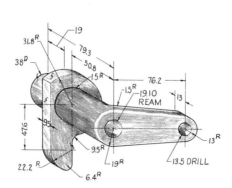

Fig. 6.41 Offset trip lever.

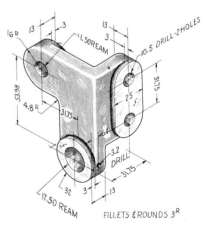

Fig. 6.42 Guide link.

removed from their original place in the pictorial assembly outward and toward the vanishing points. To prepare a sketch of this type with all of the parts shown in true geometrical perspective would result in a general picture that would be misleading and one that would be apt to confuse the nontechnical person because some parts would appear much too large or too small to be mating parts.

6.23
Pictorial Sketching on Ruled Paper. Although one should become proficient in sketching on plain white bond paper, a specially ruled paper, shown in Fig. 6.34, can be used by those who need the help of guidelines.

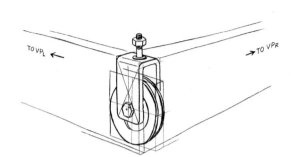

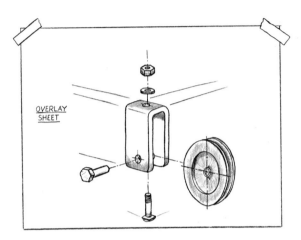

Block-in assembly in outline

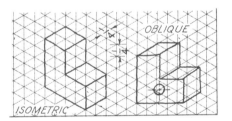

Retrace parts on overlay sheet in
exploded positions from assembly sketch

Fig. 6.33 Sketch showing the parts of a mechanism in exploded positions.

Fig. 6.34 Sketches on isometric paper.

Problems

The problems presented with this chapter have been selected to furnish practice in freehand drawing. The individual pieces that appear in pictorial form have been taken from a wide variety of mechanisms used in different fields of engineering. The student may be required to prepare complete working sketches of these parts if his instructor desires. Such an assignment, however, would presuppose an understanding of the fundamentals of dimensioning as they are presented in the beginning sections of Chapter 13.

The problems presented in Fig. 6.53–6.60 were selected to give practice in preparing pictorial sketches—isometric, oblique, and perspective. In addition to developing proficiency in sketching, these problems offer the student further opportunity to gain experience in reading drawings. Additional problems that are suitable for pictorial sketching may be found in Chapter 11.

1. (Fig. 6.35). Reproduce the one-view sketch on a sheet of sketching paper.

2–4. (Figs. 6.36–6.38). Sketch, freehand, the necessary views of the given objects as assigned. The selected length for the unit will determine the size of the views. Assume any needed dimensions that are not given in units.

5–14. (Figs. 6.39–6.48). These problems are designed to give the student further study in multiview representation and, at the same time, offer him the opportunity to apply good line technique to the preparation of sketches.

Only the necessary views on which all of the hidden lines are to be shown should be drawn.

If dimensions are to be given, ample space must be allowed between the views for their placement. The beginning sections of Chapter 13 present the basic principles of size description.

15. (Fig. 6.49). Make a complete three-view sketch of the motor base. The ribs are $\frac{3}{8}$ in. thick. At points A, four holes are to be drilled for $\frac{1}{2}$-in. bolts that are to be $2\frac{1}{2}$ in. center to center in one direction and $3\frac{1}{8}$ in. in the other. At points B four holes are to be drilled for $\frac{1}{2}$-in. bolts that fasten the motor base to a steel column. Fillets and rounds are $\frac{1}{8}$ R.

16. (Fig. 6.50). Make a three-view orthographic sketch of the motor bracket.

17. (Fig. 6.51). Make a complete three-view sketch of the tool rest and/or the tool rest bracket. The rectangular top surface of the tool rest is to be $1\frac{1}{8}$ in. above the center line of the hole for the $\frac{7}{16}$ in. bolt. The overall dimensions

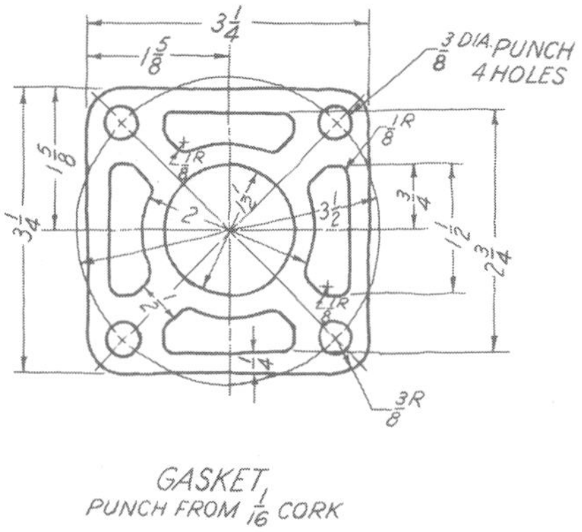

GASKET
PUNCH FROM $\frac{1}{16}$ CORK

Fig. 6.35

Fig. 6.36   Wedge block.

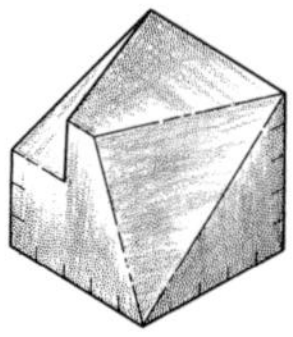

Fig. 6.37   End block.

Fig. 6.38   Corner block.

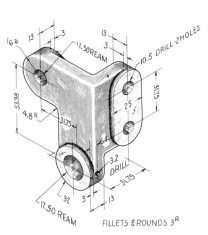

FILLETS & ROUNDS 3^R

Fig. 6.39   Shifter.

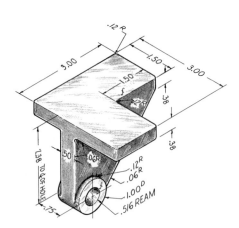

Fig. 6.40   Tool rest.

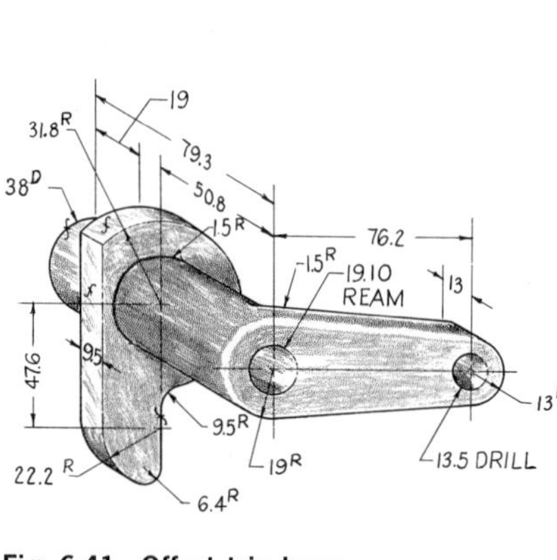

Fig. 6.41   Offset trip lever.

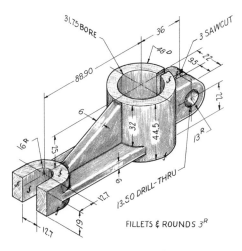

FILLETS & ROUNDS 3^R

Fig. 6.42   Guide link.

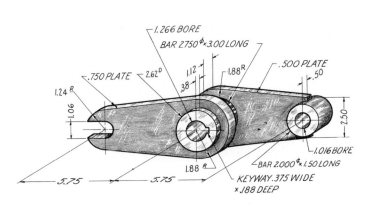

Fig. 6.43 Idler lever (weldment).

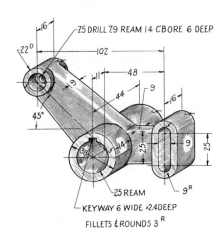

Fig. 6.44 Control bracket.

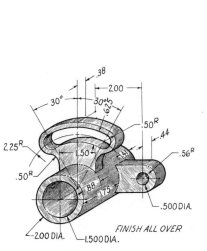

Fig. 6.45 Index guide.

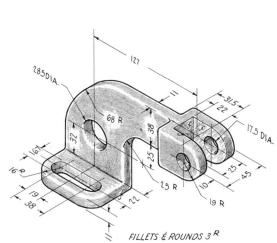

Fig. 6.46 Arm bracket.

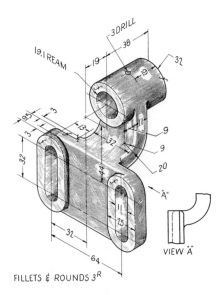

Fig. 6.47 Bearing bracket.

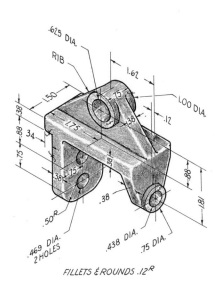

Fig. 6.48 Rear support bracket.

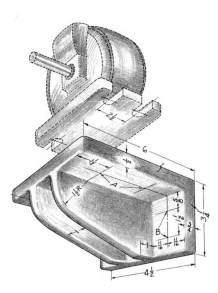

Fig. 6.49 Motor base.

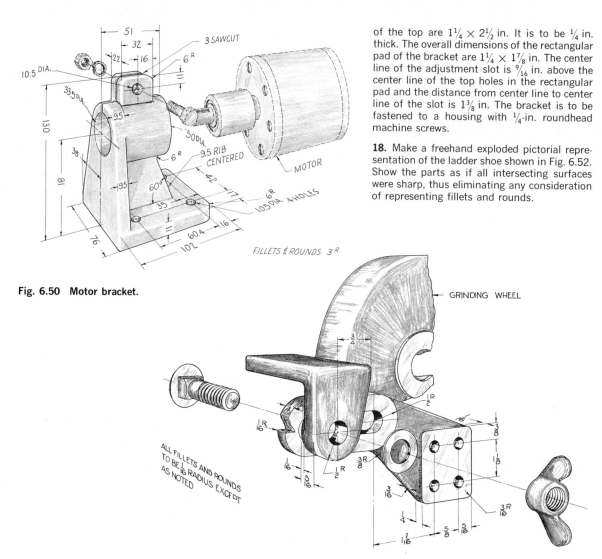

Fig. 6.50 Motor bracket.

of the top are $1\frac{1}{4} \times 2\frac{1}{2}$ in. It is to be $\frac{1}{4}$ in. thick. The overall dimensions of the rectangular pad of the bracket are $1\frac{1}{4} \times 1\frac{7}{8}$ in. The center line of the adjustment slot is $\frac{9}{16}$ in. above the center line of the top holes in the rectangular pad and the distance from center line to center line of the slot is $1\frac{3}{8}$ in. The bracket is to be fastened to a housing with $\frac{1}{4}$-in. roundhead machine screws.

18. Make a freehand exploded pictorial representation of the ladder shoe shown in Fig. 6.52. Show the parts as if all intersecting surfaces were sharp, thus eliminating any consideration of representing fillets and rounds.

Fig. 6.51 Tool rest and tool rest bracket.

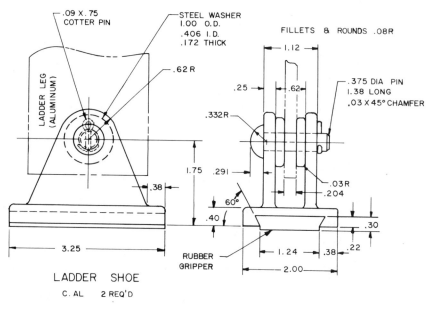

Fig. 6.52 Ladder shoe.

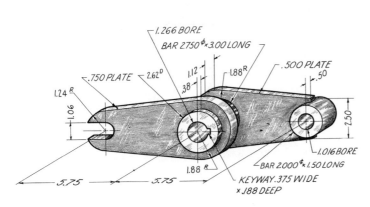

Fig. 6.43 Idler lever (weldment).

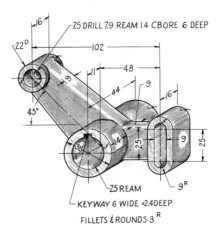

Fig. 6.44 Control bracket.

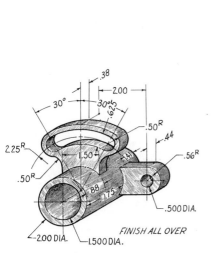

Fig. 6.45 Index guide.

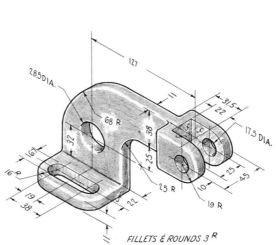

Fig. 6.46 Arm bracket.

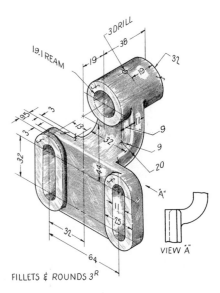

Fig. 6.47 Bearing bracket.

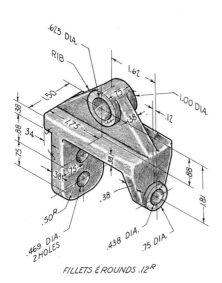

Fig. 6.48 Rear support bracket.

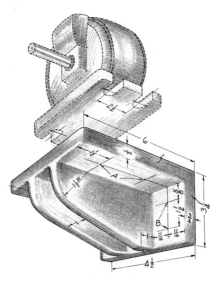

Fig. 6.49 Motor base.

*Freehand sketching
for visualization and
communication*

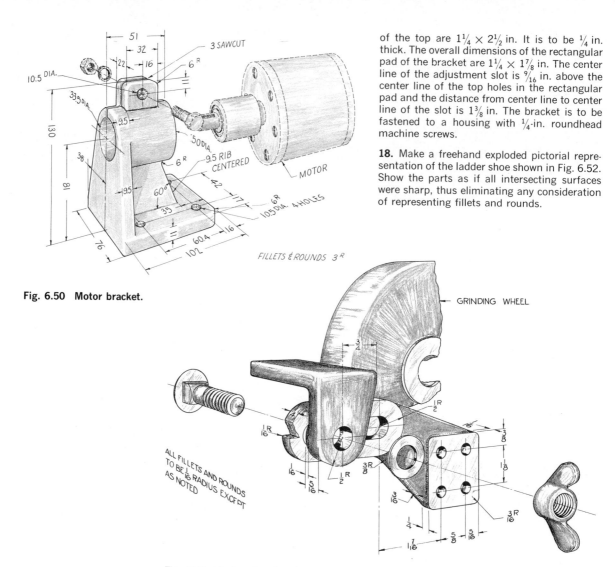

Fig. 6.50 Motor bracket.

of the top are $1\frac{1}{4} \times 2\frac{1}{2}$ in. It is to be $\frac{1}{4}$ in. thick. The overall dimensions of the rectangular pad of the bracket are $1\frac{1}{4} \times 1\frac{7}{8}$ in. The center line of the adjustment slot is $\frac{9}{16}$ in. above the center line of the top holes in the rectangular pad and the distance from center line to center line of the slot is $1\frac{3}{8}$ in. The bracket is to be fastened to a housing with $\frac{1}{4}$-in. roundhead machine screws.

18. Make a freehand exploded pictorial representation of the ladder shoe shown in Fig. 6.52. Show the parts as if all intersecting surfaces were sharp, thus eliminating any consideration of representing fillets and rounds.

Fig. 6.51 Tool rest and tool rest bracket.

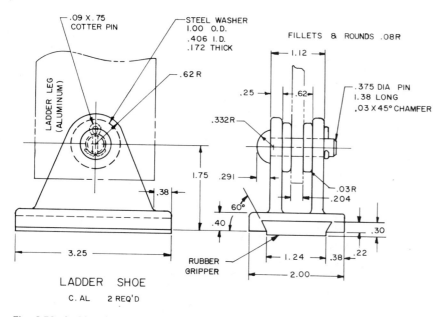

Fig. 6.52 Ladder shoe.

19–24. (Figs. 6.53–6.58). Make freehand isometric sketches of the objects as assigned.

25–28. (Figs. 6.57–6.60). Make freehand oblique sketches of the objects as assigned.

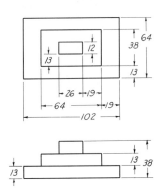

Dimensions are in millimeters

Fig. 6.53

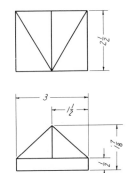

Fig. 6.54

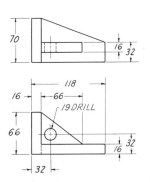

Fig. 6.55

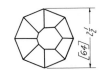

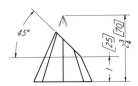

Dimensions in [ ] are in millimeters

Fig. 6.56

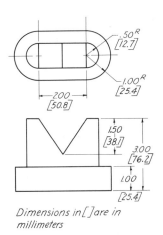

Dimensions in [ ] are in millimeters

Fig. 6.57

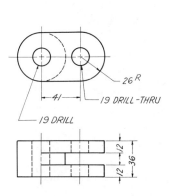

Fig. 6.58

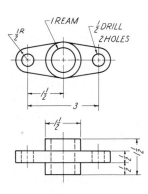

Fig. 6.59

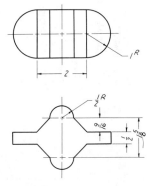

Fig. 6.60

Alcoa Seaprobe at work. The design of this recovery ship required the application of the principles of multiview drawing in the development of creative ideas. (*Courtesy Aluminum Corporation of America*)

Alcoa Seaprobe at work. The design of this recovery ship required the application of
the principles of multiview drawing in the development of creative ideas. (*Courtesy
Aluminum Corporation of America*)

19–24. (Figs. 6.53–6.58). Make freehand isometric sketches of the objects as assigned.

25–28. (Figs. 6.57–6.60). Make freehand oblique sketches of the objects as assigned.

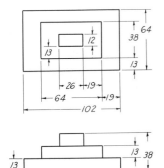

Dimensions are in millimeters

Fig. 6.53

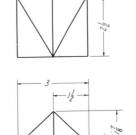

Fig. 6.54

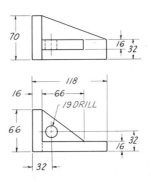

Fig. 6.55

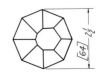

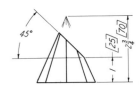

Dimensions in [ ] are in millimeters

Fig. 6.56

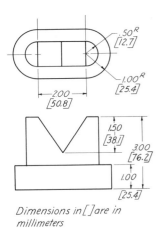

Dimensions in [ ] are in millimeters

Fig. 6.57

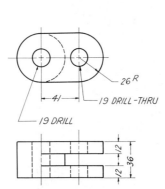

Fig. 6.58

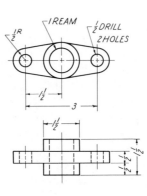

Fig. 6.59

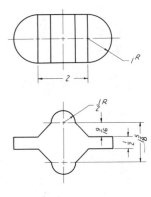

Fig. 6.60

7

Sectional views

7.1

Sectional Views (Fig. 7.1). Although the invisible features of a simple object usually may be described on an exterior view by the use of hidden lines, it is unwise to depend on a perplexing mass of such lines to describe adequately the interior of a complicated object or an assembled mechanism. Whenever a representation becomes so confused that it is difficult to read, it is customary to make one or more of the views "in section" (Fig. 7.2). A view "in section" is one obtained by imagining the object to have been cut by a cutting plane, the front portion being removed to reveal clearly the interior features. Figure 7.3 illustrates the use of an imaginary cutting plane. At this point it should be understood that a portion is shown removed only in a sectional view, not in any of the other views (Fig. 7.3).

When the cutting plane cuts an object lengthwise, the section obtained is commonly called a longitudinal section; when crosswise, it is called a cross section. It is designated as being either a full section, a half section, or a broken section. If the plane cuts entirely across the object, the section represented is known as a *full section*. If it cuts only halfway across a symmetrical object, the section is a *half section*. A *broken section* is a partial one, which is used when less than a half section is needed (Fig. 7.3).

On a completed sectional view, fine section lines are drawn across the surface cut by the imaginary plane, to emphasize the contour of the interior (see Sec. 7.8).

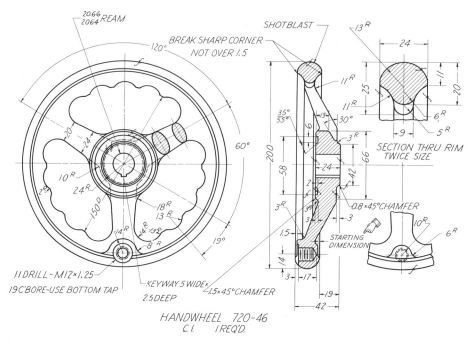

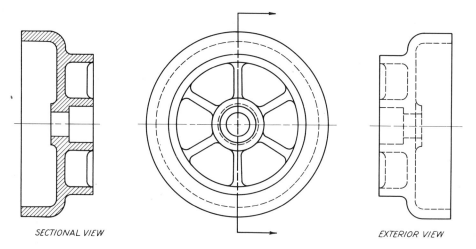

Fig. 7.1 Working drawing with sectional views. (*Courtesy Warner and Swasey Co.*)

SECTIONAL VIEW

Fig. 7.2 Sectional view.

EXTERIOR VIEW

needed in addition to a sectioned front view.

The procedure in making a full-sectional view is simple, in that the sectional view is an orthographic one. The imaginary cut face of the object simply is shown as it would appear to an observer looking directly at it from a point an infinite distance away. In any sectional view, it is considered good practice to omit all invisible lines unless such lines are necessary to clarify the representation. Even then they should be used sparingly.

7.3

Half Section. The cutting plane for a half section removes one-quarter of an object. The plane cuts halfway through to the axis or center line so that half the finished sectional view appears in section and half appears as an external view (Fig. 7.3). This type of sectional view is used when a view is needed showing both the exterior and interior construction of a symmetrical object. Good practice dictates that hidden lines be omitted from both halves of the view unless they are absolutely necessary for dimensioning purposes or for explaining the construction. Although the use of a solid object line to separate the two halves of a half section has been approved by the Society of Automotive Engineers and has been accepted by the American National Standards Institute [Fig. 7.5(*a*)], many draftsmen prefer to use a center line, as

7.2

Full Section. Since a cutting plane that cuts a full section passes entirely through an object, the resulting view will appear as illustrated in Fig. 7.3. Although the plane usually passes along the main axis, it may be offset (Fig. 7.4) to reveal important features.

A full-sectional view, showing an object's characteristic shape, usually replaces an exterior front view; however, one of the other principal views, side or top, may be converted to a sectional view if some interior feature thus can be shown to better advantage or if such a view is

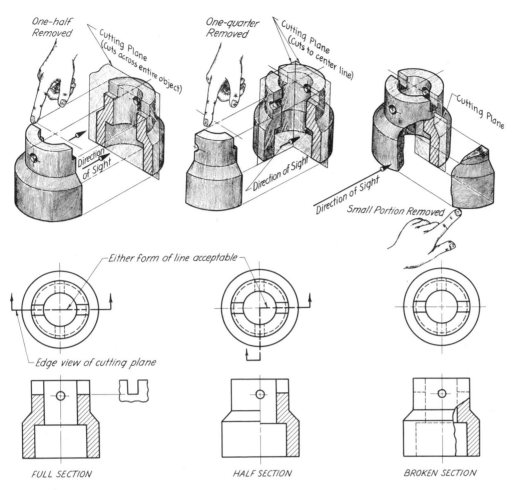

Fig. 7.3 **Types of sectional views.**

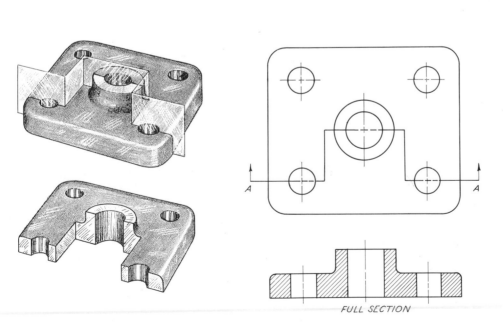

Fig. 7.4 **Offset cutting plane.**

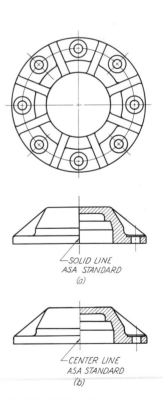

Fig. 7.5 **Half section.**

shown in Fig. 7.5(*b*). They reason that the removal of a quarter of the object is theoretical and imaginary and that an actual edge, which would be implied by a solid line, does not exist. The center line is taken as denoting a theoretical edge.

7.4

Broken Section. A broken or partial section is used mainly to expose the interior of objects so constructed that less than a half section is required for a satisfactory description (Fig. 7.6). The object theoretically is cut by a cutting plane and the front portion is removed by breaking it away. The "breaking away" gives an irregular boundary line to the section.

7.5

Revolved Section. A revolved section is useful for showing the true shape of the cross section of some elongated object,

such as a bar, or some feature of an object, such as an arm, spoke, or rib (Figs. 7.1 and 7.7).

To obtain such a cross section, an imaginary cutting plane is passed through the member perpendicular to the longitudinal axis and then is revolved through 90° to bring the resulting view into the plane of the paper (Fig. 7.8). When revolved, the section should show in its true shape and in its true revolved position, regardless of the location of the lines of the exterior view. If any lines of the view interfere with the revolved section, they should be omitted (Fig. 7.9). It is sometimes advisable to provide an open space for the section by making a break in the object (Fig. 7.7).

7.6

Removed (Detail) Sections. A removed section is similar to a revolved section,

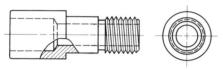

Fig. 7.6 Broken section.

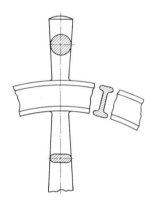

Fig. 7.7 Revolved section.*

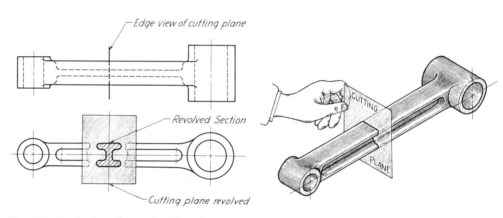

Fig. 7.8 Revolved section and cutting plane.

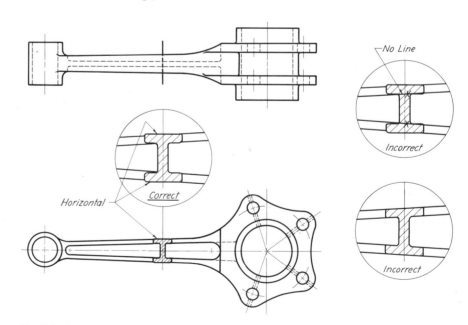

Fig. 7.9 Correct and incorrect treatment of a revolved section.

*ANSI Y14.2-1957.

except that it does not appear on an external view but instead is drawn "out of place" and appears adjacent to it (Fig. 7.10). There are two good reasons why detail sections frequently are desirable. First, their use may prevent a principal view of an object, the cross section of which is not uniform, from being cluttered with numerous revolved sections (Fig. 7.11). Second, they may be drawn to an enlarged scale in order to emphasize detail and allow for adequate dimensioning.

Whenever a detail section is used, there must be some means of identifying it. Usually this is accomplished by showing the cutting plane on the principal view and then labeling both the plane and the resulting view, as shown in Fig. 7.11.

7.7

Phantom Sections. A phantom or hidden section is a regular exterior view on which the interior construction is emphasized by crosshatching an imaginary cut surface with dotted section lines (Fig. 7.12). This type of section is used only when a regular section or a broken section would remove some important exterior detail or, in some instances, to show an accompanying part in its relative position with regard to a particular part (Fig. 7.13). Instead of using a broken line with dashes of equal length, the phantom line shown in Fig. 2.58 could have been used to represent the outline of the adjacent parts shown in Fig. 7.13.

7.8

Section Lining. Section lines are light continuous lines drawn across the imaginary cut surface of an object for the purpose of emphasizing the contour of its interior. Usually they are drawn at an angle of 45° except in cases where a number of adjacent parts are shown assembled (Fig. 7.17).

To be pleasing in appearance, these lines must be correctly executed. While on ordinary work they are spaced about .09 in. (2 mm) apart, there is no set rule governing their spacing. They simply should be spaced to suit the drawing and the size of the areas to be crosshatched. For example, on small views having small areas, the section lines may be as close as .03 in. (0.8 mm), while on large views having large areas they may be as far apart as .12 in. (3 mm). In the case of very thin plates, the cross section is shown solid black (Fig. 7.14).

The usual mistake of the beginning student is to draw the lines too close together. This, plus the unavoidable slight variations, causes the section lining to appear streaked. Although several forms of mechanical section liners are available, most draftsmen do their spacing by eye. The student is advised to do likewise, being careful to see that the initial pitch, as set by the first few lines, is maintained across the area. To accomplish this, he should check back from time to time to make sure there has been no slight gen-

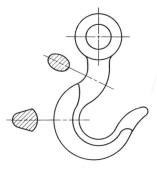

Fig. 7.10 Removed sections.*

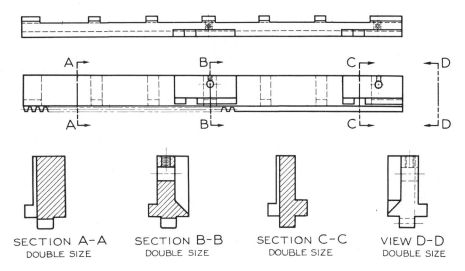

SECTION A-A
DOUBLE SIZE

SECTION B-B
DOUBLE SIZE

SECTION C-C
DOUBLE SIZE

VIEW D-D
DOUBLE SIZE

Fig. 7.11 Removed (detail) sections.*

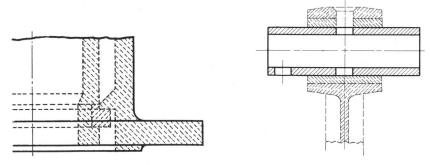

Fig. 7.12 Phantom section.

Fig. 7.13 Phantom sectioning— adjacent parts.

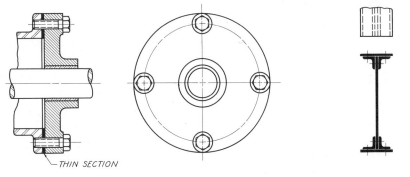

THIN SECTION

Fig. 7.14 Thin sections.

*ANSI Y14.2-1957.

133

eral increase or decrease in the spacing. An example of correct section lining is shown in Fig. 7.15(*a*), and, for comparison, examples of faulty practice may be seen in Fig. 7.15(*b*), (*c*), and (*d*). Experienced draftsmen realize that nothing will do more to ruin the appearance of a drawing than carelessly executed section lines.

As shown in Fig. 7.16, the section lines on two adjacent pieces should slope at 45° in opposite directions. If a third piece adjoins the two other pieces, as in Fig. 7.17(*a*), it ordinarily is section-lined at 30°. An alternative treatment that might be used would be to vary the spacing without changing the angle. On a sectional view showing an assembly of related parts, *all portions of the cut surface of any part must be section-lined in the same direction, for a change would lead the reader to consider the portions as belonging to different parts. Furthermore, to allow quick identification, each piece*

(and all identical pieces) in every view of the assembly drawing should be section-lined in the same direction.

Shafts, bolts, rivets, balls, and so on, whose axes lie in the plane of section, are not treated the same as ordinary parts. Having no interior construction to be shown, they are drawn in full and thus tend to make the adjacent sectioned parts stand out to better advantage (Fig. 7.18).

Whenever section lines drawn at 45° with the horizontal are parallel to part of the outline of the section (see Fig. 7.19), it is advisable to draw them at some other angle (say, 30° or 60°). Those drawn as in (*a*) and (*c*) produce an unusual appearance that is contrary to what is expected. Note the more natural effect obtained in (*b*) and (*d*) by sloping the lines at 30° and 75°.

7.9

Outline Sectioning. Very large surfaces may be section-lined around the bounding outline only, as illustrated in Fig. 7.20.

7.10

Symbolic Representation for a Cutting Plane. The symbolic lines that are used to represent the edge view of a cutting plane are shown in Fig. 7.21. The line is as heavy as an object line and is composed of either alternate long and short dashes or a series of dashes of equal length. The latter form is used in the automobile industry and has been approved by the SAE (Society of Automotive Engineers) and the American National Standards Institute. On drawings of ordinary size, when alternate long and short dashes are used for the cutting-plane line, the long dashes are .80 in. (20 mm) long, the short dashes .12 in. (3 mm) long, and the spaces .03 in. (0.8 mm) wide, depending on the size of the drawing. When drawn in pencil on manila paper, they are made with a medium pencil.

Arrowheads are used to show the direction in which the imaginary cut surface is

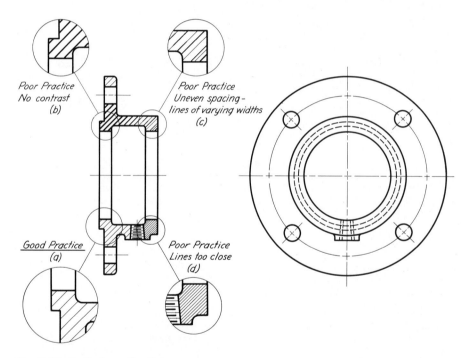

*Poor Practice
No contrast
(b)*

*Poor Practice
Uneven spacing -
lines of varying widths
(c)*

*Good Practice
(a)*

*Poor Practice
Lines too close
(d)*

Fig. 7.15 Faults in section lining.

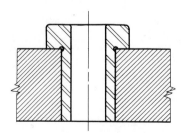

Fig. 7.16 Two adjacent pieces.

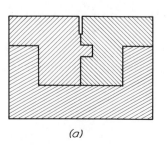

(a)

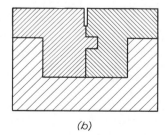

(b)

Fig. 7.17 Three adjacent pieces.

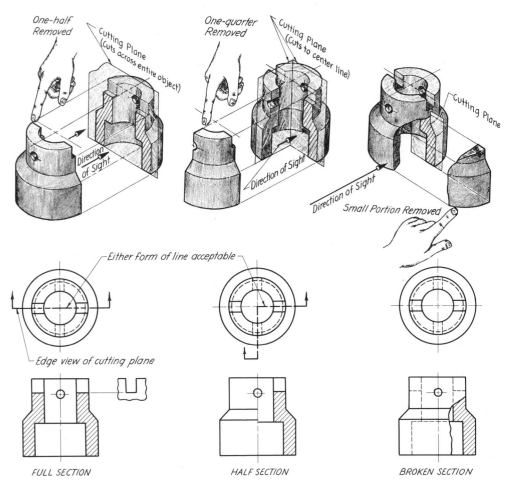

One-half Removed

Cutting Plane (Cuts across entire object)

Direction of Sight

One-quarter Removed

Cutting Plane (Cuts to center line)

Direction of Sight

Cutting Plane

Direction of Sight

Small Portion Removed

Either form of line acceptable

Edge view of cutting plane

FULL SECTION

HALF SECTION

BROKEN SECTION

Fig. 7.3 Types of sectional views.

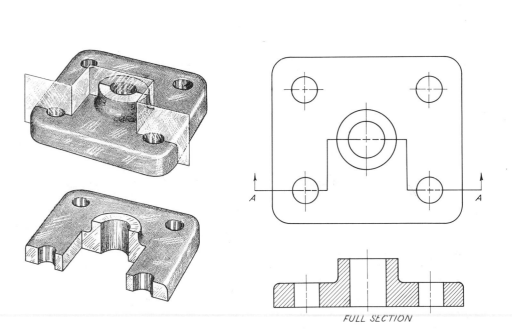

A   A

FULL SECTION

Fig. 7.4 Offset cutting plane.

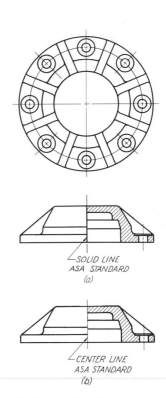

SOLID LINE ASA STANDARD
(a)

CENTER LINE ASA STANDARD
(b)

Fig. 7.5 Half section.

shown in Fig. 7.5(*b*). They reason that the removal of a quarter of the object is theoretical and imaginary and that an actual edge, which would be implied by a solid line, does not exist. The center line is taken as denoting a theoretical edge.

7.4

Broken Section. A broken or partial section is used mainly to expose the interior of objects so constructed that less than a half section is required for a satisfactory description (Fig. 7.6). The object theoretically is cut by a cutting plane and the front portion is removed by breaking it away. The "breaking away" gives an irregular boundary line to the section.

7.5

Revolved Section. A revolved section is useful for showing the true shape of the cross section of some elongated object,

such as a bar, or some feature of an object, such as an arm, spoke, or rib (Figs. 7.1 and 7.7).

To obtain such a cross section, an imaginary cutting plane is passed through the member perpendicular to the longitudinal axis and then is revolved through 90° to bring the resulting view into the plane of the paper (Fig. 7.8). When revolved, the section should show in its true shape and in its true revolved position, regardless of the location of the lines of the exterior view. If any lines of the view interfere with the revolved section, they should be omitted (Fig. 7.9). It is sometimes advisable to provide an open space for the section by making a break in the object (Fig. 7.7).

7.6

Removed (Detail) Sections. A removed section is similar to a revolved section,

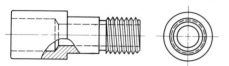

Fig. 7.6 Broken section.

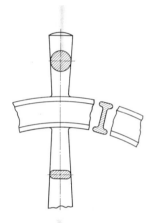

Fig. 7.7 Revolved section.*

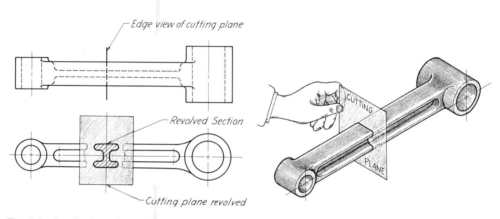

Fig. 7.8 Revolved section and cutting plane.

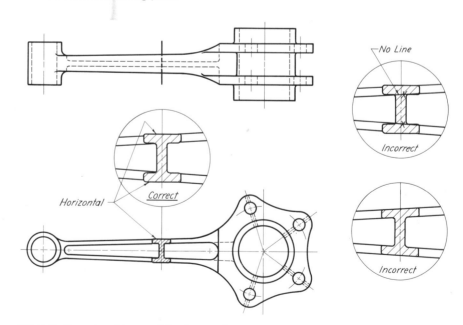

Fig. 7.9 Correct and incorrect treatment of a revolved section.

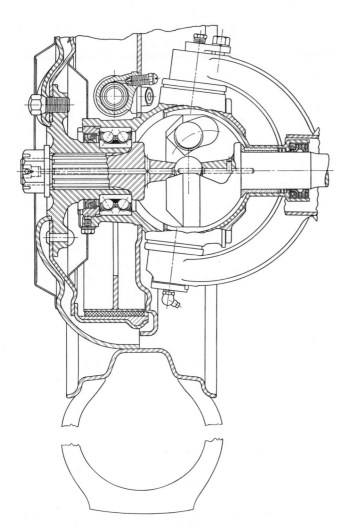

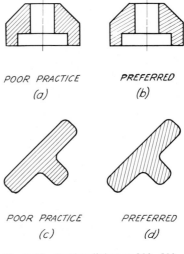

POOR PRACTICE
(a)

PREFERRED
(b)

POOR PRACTICE
(c)

PREFERRED
(d)

Fig. 7.19 Section lining at 30°, 60°, or 75°.

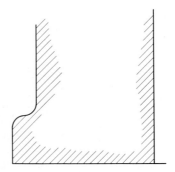

Fig. 7.20 Outline sectioning.

Fig. 7.18 Treatment of shafts, fasteners, ball bearings, and other parts. (*Courtesy New Departure Division, General Motors Corporation*)

viewed, and reference letters are added to identify it (Fig. 7.22).

Whenever the location of the cutting plane is obvious, it is common practice to omit the edge-view representation, particularly in the case of symmetrical objects. But if it is shown, and coincides with a center line, it takes precedence over the center line.

7.11
Summary of the Practices of Sectioning

1. A cutting plane may be offset in order to cut the object in such a manner as to reveal an important detail that would not be shown if the cutting plane were continuous (Fig. 7.4).

2. All visible lines beyond the cutting plane for the section are usually shown.

3. Invisible lines beyond the cutting plane for the section are usually not

shown, unless they are absolutely necessary to clarify the construction of the piece. In a half section, they are omitted in the unsectioned half, and either a center line or a solid line is used to separate the two halves of the view (Figs. 7.3 and 7.5).

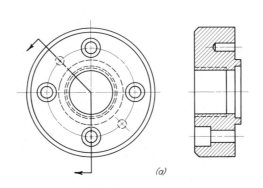

(a)

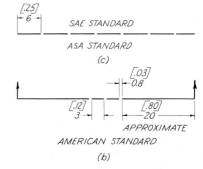

Fig. 7.21 Cutting plane lines (AN Standard).

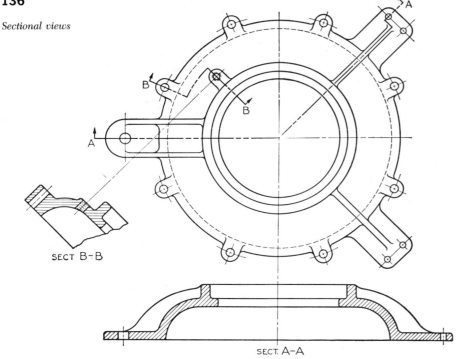

SECT B-B

SECT. A-A

Fig. 7.22 Sectional view.

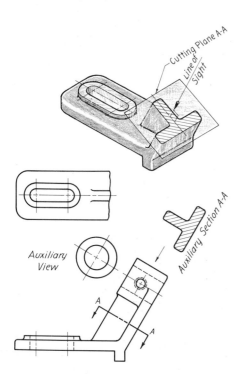

Fig. 7.23 Auxiliary section.

4. On a view showing assembled parts, the section lines on adjacent pieces are drawn in opposite directions at an angle of 45° (Fig. 7.16).

5. On an assembly drawing, the portions of the cut surface of a single piece in the same view or different views always should be section-lined in the same direction, with the same spacing (Fig. 7.18).

6. The symbolic line indicating the location of the cutting plane may be omitted if the location of the plane is obvious (Fig. 7.1).

7. On a sectioned view showing assembled pieces, an exterior view is preferred for shafts, rods, bolts, nuts, rivets, and so forth, whose axes are in the plane of section (Fig. 7.18).

7.12

Auxiliary Sections. A sectional view, projected on an auxiliary plane, is sometimes necessary to show the shape of a surface cut by a plane or to show the cross-sectional shape of an arm, rib, and so forth, inclined to any two or all three of the principal planes of projection (Fig. 7.23). When a cutting plane cuts an object, as in Fig. 7.23, arrows should show the direction in which the cut surface is viewed. Auxiliary sections are drawn by the usual method for drawing auxiliary views. When the bounding edge of the section is a curve, it is necessary to plot enough points to obtain a smooth one. Section 8.11 explains in detail the method for constructing the required view. A section view of this type usually shows only the inclined cut surface.

7.13

Conventional Sections. Sometimes a less confusing sectioned representation is obtained if certain of the strict rules of projection are violated, as explained in Chapter 5. For example, an unbalanced and confused view results when the sectioned view of the pulley shown in Fig. 7.24 is drawn in the true projection, as in (*a*). It is better practice to preserve symmetry by showing the spokes as if they were aligned into one plane, as in (*c*). Such treatment of unsymmetrical features is not misleading, since their actual arrangement is revealed in the circular view. The spokes are not sectioned in the preferred view. If they were, the first impression would be that the wheel had a solid web (*b*). See also Fig. 7.25.

When there are an odd number of holes in a flange, as is the case with the part in Fig. 7.26, they should be shown aligned in the sectioned view to reveal their true location with reference to the rim and the axis of the piece. To secure the so-called aligned section, one usually considers the

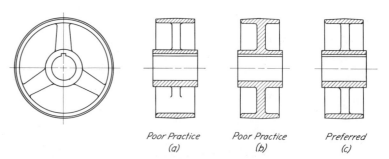

Poor Practice
(a)

Poor Practice
(b)

Preferred
(c)

Fig. 7.24 Conventional treatment of spokes in section.

cutting plane to be bent to pass through the angled hole, as shown in the pictorial drawing. Then, the bent portion of the plane (with the hole) is imagined to be revolved until it is aligned with the other portion of the cutting plane. As straightened out, the imaginary continuous plane produces the preferred section view shown in (*a*).

Figure 7.27 shows another example of conventional representation. The sectional view is drawn as though the upper projecting lug had been swung until the portion of the cutting plane through it formed a continuous plane with the other portion (Sec. 5.42). It should be noted that the hidden lines in the sectioned view are necessary for a complete description of the construction of the lugs.

7.14

Ribs in Section. When a machine part has a rib cut by a plane of section (Fig. 7.28), a "true" sectional view taken through the rib would prove to be false and misleading, because the crosshatching on the rib would cause the object to appear "solid." The preferred treatment is to omit arbitrarily the section lines from the rib, as illustrated by Fig. 7.28(*a*). The resulting sectional view may be considered the view that would be obtained if the plane were offset to pass just in front of the rib (*b*).

An alternative conventional method, approved but not as frequently used, is illustrated in Fig. 7.29. This practice of omitting alternate section lines some-

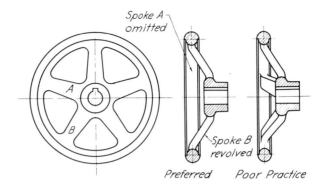

Fig. 7.25 Spokes in section.*

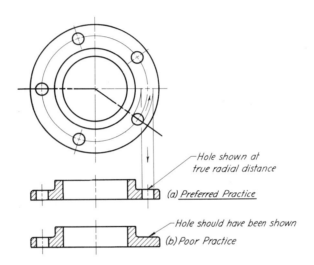

Hole shown at true radial distance

(a) *Preferred Practice*

Hole should have been shown

(b) *Poor Practice*

Fig. 7.26 Drilled flanges.

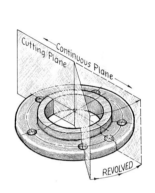

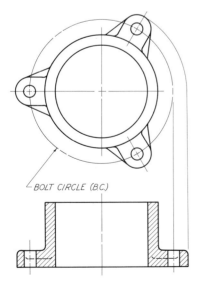

BOLT CIRCLE (B.C.)

Fig. 7.27 Revolution of a portion of an object.

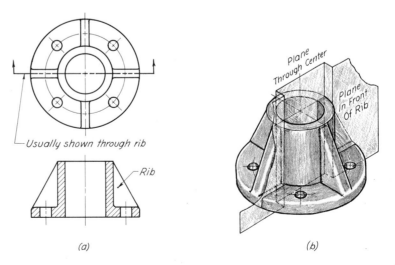

Usually shown through rib

Rib

(a)

(b)

Fig. 7.28 Conventional treatment of ribs in section.

* ANSI Y14.2-1957.

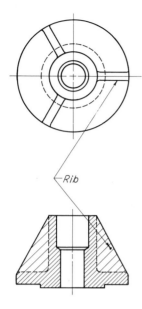

Fig. 7.29 Alternative treatment of ribs in section.

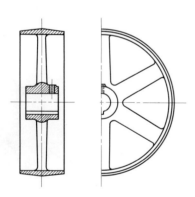

Fig. 7.30 Half view.

times is adopted when it is necessary to emphasize a rib that might otherwise be overlooked.

7.15

Half Views. When the space available is insufficient to allow a satisfactory scale to be used for the representation of a symmetrical piece, it is considered good practice to make one view a half view, as shown in Fig. 7.30. The half view, however, must be the top or side view and not the front view, which shows the characteristic contour. The half view should be the rear half.

7.16

Material Symbols. The section-line symbols recommended by the American National Standards Institute for indicating various materials are shown in Fig. 7.31. Code section lining ordinarily is not used on a working (detail) drawing of a separate part. It is considered unnecessary to indicate a material symbolically when its exact specification must be given as a note. For this reason, and in order to save time as well, the easily drawn symbol for cast iron is commonly used on detail drawings for all materials. Contrary to this general practice, however, some few chief draftsmen insist that symbolic section lining be used on all detail drawings prepared under their supervision.

Code section lining usually is employed on an assembly section showing the various parts of a unit in position, because a distinction between the materials causes the parts to "stand out" to better advantage. Furthermore, a knowledge of the type of material of which an individual part is composed often helps the reader to identify it more quickly and understand its function.

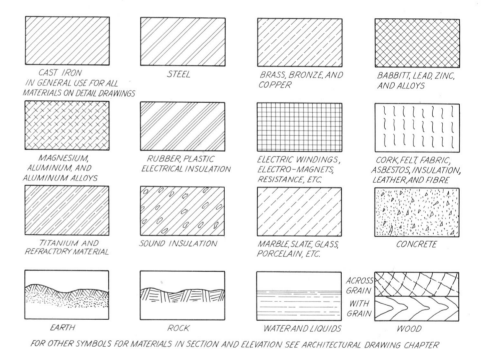

Fig. 7.31 Material symbols (AN Standard).

Problems

The following problems were designed to emphasize the principles of sectioning. Those drawings that are prepared from the pictorials of objects may be dimensioned if the elementary principles of dimensioning (Chapter 13) are carefully studied.

Metric: When a problem has been dual-dimensioned and the metric values represent converted inch values, use the nearest full millimeter value for a dimension except in the case of a critical dimension or a machined hole. For drilled holes see Appendix Tables 35 and 36. For a dimension that will be divided by two, the dimension may be taken to the nearest even number of millimeters.

1. (Fig. 7.32). Reproduce the top view and change the front view to a full section view in accordance with the indicated cutting plane.

2. (Fig. 7.33). Reproduce the top view and change the front and side views to sectional views that will be in accordance with the indicated cutting planes.

3. (Fig. 7.34). Draw a front view of the pulley (circular view) and a side view in full section.

4. (Fig. 7.35). Reproduce the top view of the rod support and draw the front view in full section. Read Sec. 5.43 before starting to draw.

5. (Fig. 7.36). Draw a front view of the V pulley (circular view) and a side view in full section.

6. (Fig. 7.37). Reproduce the two view of the hand wheel and change the right-side view to a full section.

7. (Fig. 7.38). Reproduce the top view of the control housing cover and convert the front view to a full section.

8. (Fig. 7.39). Reproduce the circular front view of the pump cover and convert the right-side view to a full section.

9. (Fig. 7.40). Reproduce the front and top views of the steady brace. Complete the top view and draw the required side view and auxiliary section. Since this is a structural drawing, the figure giving the value of a distance appears above an unbroken dimension line in accordance with the custom in this field of engineering. The slope (45°) of the inclined member to which the plates are welded is indicated by a slope triangle with 12-in. legs.

This problem has been designed to make it necessary for the student to test his power of visualization if he is to determine the shape of the inclined structural member. Good judgment must be exercised in determining the location of the third hole in each plate. All hidden object lines should be shown.

10–19. (Figs. 7.41–7.50). These problems may be dimensioned, as are working drawings. For each object, the student should draw all the views necessary for a working drawing of the part. Good judgment should be exercised in deciding whether the sectional view should be a full section or a half section. After the student has made his decision, he should consult his class instructor.

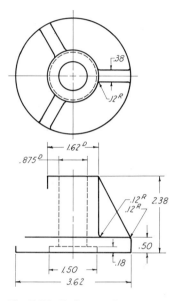

Fig. 7.35 **Rod support.**

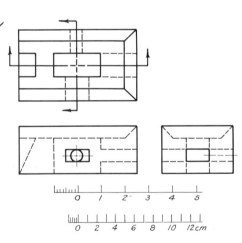

Fig. 7.33 **Mutilated block.**

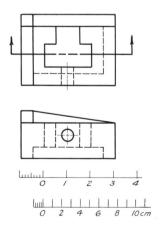

Fig. 7.32 **Mutilated block.**

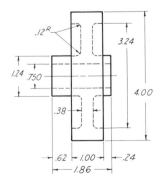

Fig. 7.34 **Pulley.**

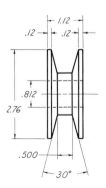

Fig. 7.36 **V pulley.**

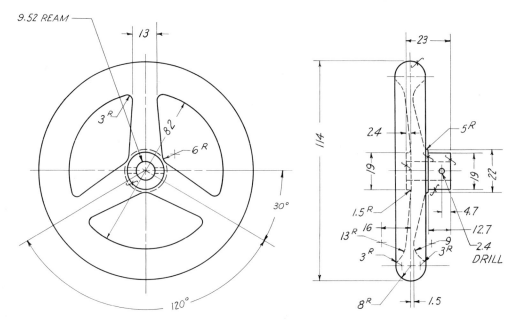

Fig. 7.37 Hand wheel.

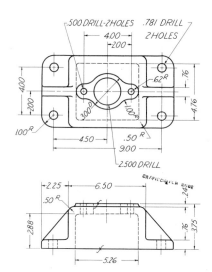

Fig. 7.38 Control housing cover.

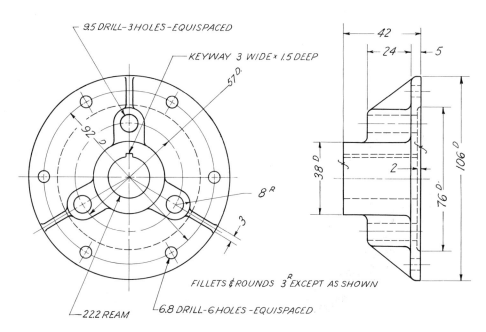

Fig. 7.39 Pump cover.

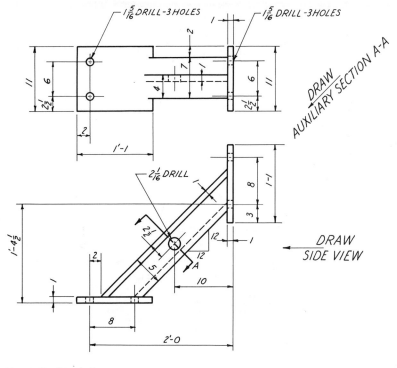

$1\frac{5}{16}$ DRILL-3 HOLES

$1\frac{5}{16}$ DRILL-3 HOLES

DRAW AUXILIARY SECTION A-A

$2\frac{1}{16}$ DRILL

DRAW SIDE VIEW

Fig. 7.40 Steady brace.

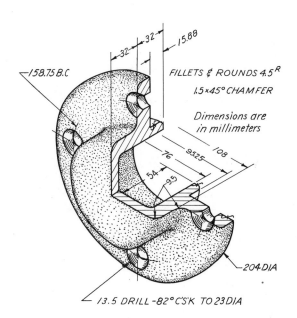

158.75 B.C

FILLETS & ROUNDS 4.5^R

1.5 × 45° CHAMFER

Dimensions are in millimeters

204 DIA

13.5 DRILL -82° C'S'K TO 23 DIA

Fig. 7.41 Cover.

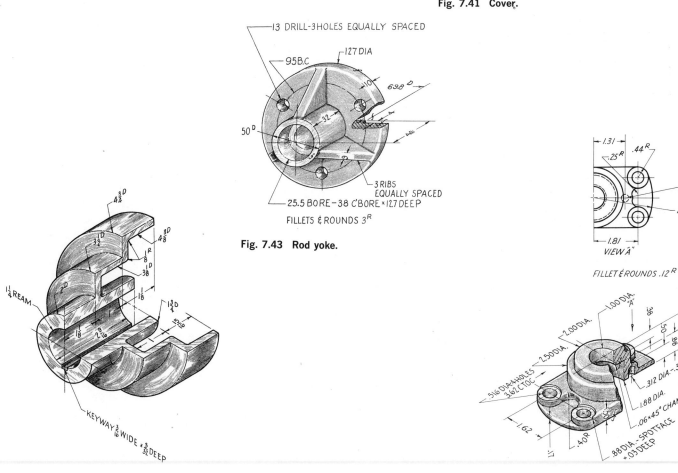

13 DRILL-3 HOLES EQUALLY SPACED

127 DIA

95 B.C

63.8 D

3 RIBS EQUALLY SPACED

25.5 BORE—38 C'BORE × 12.7 DEEP

FILLETS & ROUNDS 3^R

Fig. 7.43 Rod yoke.

1.31

.25^R

.44^R

.30^R

R

1.81

VIEW "A"

FILLET & ROUNDS .12^R

$4\frac{3}{4}$D

$3\frac{1}{2}$D

$4\frac{3}{8}$D

$1\frac{1}{8}$R

$3\frac{3}{8}$D

$1\frac{1}{4}$ REAM

$1\frac{1}{8}$

KEYWAY $\frac{3}{16}$ WIDE × $\frac{3}{32}$ DEEP

Fig. 7.42 Cone pulley.

1.00 DIA. "A"

2.00 DIA.

2.50 DIA.

.312 DIA. - .34 DEEP

1.88 DIA.

.06 × 45° CHAMFER

5/16 DIA-4 HOLES 3.62 C TO C

1.62

.40 R

.88 DIA.- SPOTFACE + .03 DEEP

Fig. 7.44 Control housing cover.

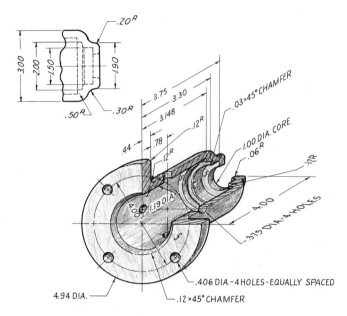

Fig. 7.45 Centering bearing.

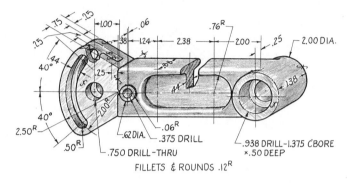

Fig. 7.46 Slotted guide link.

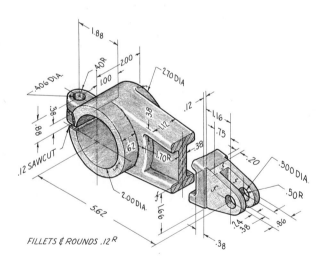

Fig. 7.47 Shifter link.

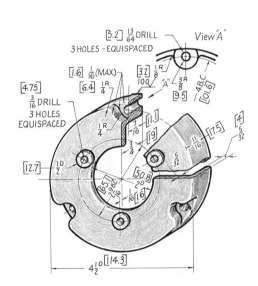

Fig. 7.48 Cover. (Dimension values shown in [ ] are in millimeters.)

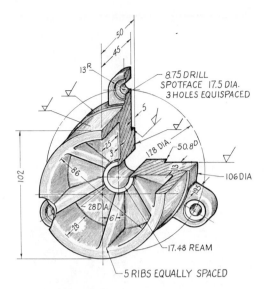

Fig. 7.49 End guide.

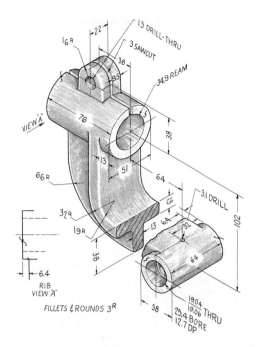

Fig. 7.50 Hanger bracket.

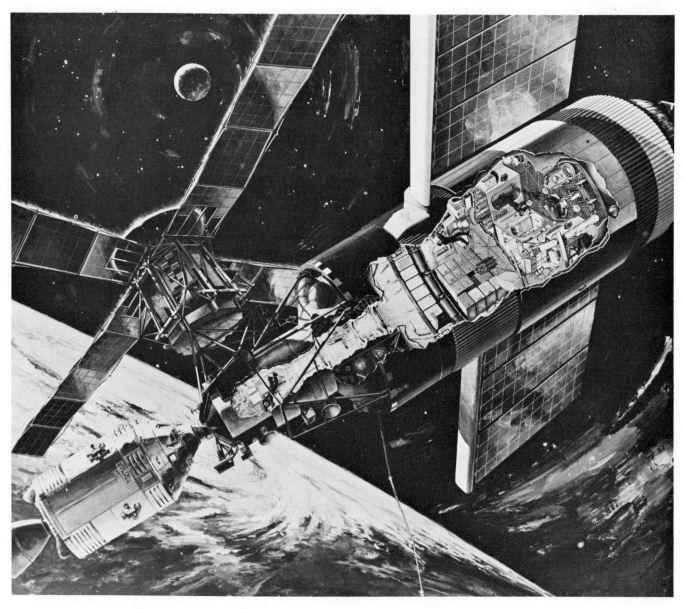

Skylab—manned orbital scientific space station—made extensive use of the hardware and technological knowledge developed and acquired during previous space missions. The development of the needed hardware required innumerable multiview drawings; many having auxiliary views. Skylab was developed and placed into orbit to gain added knowledge of manned earth-orbital operations and to conduct selected scientific and medical investigations (*Courtesy National Aeronautics and Space Administration*)

Auxiliary views

A □ Primary auxiliary views

8.1

Introduction. When it is desirable to show the true size and shape of an irregular surface, which is inclined to two or more of the coordinate planes of projection, a view of the surface must be projected on a plane parallel to it. This imaginary projection plane is called an *auxiliary plane*, and the view obtained is called an *auxiliary view* (Fig. 8.1).

The theory underlying the method of projecting principal views applies also to auxiliary views. In other words, an auxiliary view shows an inclined surface of an object as it would appear to an observer stationed an infinite distance away (Fig. 8.2).

8.2

Use of Auxiliary Views. In commercial drafting, an auxiliary view ordinarily is a partial view showing only an inclined surface. The reason for this is that a projection showing the entire object adds very little to the shape description. The added lines are likely to defeat the intended purpose of an auxiliary view. For example, a complete drawing of the casting in Fig. 8.3 must include an auxiliary view of the inclined surface in order to show the true shape of the surface and the location of the holes. Compare the views in (*a*) and (*b*) and note the confused appearance of the view in (*b*). In technical schools, some instructors require that an auxiliary view show the entire object, including all invisible lines. Such a requirement, though im-

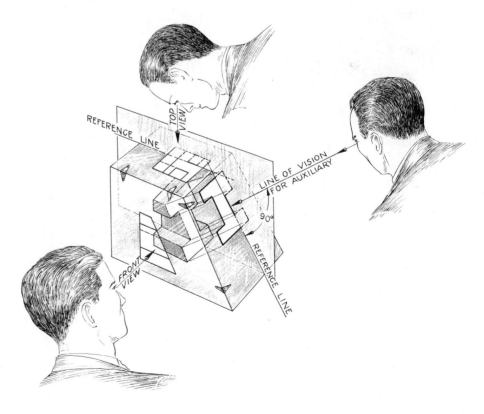

Fig. 8.1 Theory of projecting an auxiliary view.

practical commercially, is justified in the classroom, for the construction of a complete auxiliary view furnishes excellent practice in projection.

A partial auxiliary view often is needed to complete the projection of a foreshortened feature in a principal view. This second important function of auxiliary views is illustrated in Fig. 8.16 and explained in Sec. 8.13.

8.3
Types of Auxiliary Views. Although auxiliary views may have an infinite number of positions in relation to the three principal planes of projection, primary auxiliary views may be classified into three general types in accordance with position relative to the principal planes. Figure 8.4 shows the first type, where the auxiliary plane is perpendicular to the frontal plane and inclined to the horizontal plane of projection. Here the auxiliary view and top view have one dimension that is common to both: the depth. Note that the auxiliary plane is hinged to the frontal plane and that the auxiliary view is projected from the front view.

In Fig. 8.5 the auxiliary plane is perpendicular to the horizontal plane and inclined to the frontal and profile planes of projection. The auxiliary view is projected from the top view, and its height is the same as the height of the front view.

The third type of auxiliary view, as shown in Fig. 8.6, is projected from the side view and has a common dimension with both the front and top views. To construct it, distances may be taken from either the front or top view.

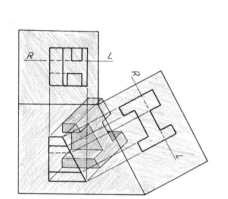

Fig. 8.2 Auxiliary view.

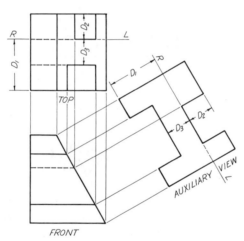

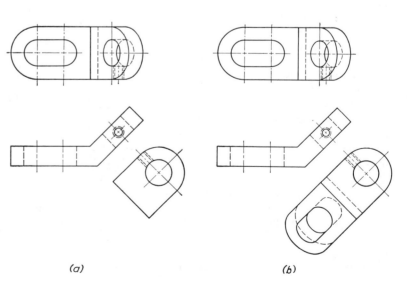

(a)   (b)

Fig. 8.3 Partial and complete auxiliary views.

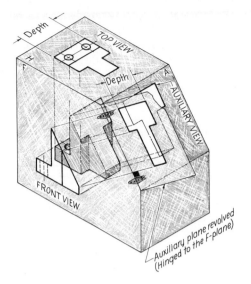

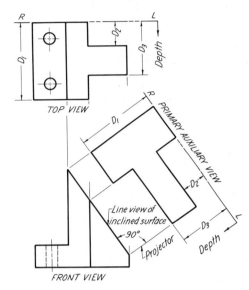

Transfer the D distances in the top, taken from R-L in the direction of the arrow, to the auxiliary view. Both of these views show related distances that are equal in the direction of depth. (See pictorial drawing)

Fig. 8.4 Auxiliary view projected from front view.

Height distances (D_1 and D_2) for the auxiliary view are equal to the height distances (D_1 and D_2) in the front view.

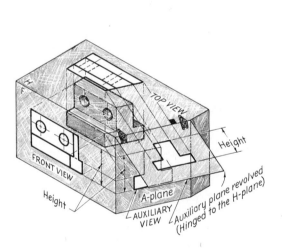

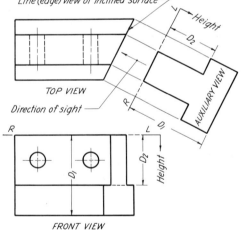

Fig. 8.5 Auxiliary view projected from top view.

All three types of auxiliary views are constructed similarly. Each is projected from the view that shows the slanting surface as a line, and the distances for the view are taken from the other principal view that has a common dimension with the auxiliary. A careful study of the three illustrations will reveal the fact that the inclined auxiliary plane is always hinged to the principal plane to which it is perpendicular.

8.4
Symmetrical and Unsymmetrical Auxiliary Views. Since auxiliary views are either symmetrical or unsymmetrical about a center line or reference line, they may be termed (1) symmetrical, (2) unilateral, or

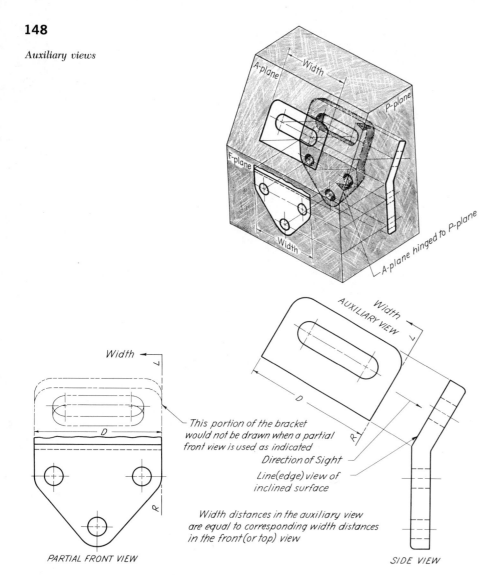

Fig. 8.6 Auxiliary view projected from side view.

This portion of the bracket would not be drawn when a partial front view is used as indicated

Direction of Sight

Line(edge) view of inclined surface

Width distances in the auxiliary view are equal to corresponding width distances in the front (or top) view

PARTIAL FRONT VIEW

SIDE VIEW

AUXILIARY VIEW

(3) bilateral, according to the degree of symmetry. A symmetrical view is drawn symmetrically about a center line, the unilateral view entirely on one side of a reference line, and the bilateral view on both sides of a reference line.

8.5

To Draw a Symmetrical Auxiliary View. When an inclined surface is symmetrical, the auxiliary view is "worked" from a center line (Fig. 8.7). The first step in drawing such a view is to draw a center line parallel to the inclined line that represents an edge view of the surface. If the object is assumed to be enclosed in a glass box, this center line may be considered the line of intersection of the auxiliary plane and an imaginary vertical center plane. There are professional draftsmen who, not acquainted with the glass box, proceed without theoretical explanation. Their method is simple to draw a working center line for the auxiliary view and a corresponding line in one of the principal views.

Although theoretically, this working center line may be drawn at any distance from the principal view, actually it should be so located to give the whole drawing a balanced appearance. If not already shown, it also must be drawn in the principal view showing the true width of the inclined surface.

The next step is to draw projection lines from each point of the sloping face, remembering that the projectors make an angle of 90° with the inclined line representing the surface. With the projectors drawn, the location of each point in the auxiliary can be established by setting the dividers to each point's distance from the center line in the principal view and transferring the distance to the auxiliary view. For example, point X is projected to the auxiliary by drawing a projector from point X in the front view perpendicular to the center line. Since its distance from the center line in the top view is the same as it is from the center line in the auxiliary view, the point's location along the projector may be established by using the distance taken from the top view. In the case of point X, the distance is set off from the center line toward the front view. Point Y is set off from the center line away from the front view. A careful study of Fig. 8.7 reveals the fact that if a point lies between the front view and the center line of the top view, it will lie between the front view and the center line of the auxiliary view, and, conversely, if it lies away

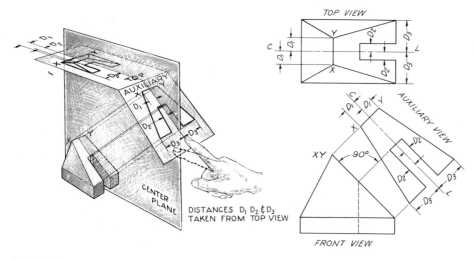

Fig. 8.7 Symmetrical auxiliary view of an inclined surface.

DISTANCES D_1 D_2 & D_3 TAKEN FROM TOP VIEW

CENTER PLANE

TOP VIEW

AUXILIARY VIEW

FRONT VIEW

from the front view with reference to the center line of the top view, it will lie away from the front view with reference to the center line of the auxiliary view.

8.6

Unilateral Auxiliary Views. When constructing a unilateral auxiliary view, it is necessary to work from a reference line that is drawn in a manner similar to the working center line of a symmetrical view. The reference line for the auxiliary view may be considered to represent the line of intersection of a reference plane, coinciding with an outer face, and the auxiliary plane (Fig. 8.8). The intersection of this plane with the top plane establishes the reference line in the top view. All the points are projected from the edge view of the surface, as in a symmetrical view, and it should be noted in setting them off that they all fall on the same side of the reference line.

Figure 8.9 shows an auxiliary view of an entire object. In constructing such a view, it should be remembered that the projectors from all points of the object are perpendicular to the auxiliary plane, since the observer views the entire figure by looking directly at the inclined surface. The distances perpendicular to the auxiliary reference line were taken from the front view.

8.7

Bilateral Auxiliary Views. The method of drawing a bilateral view is similar to that of drawing a unilateral view, the only difference being that in a bilateral view the inclined face lies partly on both sides of the reference plane, as shown in Fig. 8.10.

8.8

Curved Lines in Auxiliary Views. To draw a curve in an auxiliary view, the draftsman must plot a sufficient number of points to ensure a smooth curve (Fig. 8.11). The points are projected first to the inclined line representing the surface in the front view and then to the auxiliary view. The distance of any point from the center line in the auxiliary view is the same as its distance from the center line in the end view.

8.9

Projection of a Curved Boundary. In Fig. 8.11, the procedure is illustrated for plotting the true size and shape of an inclined surface bounded by a curved outline. A similar procedure can be followed to plot a

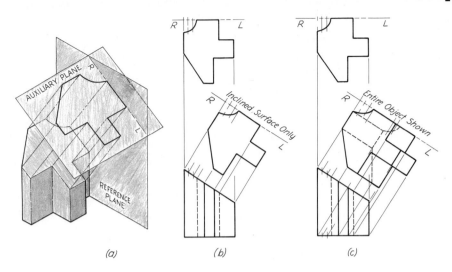

(a)   (b)   (c)

Fig. 8.8 Unilateral auxiliary view.

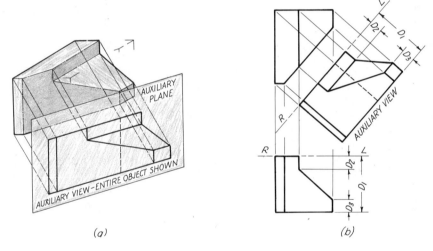

(a)   (b)

Fig. 8.9 Auxiliary view of an object.

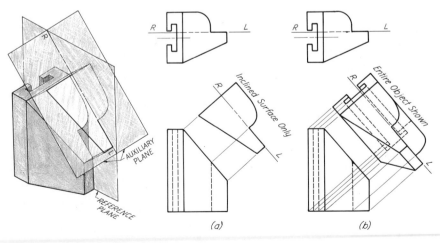

(a)   (b)

Fig. 8.10 Bilateral auxiliary view.

150

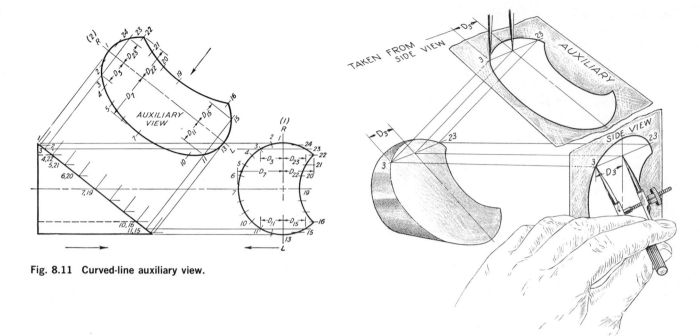

Fig. 8.11 Curved-line auxiliary view.

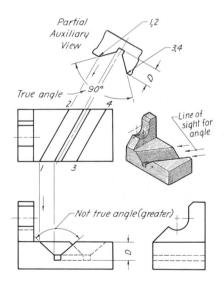

Fig. 8.12 Plotted boundary curves.

curve, such as the one on the left end of the object shown in Fig. 8.12, in an auxiliary view. Since the points on the curved outline of the vertical surface are being viewed from the same direction as those on the curved boundary of the inclined surface, the projectors from points on both surfaces will be parallel. From the pictorial drawing, it can be observed that points A and A' and B and B' lie on elements of the cylindrical surface. Also it should be noted that A and A' are the same distance from the reference plane as are B and B'. It is for this reason that

point A' in the auxiliary view is at the intersection of the projector from A' in the front view and a line through A in the auxiliary view, drawn parallel to the reference line RL.

8.10
Dihedral Angles. Frequently, an auxiliary view may be needed to show the true size of a dihedral angle—that is, the true size of the angle between two planes. In Fig. 8.13, it is desirable to show the true size of the angle between the planes forming the V-slot by means of a partial auxiliary view, as shown. The direction of sight (see pictorial) must be taken parallel to the edge lines 1–2 and 3–4 so that these lines will appear as points and the surfaces forming the dihedral angle will project as line views in the auxiliary view. The reference line for the partial auxiliary view would necessarily be drawn perpendicular to the lines 1–2 and 3–4 in the top view. Since the plane on which the auxiliary view is projected is a vertical one, height dimensions were used—that is, distances in the direction of the dimension D in the auxiliary view were taken from the front view.

8.11
To Construct an Auxiliary View, Practical Method. The usual steps in constructing an auxiliary view are shown in Fig. 8.14. The illustration should be studied carefully, as each step is explained in the drawing.

Fig. 8.13 To determine the true dihedral angle between inclined surfaces.

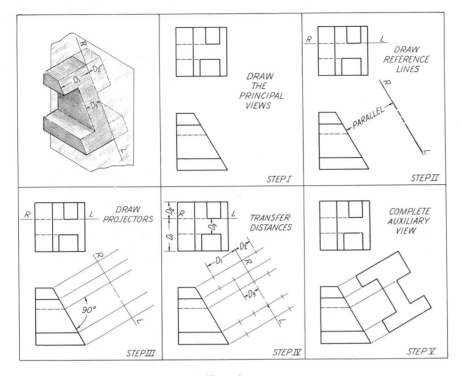

STEP I — DRAW THE PRINCIPAL VIEWS

STEP II — DRAW REFERENCE LINES — PARALLEL — R L

STEP III — DRAW PROJECTORS — 90°

STEP IV — TRANSFER DISTANCES

STEP V — COMPLETE AUXILIARY VIEW

Fig. 8.14 Steps in constructing an auxiliary view.

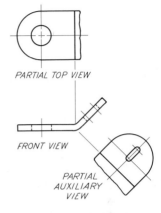

PARTIAL TOP VIEW

FRONT VIEW

PARTIAL AUXILIARY VIEW

Fig. 8.15 Partial views.

8.12

Auxiliary and Partial Views. Often the use of an auxiliary view allows the elimination of one of the principal views (top or side) or makes possible the use of a partial principal view. The shape description furnished by the partial views shown in Fig. 8.15 is sufficient for a complete understanding of the shape of the part. The use of partial views simplifies and drawing, saves valuable drafting time, and tends to make the drawing easier to read.

A break line is used at a convenient location to indicate an imaginary break for a partial view.

8.13

Use of an Auxiliary View to Complete a Principal View. As previously stated, it is frequently necessary to project a foreshortened feature in one of the principal views from an auxiliary view. In the case of the object shown in Fig. 8.16, the foreshortened projection of the inclined face in the top view can be projected from the auxiliary view. The elliptical curves are plotted by projecting points from the auxiliary view to the front view and from there to the top view. The location of these points in the top view with respect to the center line is the same as their location in the auxiliary view with respect to the auxiliary center line. For example, the distance

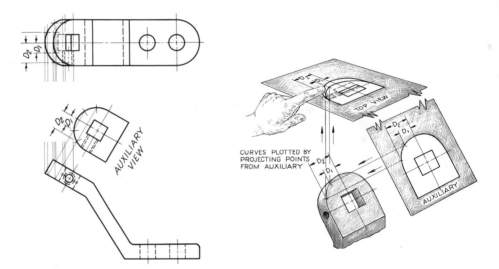

AUXILIARY VIEW

CURVES PLOTTED BY PROJECTING POINTS FROM AUXILIARY

TOP VIEW

AUXILIARY

Fig. 8.16 Use of auxiliary to complete a principal view.

D_1 from the center line in the top view is the same as the distance D_1 from the auxiliary center line in the auxiliary view.

The steps in preparing an auxiliary view and using it to complete a principal view are shown in Fig. 8.17.

8.14

Line of Intersection. It is frequently necessary to represent a line of intersection between two surfaces when making a

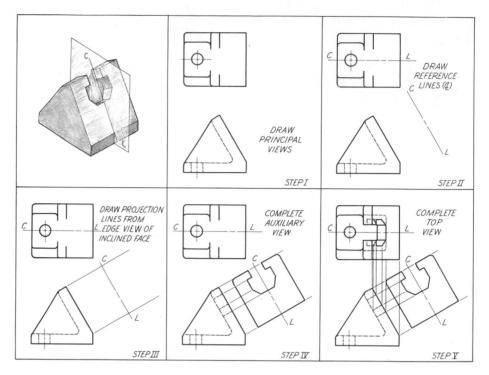

Fig. 8.17 Steps in preparing an auxiliary view and completing a principal view.

multiview drawing involving an auxiliary view. Figure 8.18 shows a method for drawing the line of intersection on a principal view. In this case the scheme commonly used for determining the intersection involves the use of elements drawn on the surface of the cylindrical portion of the part, as shown on the pictorial drawing. These elements, such as AB, are common to the cylindrical surface. Point B, where the element pierces the flat surface, is a point that is common to both surfaces and therefore lies on the line of intersection.

On the orthographic views, element AB appears as a point on the auxiliary view and as a line on the front view. The loca-

tion of the projection of the piercing point on the front view is visible upon inspection. Point B is found in the other principal view by projecting from the front view and setting off the distance D taken from the auxiliary view. The distance D of point B from the center line is a true distance for both views. The center line in the auxiliary view and side view can be considered as the edge view of a reference plane or datum plane from which measurements can be made.

8.15

True Length of a Line. The true length of an oblique line may be determined either by means of an auxiliary view or by revo-

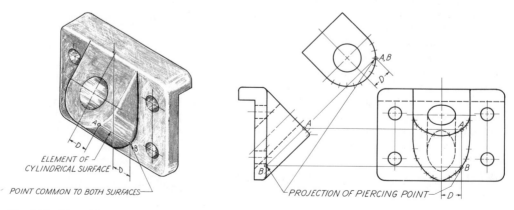

Fig. 8.18 Line of intersection.

lution of the line. Separate discussions of the procedure to be followed in the application of these methods are given in Chapters 5 and 9. To determine the true length of a line by revolution, see Sec. 5.29. To find the true length through the use of an auxiliary view, read Sec. 9.7.

B □ Secondary auxiliary views

8.16

Secondary (Oblique) Auxiliary Views. Frequently an object will have an inclined face that is not perpendicular to any one of the principal planes of projection. In such cases it is necessary to draw a primary auxiliary view and a secondary auxiliary or oblique view (Fig. 8.19). The primary auxiliary view is constructed by projecting the figure on a primary auxiliary plane that is perpendicular to the inclined surface and one of the principal planes. This plane may be at any convenient location. In the illustration, the primary auxiliary plane is perpendicular to the frontal plane. Note that the inclined face appears as a straight line in the primary auxiliary view. Using this view as a regular view, the secondary auxiliary view may be projected on a plane parallel to the inclined face. Figure 8.19(*b*) shows a practical application of the theoretical principles shown pictorially in (*a*).

It is suggested that the student read Sec. 9.14 in which the procedure for drawing the normal (true-shape) view of an oblique surface is presented step by step.

Figure 8.20 shows the progressive steps in preparing and using a secondary auxiliary view of an oblique face to complete a principal view. Reference planes have been used as datum planes from which to take the necessary measurements. Step II shows the partial construction of the primary auxiliary view in which the inclined surface appears as a line. Step III shows the secondary auxiliary view projected from the primary view and completed, using the known measurements of the lug. The primary auxiliary view is finished by projecting from the secondary auxiliary view. Step IV illustrates the procedure for projecting from the secondary auxiliary view to the top view through the primary auxiliary in order to complete the foreshortened view of the lug. It should be noted that distance D_1 taken from refer-

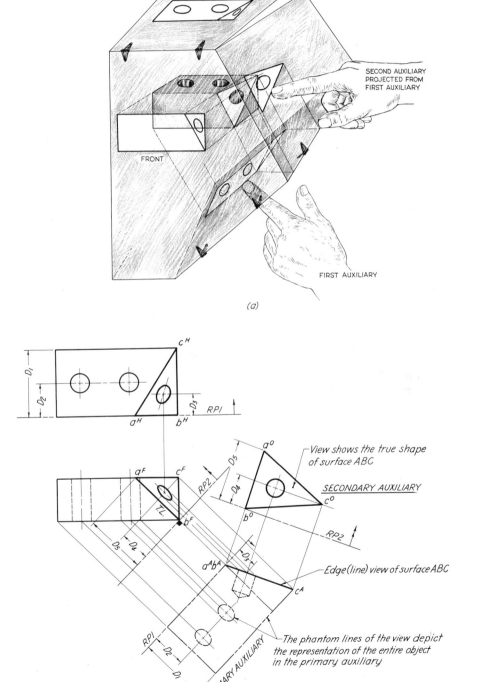

(a)

(b)

Fig. 8.19 Secondary auxiliary view of an oblique face.

ence R_2P_2 in the secondary auxiliary is transferred to the top view because both views show the same width distances in true length. A sufficient number of points should be obtained to allow the use of an irregular curve. Step V shows the projec-

tion of these points on the curve to the front view. In this case the measurements are taken from the primary auxiliary view because the height distances from reference plane R_1P_1 are the same in both views.

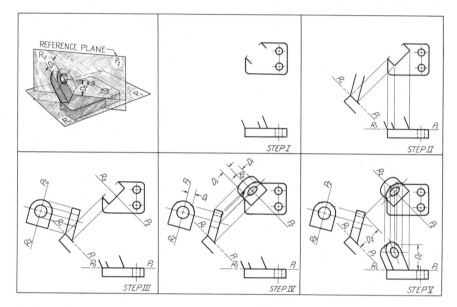

Fig. 8.20 Steps in drawing a secondary auxiliary view and using it to complete a principal view.

Problems

The problems shown in Fig. 8.21 are designed to give the student practice in constructing auxiliary views of the inclined surfaces of simple objects formed mainly by straight lines. They will provide needed drill in projection if, for each of the objects in Fig. 8.21, an auxiliary is drawn showing the entire object. Complete drawings may be made of the objects shown in Figs. 8.22–8.30. If the views are to be dimensioned, the student should adhere to the rules of dimensioning given in Chapter 13 and should not take too seriously the locations for the dimensions on the pictorial representations.

Metric: When a problem has been dual-dimensioned and the metric values represent converted inch values, use the nearest full millimeter value for a dimension except in the case of a critical dimension or a machined hole. For drilled holes see Appendix Tables 35 and 36. For a dimension that will be divided by two, the dimension may be taken to the nearest even number of millimeters.

1. (Fig. 8.21). Using instruments, reproduce the given views of an assigned object and draw an auxiliary view of its inclined surface.

2. (Fig. 8.22). Draw the views that would be necessary on a working drawing of the dovetail bracket.

3. (Fig. 8.23). Draw the necessary views of the anchor bracket. Make partial views for the top and end views.

4. (Fig. 8.24). Draw the views that would be necessary on a working drawing of the feeder bracket.

5. (Fig. 8.25). Draw the necessary views of the anchor clip. It is suggested that the top view be a partial one and that the auxiliary view show only the inclined surface.

6. (Fig. 8.26). Draw the views that would be necessary on a working drawing of the angle bracket. Note that two auxiliary views will be required.

7. (Fig. 8.27). Draw the necessary views of the offset guide. It is suggested that partial views be used, except in the view where the inclined surface appears as a line.

8. (Fig. 8.28). Draw the views that would be needed on a working drawing of the gear cover. The opening on the inclined face is circular.

9. (Fig. 8.29). Draw the necessary views of the cutoff clip.

10. (Fig. 8.30). Draw the necessary views of the ejector clip.

11. (Fig. 8.31). Draw the views as given. Complete the top view.

12. (Fig. 8.32). Draw the views as given. Complete the auxiliary view and the front view.

13. (Fig. 8.33). Draw the views that would be necessary on a working drawing of the 45° elbow.

14. Make a multiview drawing of the airplane engine mount shown in Fig.. 8.34. The engine mount is formed of three pieces of steel plate welded to a piece of steel tubing. The completed drawing is to consist of four views. It is suggested that the front view be the view obtained by looking along and parallel to the axis of the

Fig. 8.21

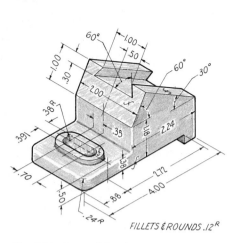

Fig. 8.22   Dovetail bracket.

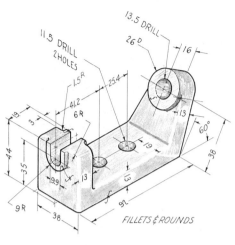

Fig 8.23   Anchor bracket.

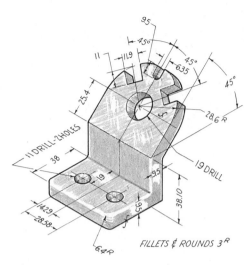

Fig. 8.24   Feeder bracket.

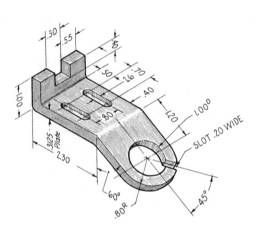

Fig. 8.25   Anchor clip.

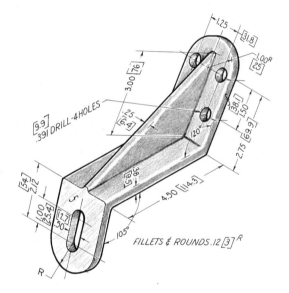

Fig. 8.26   Angle bracket. (Dimension values shown in [ ] are in millimeters.)

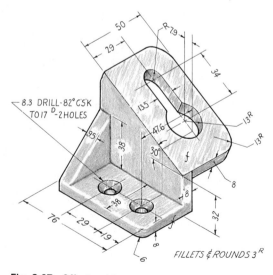

Fig. 8.27   Offset guide.

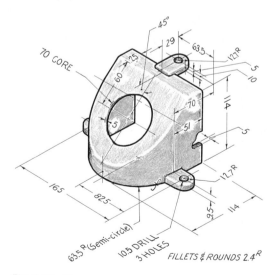

Fig. 8.28   Gear cover.

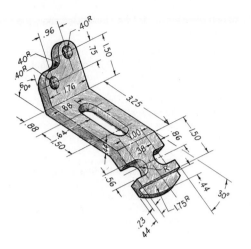

Fig. 8.29 Cut-off clip.

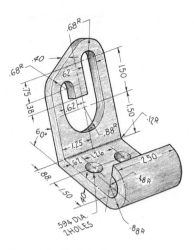

Fig. 8.30 Ejector clip.

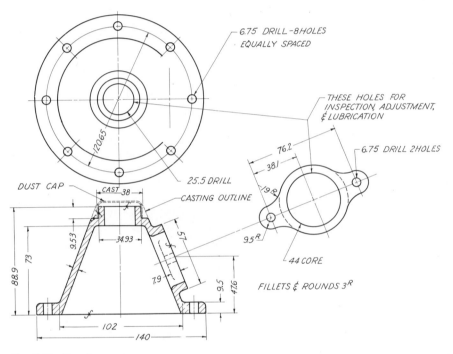

6.75 DRILL - 8 HOLES
EQUALLY SPACED

THESE HOLES FOR
INSPECTION, ADJUSTMENT,
& LUBRICATION

6.75 DRILL 2 HOLES

25.5 DRILL

DUST CAP

CASTING OUTLINE

44 CORE

FILLETS & ROUNDS 3ᴿ

Fig. 8.31 Housing cover.

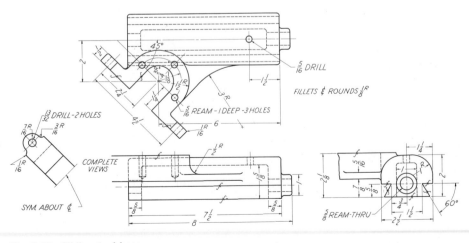

5/16 DRILL

FILLETS & ROUNDS 1/8 ᴿ

5/16 REAM - 1 DEEP - 3 HOLES

13/32 DRILL - 2 HOLES

COMPLETE
VIEWS

SYM. ABOUT ₵

5/8 REAM-THRU

Fig. 8.32 Sliding tool base.

tube. The remaining views that are needed are an auxiliary view showing only the inclined lug, a side view that should be complete with all hidden lines shown, and a partial top view with the inclined lug omitted.

15. (Fig. 8.35). Draw the views given and add the required primary and secondary auxiliary views.

16. (Fig. 8.36). Draw the necessary views of the tool holder.

17. (Fig. 8.37). Draw the necessary views of the angle block. One view should be drawn to show the true size and shape of the oblique surface.

18. (Fig. 8.38). Draw the necessary views of the locating slide.

19. (Fig. 8.39). Using instruments, draw a secondary auxiliary view that will show the true size and shape of the inclined surface of an as-

signed object. The drawing must also show the given principal views.

20. (Fig. 8.40). Draw the layout for the support anchor as given and then, using the double-auxiliary-view method, complete the views as required.

The plate and cylinder are to be welded. Since the faces of the plate show as oblique surfaces in the front and top views, double auxiliary views are necessary to show the thickness and the true shape.

Start the drawing with the auxiliary views that are arranged horizontally on the paper, then complete the principal views. The inclined face of the cylinder will show as an ellipse in top and front views; but do not show this in the auxiliary view that shows the true shape of the square plate.

How would you find the view that shows the true angle between the inclined face and the axis of the cylinder?

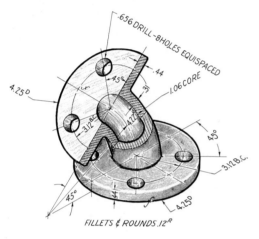

Fig. 8.33 45° elbow.

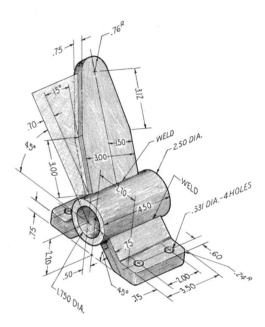

Fig. 8.34 Airplane engine mount.

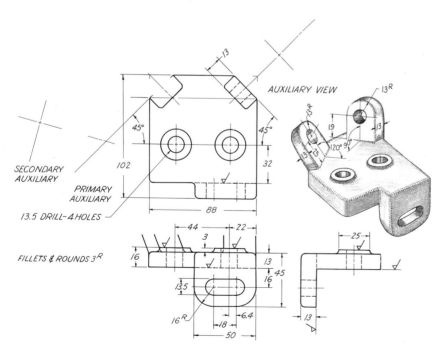

Fig. 8.35 Cross anchor.

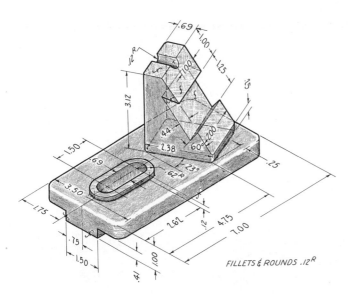

Fig. 8.36 Tool holder.

FILLETS & ROUNDS .12^R

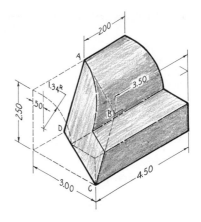

Fig. 8.37 Angle block.

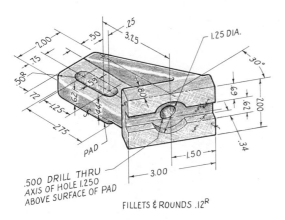

.500 DRILL THRU
AXIS OF HOLE 1.250
ABOVE SURFACE OF PAD

FILLETS & ROUNDS .12^R

Fig. 8.38 Locating slide.

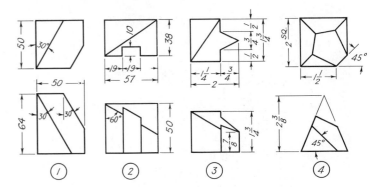

Fig. 8.39

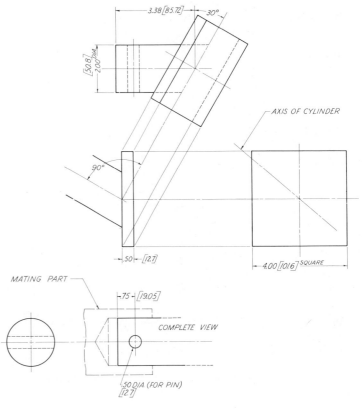

AXIS OF CYLINDER

MATING PART

COMPLETE VIEW

Fig. 8.40 Support anchor. (Dimension values shown in [ ] are in millimeters.)

The partially completed underground Judiciary Square Station on the 98-mile (157.7-km) Metro System in our nation's capitol. When finished 47.9 miles (77.1 km) of the line will be underground, and 50.6 miles (81.4 km) will be at the surface or on aerial ways. The system could not have been designed successfully without the application of graphic methods. (*Courtesy Washington Metropolitan Area Transit Authority*)

Basic spatial geometry for design and analysis

A □ Basic descriptive geometry

9.1

Introduction. On many occasions, problems arise in engineering design that may be solved quickly by applying the basic principles of orthographic projection. If one thoroughly understands the solution for each of the problems presented, he should find it easy, at a later time, to analyze and solve almost any of the practical problems he may encounter.

It should be pointed out at the very beginning that to solve most types of problems one must apply the principles and methods used to solve a few basic problems, such as: (1) to find the true length of a line, (2) to find the point projection of a line, and (3) to find the true size and shape of a surface (Fig. 9.1). To find information such as the angle between surfaces, the angle between lines, or the clearance between members of a structure, one must use, in proper combination, the methods of solving these basic problems. Success in solving problems by projection depends largely on the complete understanding of the principles of projection, the ability to visualize space conditions, and the ability to analyze a given situation. Since the ability to analyze and to visualize are of utmost importance in engineering design, the student is urged to develop these abilities by resisting the temptation to memorize step procedures.

9.2

Projection of a Point. Figure 9.2(*a*) shows the projection of point *S* on the three

Fig. 9.1 The frame of this satellite could not have been designed without the application of descriptive geometry methods. (*Courtesy TRW Systems Group*)

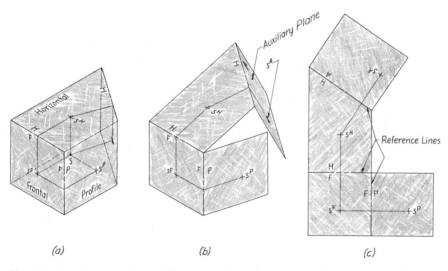

(a) *(b)* *(c)*

Fig. 9.2 Frontal, horizontal, profile, and auxiliary views of a point S.

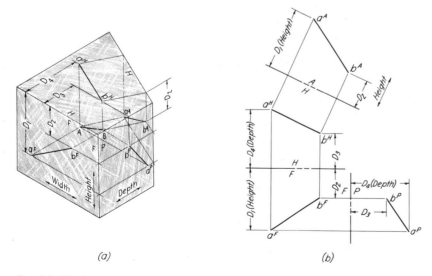

(a) *(b)*

Fig. 9.3 The line in space and in successive views.

principal planes of projection and a supplementary plane A. The notation used is as explained in Sec. 5.8. Point s^F is the view of point S on the frontal plane, s^H is the view of S on the horizontal plane, and s^P is its view on the profile plane. For convenience and ease in recognizing the projected view of a point on a supplementary plane, the supplementary planes are designated as A-planes and O-planes. A (auxiliary)-planes are always perpendicular to one of the principal planes. Point s^A is the view of S on the A-plane. The view of S on an O (oblique)-plane would be designated s^O. Additional O-planes are identified as O_1, O_2, O_3, etc., in the order that they follow the first O-plane.

Since it is necessary to represent on one plane (the working surface of our drawing paper) the views of point S that lie on mutually perpendicular planes of projection, the planes are assumed to be hinged so that they can be revolved, as shown in Fig. 9.2(*b*), until they are in a single plane, as in (*c*). The lines about which the planes of projection are hinged are called *reference lines*. A reference line is identified by the use of capital letters representing the adjacent planes, as FH, FP, FA, HA, AO, and so forth (see Fig. 9.2).

It is important to note in (*c*) that the projections s^F and s^H fall on a vertical line, s^F and s^P lie on a horizontal line, and s^H and s^A lie on a line perpendicular to the reference line HA. In each case this results from the fact that point S and its projections on adjacent planes lie in a plane perpendicular to the reference line for those planes [Fig. 9.2(*a*)]. This important principal of projection determines the location of views when the relationship of lines and planes form the problem.

9.3

Projection of a Straight Line. Capital letters are used for designating the end points of the actual line in space. In the projected views, these points are identified as shown in Fig. 9.3. The student should read Sec. 5.8, which presents the principles of multiview drawing. In particular, he should study the related illustration, which shows some typical line positions.

9.4

Projection of a Plane in Space. Theoretically, a plane is considered to be flat and unlimited in extent. A plane can be delineated graphically by: (1) two intersecting lines, (2) a line and a point not on the line, (3) two parallel lines, and, lastly, (4) three points not on a straight line. For graphical purposes and to facilitate the solution of space problems as presented in this chapter, planes will be bounded and usually triangular. The space picture and multiview representation of a plane ABC are given in Fig. 9.4. The pictorial at the left in (*a*) shows the projected views on the principal planes of projection. In (*b*) the planes are shown being opened outward to be in the plane of the paper, as in (*c*). It should be noted that plane ABC projects as a line (edge view) on the auxiliary plane. The three points A, B, and C of the plane are projected in the same manner as the single point S in Fig. 9.2 and are identified similarly (a^F, a^H, a^P, a^A, etc.).

9.5

Parallel Lines. Any two lines in space must be either (1) parallel, (2) intersecting, or (3) nonintersecting and nonparallel (called *skew lines*). Figure 9.18 shows intersecting lines, while Fig. 9.17 shows skew lines. Parallel lines are shown in Fig. 9.5.

It might be stated as a rule of projection, with one exception, that when two lines are parallel their projections will be parallel in every view (Fig. 9.5). In other words the lines will appear to be parallel in every view in which both appear. This is true even though in specific views they may appear as points or their projections may coincide. In either case they are still parallel because both conditions indicate that the lines have the same direction. The exception that has been mentioned occurs when the F- and H-projections of two inclined profile lines are shown. For proof of parallelism a supplementary view should be drawn, which may or may not be the profile view.

The true or shortest distance between two parallel lines can be determined on the view that will show these lines as points. The true distance can be measured between the points (Fig. 9.5).

9.6

To Determine the True Length of a Line. An observer can see the true length of a line when he looks in a direction perpendicular to it. It is suggested that the student hold a pencil before him and move it into the following typical line positions to observe the conditions under which the pencil, representing a line, appears in true length.

    1. *Vertical line.* The vertical line is perpendicular to the horizontal and will

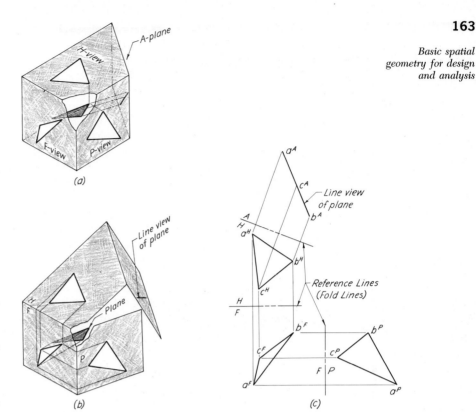

Fig. 9.4 **The plane in space and in successive views.**

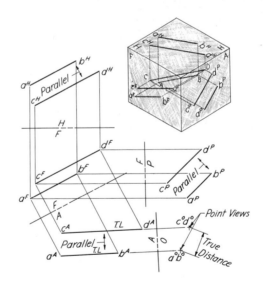

Fig. 9.5 **Parallel lines.**

therefore appear as a point in the H (top)-view. It will appear in true length in the F (frontal)-view, in true length in the P (profile)-view, and in true length in any auxiliary view that is projected on an auxiliary plane that is perpendicular to the horizontal plane of projection.

    2. *Horizontal line.* The horizontal line will appear in true length when viewed from above because it is parallel to the *H*-plane of projection and its end points are theoretically equidistant from an observer looking downward.

3. *Inclined line.* The inclined line will show true length in the *F*-view or *P*-view, for by definition (Sec. 5.8) an inclined line is one that is parallel to either the *F*-plane or the *P*-plane of projection. However, it cannot be parallel to both planes of projection at the same time.

4. *Oblique line.* The oblique line will not appear in true length in any of the principal views because it is inclined to all of the principal planes of projection. It should be apparent, in viewing the pencil alternately from the directions used to obtain the principal views, namely, from the front, above, and side, that one end of the pencil is always

farther away from the observer than the other. Only when looking directly at the pencil from such a position that the end points are equidistant from the observer can the true length be seen. On a drawing, the true length projection of an oblique line will appear in a supplementary *A* (auxiliary)-view on a plane that is parallel to the line.

9.7
To Determine the True Length of an Oblique Line. In order to find the true length of an oblique line, it is necessary to select an auxiliary plane of projection that will be parallel to the line (Figs. 9.6 and 9.7).

> **Given:** The *F* (frontal)-view $a^F b^F$ and the *H* (top)-view $a^H b^H$ of the oblique line *AB* (Fig. 9.6).
>
> **Solution:** (1) Draw the reference line *HA* parallel to the projection $a^H b^H$. The *A*-plane for this reference line will be parallel to *AB* and perpendicular to the *H*-plane (see pictorial drawing). (2) Draw lines of projection from points a^H and b^H perpendicular to the reference line. (3) Transfer height measurements from the *F*-view to the *A*-view to locate a^A and b^A. In making this transfer of measurements, the students should attempt to visualize the space condition for the line and understand that, since the *F*-view and *A*-view both show height and because the planes of projection for these views are perpendicular to the *H*-plane, the perpendicular distance D_1 from the reference line *HF* to point a^F must be the same as the distance D_1 from the reference line *HA* to point a^A.

The projection $a^A b^A$ shows the true length of the line *AB*.

It was not necessary to use an auxiliary plane perpendicular to the *H*-plane to find the true length of line *AB* in Fig. 9.6. The auxiliary plane could just as well have been perpendicular to either the *F*- or *P*-planes. Figure 9.7 shows the use of an auxiliary plane perpendicular to the frontal plane to find the true length of the line. In this case the auxiliary view has depth distances in common with top view, as indicated.

9.8
Perpendicular Lines. Lines that are perpendicular in space will have their projections perpendicular in any view that shows either or both of the lines in true length. A second rule of perpendicularity might be that, when a line is perpendicular to a

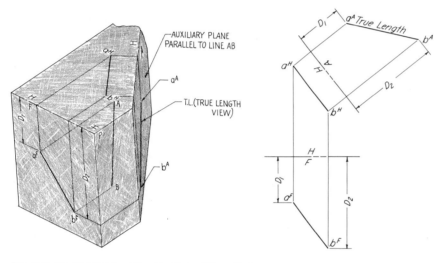

Fig. 9.6 To find the true length of an oblique line.

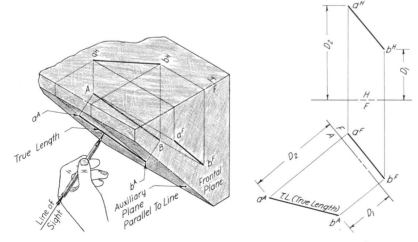

Fig. 9.7 To find the true length of a line.

plane, it will be perpendicular to every line in that plane. A careful study of Fig. 9.8 will verify these rules. For instance, it should be noted that the lines AB and CD lie in a plane that is outlined with broken lines and that $e^H f^H$ is perpendicular to $a^H b^H$ because $a^H b^H$ shows the true length of the line AB. In the A-view, we see that $e^A f^A$ is perpendicular to the line view of the plane and is therefore perpendicular to both AB and CD. The O-view shows the true shape (TSP) of the plane and the line EF as a point. This again verifies the fact that EF is perpendicular to the plane and to lines AB and CD. Otherwise line EF would not appear as a point. Note also that the O-view shows the true length (TL) of AB and CD.

9.9

Bearing of a Line. The bearing of a line is the horizontal angle between the line and a north–south line. A bearing is given in degrees with respect to the meridian and is measured from 0° to 90° from either north (N) or south (S). The bearing reading indicates the quadrant in which the line is located by use of the letters N and E, S and E, S and W, or N and W, as N 48° E or S 54° 40′ W. The bearing of a line is measured in the H-view (Fig. 9.9).

9.10

Point View of a Line (Fig. 9.10). It was pointed out in the first section of this chapter that the solutions of many types of problems depend on an understanding of a few basic constructions. One of these basic constructions involves the finding of the view showing the point view or point projection of a line. For instance, this construction is followed when it is necessary to determine the dihedral angle between two planes, for the true size of the angle will appear in the view that shows the line common to the two planes as a point.

A line will show as a point on a projection plane that is perpendicular to the line. The observer's direction of sight must be along and parallel to the line. When a line appears in true length on one of the principal planes of projection, only an auxiliary view is needed to show the line as a point. However, in the case of an oblique line both an auxiliary and an oblique view are required, for a point view must always follow a true-length view. In other words, the plane of projection for the view showing the line as a point must be adjacent to

the plane for the true view and be perpendicular to it.

Given: The F-view $a^F b^F$ and the H-view $a^H b^H$ of the oblique line AB (Fig. 9.10).

Solution: (1) Draw the view showing the TL (true length) of AB. This is an auxiliary view drawn, as explained in Sec. 9.7. (2) Draw reference line AO perpendicular to the true-length projection $a^A b^A$. This reference line is for an O-plane that is perpendicular to the A-plane. (3) Draw a projection line from $a^A b^A$ and transfer the distance D_3 from the H-view to the O-view. It should be noted from the pictorial drawing that the distance D_3 is common to both of these views, and that points a^O and b^O coincide to give a point or end view of line AB.

9.11

To Find the Shortest Distance from a Point to a Line. The shortest distance between a given point and a given straight line must be measured along a perpendicular

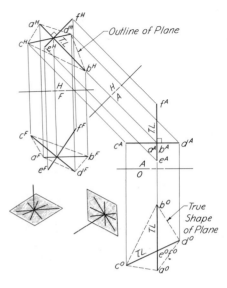

Fig. 9.8 Perpendicular lines.

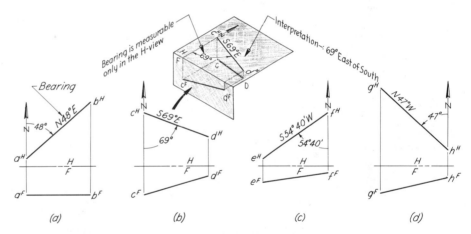

Fig. 9.9 Bearing of a line.

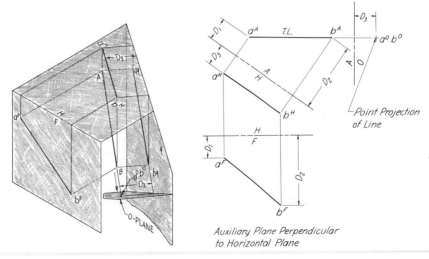

Fig. 9.10 Point view of a line.

drawn from the point to the line. Since lines that are perpendicular will have their projections show perpendicular in any view showing either or both lines in true length (Sec. 9.8), the perpendicular must be drawn in the view showing the given line in true length.

Given: The *F*- and *H*-views of the line *AB* and point *C* (Fig. 9.11).

Solution: (1) Draw the *A* (auxiliary)-view showing the true-length view $a^A b^A$ of line *AB* and view c^A of point *C*. (2) Draw $c^A x^A$ perpendicular to $a^A b^A$. Line $c^A x^A$ is a view of the required perpendicular from point *C* to its juncture with line *AB* at point *X*. (3) Draw reference line *AO* parallel to $c^A x^A$. This reference line locates an *O*-plane, which will be parallel to the perpendicular *CX* and perpendicular to the *A*-plane. The *O*-view will show the true length of *CX*. Line *CX* does not show true length in any of the other views.

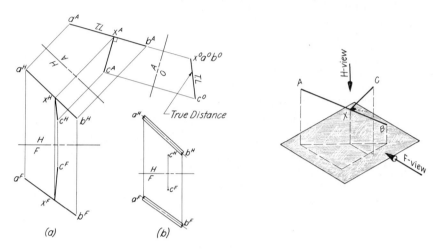

Fig. 9.11 To find the shortest distance from a point to a line.

9.12
Principal Lines of a Plane (Fig. 9.12). Those lines that are parallel to the principal planes of projection are the *principal lines of a plane*. A principal line may be either a horizontal line, a frontal line, or a profile line. Principal lines are true-length lines and one such line may be drawn in any plane to appear true length in any one of the principal views, as desired. This is an important principle that is the basis for the solution of many problems involving lines and planes.

9.13
To Obtain the Edge View of a Plane. When a plane is vertical, an edge view of it will be seen from above and it will be represented by a line in the top view. Should a plane be horizontal, it will appear as an edge in the frontal view. However, planes are not always vertical or horizontal; frequently they are inclined or oblique to the principal planes of projection.

Finding the edge view of a plane is a basic construction that is used to determine the slope of a plane (dip), to determine clearance, and to establish perpendicularity. The method presented here is part of the construction used to obtain the true size and shape of a plane, to determine the angle between a line and plane, and to establish the location at which a line pierces a plane.

The edge view of an oblique plane can be obtained by viewing the plane with direction of sight parallel to it. The edge view will then appear in an auxiliary view. When the auxiliary view shows height, the slope of the plane is shown (Fig. 9.13).

Given: The *F*- and *H*-views of plane *ABC*.

Solution: (1) Draw the horizontal line *AX* in the plane. Because *AX* is parallel to the horizontal, $a^F x^F$ will be horizontal and must be drawn before the position of $a^H x^H$ can be established. (2) Draw refer-

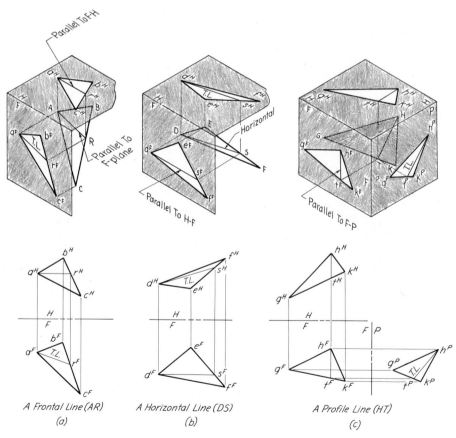

| *A Frontal Line (AR)* | *A Horizontal Line (DS)* | *A Profile Line (HT)* |
| (a) | (b) | (c) |

Fig. 9.12 Location of a principal line in a plane.

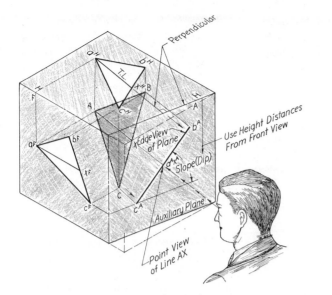

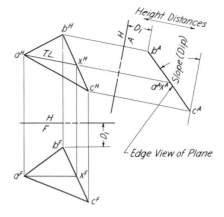

Fig. 9.13 To find the edge view of a plane.

ence line HA perpendicular to $a^H x^H$. (3) Construct the A-view, which will show the plane ABC as a straight line. Since the F-view and A-view have height as a common dimension, the distances used in constructing the A-view were taken from the F-view. It should be noted by the reader that an edge view found by projecting from the front view will show the angle that the plane makes with the F-plane. Similarly, the edge view found by projecting from the side view will show the angle with the P-plane.

9.14
To Find the True Shape (TSP) of an Oblique Plane.

Finding the true shape of a plane by projection is another of the basic constructions that the student must understand, for it is used to determine the solution of two of the problems that are to follow. In a way, the construction shown in Fig. 9.14 is a repetition of that shown in Fig. 8.19. However, repetition in the form of another presentation should help even those students who feel they understand the method for finding the true shape of an oblique surface of an object.

To see the true size and shape of an oblique plane an observer must view it with a line of sight perpendicular to it. To do this he must, as the first step in the construction, obtain an edge view of the plane. An O-view taken from the A-view will then show the true shape of the plane.

Given: The F- and H-views of plane ABC.

Solution: (1) Draw a frontal line CD in plane ABC. (2) Draw reference line FA

perpendicular to $c^F d^F$ and construct the A-view showing the edge view of the plane. The auxiliary view has depth in common with the H-view. (3) Draw reference line AO parallel to the edge view in the A-view and construct the O-view, which will show the true size and shape of plane ABC. The needed distances from the reference line to points in the O-view are found in the F-view.

9.15
To Find the Piercing Point of a Line and a Plane.

Determining the location of the point where a line pierces a plane is an-

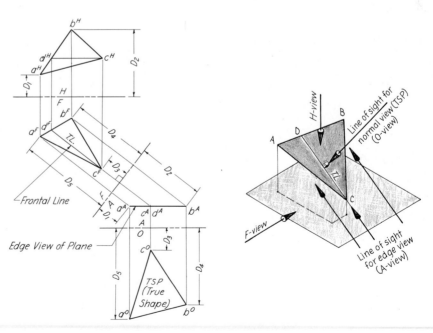

Fig. 9.14 To find the true shape of an oblique plane.

other fundamental operation with which one must be familiar. A line, if it is not parallel to a plane, will intersect the plane at a point that is common to both. In a view showing the plane as an edge, the piercing point appears where the line intersects (cuts) the edge view. This method is known as the edge-view method to distinguish it from the cutting-plane method, which may be found in Chapter 10.

The simple cases occur when the plane appears as an edge in one of the principal views. The general case, for which we use an auxiliary plane, is given in Fig. 9.15.

Given: Plane *ABC* and line *ST*.

Solution: (1) Draw the horizontal line *BX* in *ABC*. (2) Draw reference line *HA* perpendicular to $b^H x^H$ and construct the *A*-view showing an edge view of the plane as line $a^A b^A c^A$ and the view of the line $s^A t^A$. Point p^A where the line cuts the edge view of the plane is the *A*-view of the piercing point. (3) Project point *P* back from the *A*-view first to the *H*-view and then to the *F*-view.

9.16
To Determine the Angle Between a Line and a Given Plane. The true angle between a given line and a given plane will be seen in the view that shows the plane as an edge (a line) and the line in true length. The solution shown in Fig. 9.16 is based on this premise. The solution as presented might be called the *edge-view method.*

Given: The *F*- and *H*-views of plane *ABC* and line *ST*.

Solution: (1) Draw the frontal line *BX* in the plane *ABC*. (2) Draw reference line *FA* perpendicular to $x^F b^F$ and construct the *A*-view. This view will show plane *ABC* as line $a^A b^A c^A$; however, since this view does not show *ST* in true length, the true angle is not shown. (3) Draw reference line *AO* parallel to the edge view of the plane and construct the *O*-view that will show line *ST* viewed obliquely and plane *ABC* in its true size and shape. (4) Draw reference line OO_1 parallel to $s^O t^O$ and construct the second oblique view.

Line $s^{O_1} t^{O_1}$ will show the true length of *ST* in this second oblique view. Plane *ABC* will be seen again as an edge (line), for it now appears on an adjacent view taken perpendicular to the view showing

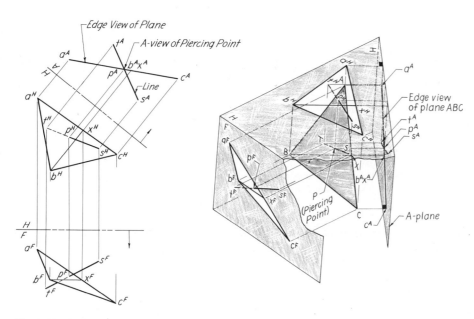

Fig. 9.15 To find the piercing point of a line and a plane.

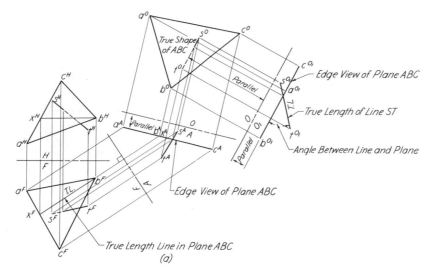

Fig. 9.16 To find the angle between a line and a plane.

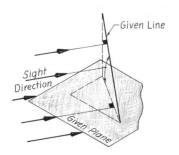

True angle can be seen in a view which shows the plane as an edge and the line in true length

(b)

true shape (TSP). The required angle can now be measured between the true-length view of line ST and the edge view of plane ABC.

In the illustration of Fig. 9.16 three supplementary views were required to obtain the true angle. If the plane had appeared as an edge in one of the given views, only two supplementary views would have been needed; if it had appeared in true shape in a given view only one additional view, properly selected to show the plane as an edge and the line in true length, would be needed.

9.17
To Determine the Angle Between Two Nonintersecting (Skew) Lines.

The angle between two nonintersecting lines is measurable in a view that shows both lines in true length. The analysis and construction that might be followed to obtain the needed view are shown in Fig. 9.17.

> **Given:** The two nonintersecting lines AB and CD.
>
> **Solution:** (1) Draw the reference line HA parallel to $a^H b^H$ and construct the A-view that shows AB in true length and the view of the line CD. (2) Draw reference line AO perpendicular to $a^A b^A$ and draw the O-view in which the line AB will appear as a point ($a^o b^o$). (3) Draw reference line OO_1, parallel to $c^o d^o$ and construct the O-view showing the true length of CD. Line AB must also show true length in this view since AB, showing as a point in the O-view, is parallel to the OO_1-plane. With both lines shown in true length in the O_1-view, the required angle may be measured in this view.

9.18
To Determine the True Angle Between Two Intersecting Oblique Lines.

Since, as previously stated, two intersecting lines establish a plane, the true angle between the intersecting lines may be seen in a true-shape view of a plane containing the lines (Fig. 9.14). In Fig. 9.18 line AC completes plane ABC containing the given lines AB and BC. It is necessary to find the true angle between AB and BC.

> **Solution:** (1) Draw the frontal line XC. (2) Draw reference line FA perpendicular to $c^F x^F$ and construct the A-view showing an edge view of plane ABC. (3) Draw reference line AO parallel to the edge view $a^A c^A b^A$ and construct the O-view, which will show the TSP of plane ABC. In this view it is desirable to show only the given lines. The true angle between AB

and BC is shown by $a^o b^o c^o$. As a practical application this method might be used to determine the angle between two adjacent sections of bent rod, as shown in (b).

9.19
Distance Between Two Parallel Planes.

When two planes are parallel their edge views will appear as parallel lines in the same view. The clearance or perpendicular distance between them can be measured in this view (Fig. 9.19). The existence of planes as parallel lines in a view is another proof that they are parallel.

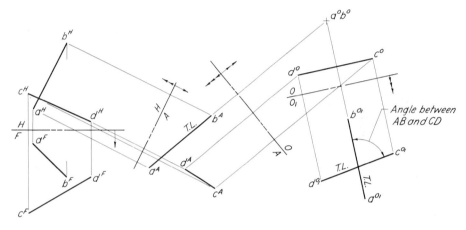

Fig. 9.17 To find the angle between two skew lines.

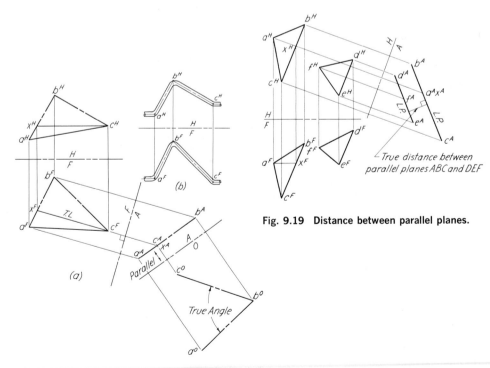

Fig. 9.18 To find the true angle between two intersecting lines.

Fig. 9.19 Distance between parallel planes.

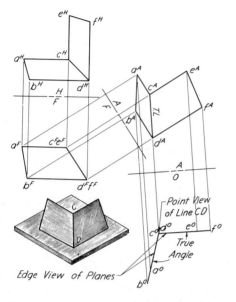

Fig. 9.20 To find the dihedral angle between two planes.

9.20

To Find the Dihedral Angle Between Two Planes. The angle between two planes is known as a *dihedral angle*. The true size of this angle between intersecting planes may be seen in a plane that is perpendicular to both. For this condition as set forth, the intersecting planes will appear as edges and the line of intersection of the two planes as a point. The true angle may be measured between the edge views of the planes (Fig. 9.20).

Given: The intersecting planes $ABCD$ and $CDEF$. The line of intersection is line CD, as shown in the pictorial drawing.

Solution: (1) Draw reference line FA parallel to $c^F d^F$ and construct the A-view. This view will show CD in true length ($c^A d^A$). (2) Draw reference line AO perpendicular to $c^A d^A$ and construct the adjacent O-view. Since this view was taken looking along line CD, points C and D are coincident and appear as a single point identified as $c^O d^O$. The intersecting planes show as edge views and the true angle between the given planes may be measured between these edge-view lines. When two planes are given that do not intersect, the dihedral angle may be found after the line of intersection has been determined.

9.21

To Find the Shortest Distance Between Two Skew Lines. As was stated in Sec. 9.5, any two lines that are not parallel and do not intersect are called *skew lines*. The shortest distance between any such lines must be measured along one line and only one line that can be drawn perpendicular

to both. This common perpendicular can be drawn in a view that is taken to show one line as a point. Its projection will be perpendicular to the view of the other line and will show in true length (Fig. 9.21).

Given: The F- and H-views of two skew lines AB and CD.

Solution: (1) Draw an A-view adjacent to the H-view to show line AB in true length ($a^A b^A$). Line CD should also be shown in this same view ($c^A d^A$). (2) Draw reference line AO perpendicular to $a^A b^A$ and draw the O-view in which line AB will appear as a point ($a^O b^O$). It is in this view that the exact location of the required perpendicular can be established. (3) Draw the line $e^O f^O$ through point $a^O b^O$ perpendicular to $c^O d^O$. The shortest distance between the skew lines now appears in true distance as the length of $e^O f^O$. Although the CD does not appear in true length in the O-view, $e^O f^O$ does, and hence $c^O d^O$ and $e^O f^O$ will appear perpendicular. (4) Complete the A-view by first locating point e^A on $c^A d^A$ and then draw $e^A f^A$ parallel to reverence line AO. (5) Locate points E and F in the H- and F-views remembering that point E is located on line CD and point F on line AB.

In engineering design an engineer frequently has to locate and find the length of the shortest line between two skewed members in order to determine clearance or the length of a connecting member. In underground construction work one might use this method to locate a connecting tunnel.

In Fig. 9.21(b) and (c) we see the two rods for which the clearance distance was determined in (a). Lines AB and CD represent the center lines of the rods.

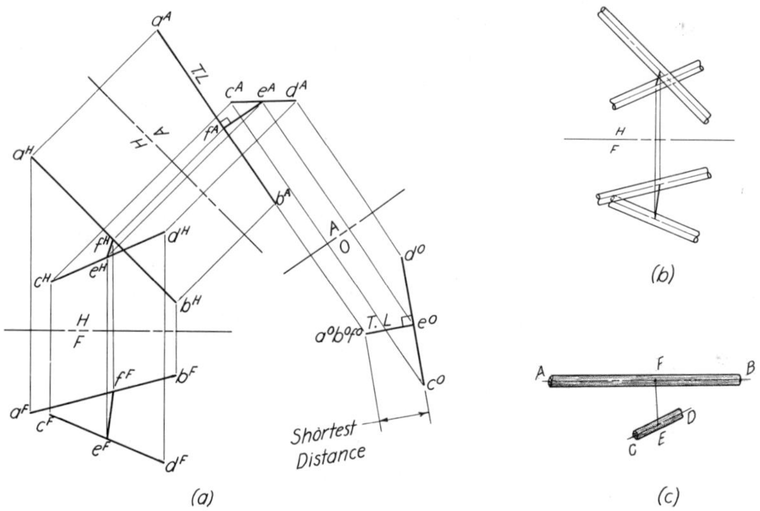

B □ Vector geometry

9.22

Vector Methods. In order to be successful in solving some types of problems that arise in design, a well-trained technologist should have a working knowledge of *vector geometry*. The methods presented in this chapter should furnish the student with some background knowledge for solving force problems as they appear in the study of mechanics, strength of materials, and design. Through discreet use of the methods of vector geometry, as well as mathematical methods, it is possible to solve engineering problems quickly within a fully acceptable range of accuracy. Since any quantity having both magnitude and

(a)

(b)

(c)

Fig. 9.21 To find the shortest distance between two skew lines.

direction may be represented by a fixed or rotating vector, vector operations are commonly used for problems in the design of frame structures, problems dealing with velocities in mechanisms, and for problems arising in the study of electrical properties. Because a student in a beginning course in engineering graphics should have basic principles rather than specialized cases presented to him, the methods given in this chapter for solving both two-dimensional and three-dimensional force problems deal mainly with static structures or, in other words, structures with forces acting so as to be in equilibrium. In a study of physics, graphical methods are useful for the composition and resolution of forces.

It is hoped that as a student progresses through his other undergraduate courses, he will desire to learn more about the use of vector methods for solving problems, and that he will become able to recognize the cases when he may have a choice between a graphical and an algebraic method. The graphical method is the better for many cases because it is much quicker and can be checked more easily.

An example of a vector addition is shown in Fig. 9.22. An airplane is flying north with a cross wind from the west. If the speed of the plane is 250 km per hr (250 kilometers per hour) and the wind is blowing toward the east at 98 km per hr, the plane will be flying NE (northeast) at 270 km per hr. Vectors can be used for problems of this type because forces acting on a body have both magnitude and direction.

9.23

Force. In our study of vector methods, a force may be defined as a cause which tends to produce motion in an object.

A force has four characteristics which determine it. First, a force has "magnitude." The value of this magnitude is expressed in terms of some standard unit. Second, a force has a "line of action." This is the line along which the force acts. Third, a force has "direction." This is the direction in which it tends to move the object upon which it acts. Fourth, and last, a force has a "point of application." This is the place at which it acts upon the object, often assumed to be a point at the center of gravity.

When values are given for magnitude, it is essential that consistent units be used. The British gravitational system uses feet, pounds, and seconds as basic units while

the metric(SI) absolute system, now being adopted in the U.S., uses the kilogram, meter, and second as units of mass, length, and time. Since the student will find it necessary to be familiar with both systems during this period of change to SI, both systems have been used in this chapter.

The SI unit of force (Fig. 9.23), called a newton (N), is the force that must be applied to a mass of one kilogram (kg) to produce an acceleration of one meter per second per second. The weight of a body in newtons is equal to its mass in kilograms multiplied by 9.81m/s^2. See Appendix Table 7.

Conversion Factors

English to SI	SI to English
Force 1 lb = 4.448 N	1 N = 0.2248 lb
= 4.448 kg·m/s²	

9.24

Vector. A force can be represented graphically by a straight line segment with an arrowhead at one end. Such an arrow when used for this purpose is known as a *vector* (Fig. 9.23). The position of the body of the arrow represents the line of action of the force while the arrowhead points out the direction. The magnitude is represented to some selected scale by the overall length of the arrow itself.

When a force acts in a two-dimensional plane, only one view of the vector is needed. However, if the force is in space, two views of the vector must be given.

9.25

Addition of Vector Forces—Two Forces. For a thorough understanding of the principles of vector addition, two simple examples will be considered first.

If one of two men who find it necessary to move a supply cabinet pushes on it with 60 lb of force while the other pushes in the same direction with 40 lb of force, the total force exerted to move the cabinet is 100 lb. The representation of two or more such forces in the manner shown in Fig. 9.24 amounts to a vector addition. Should these men be in a prankish mood and decide to push in opposite directions, as illustrated in Fig. 9.25, the cabinet might move provided the 20-lb resultant force were sufficient to overcome friction. The 20-lb resultant comes from a graphical addition.

Now let it be supposed that force A represented by F_A and force B represented by F_B in Fig. 9.26 act from a point

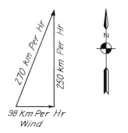

Fig. 9.22 Vector problem.

Fig. 9.23 Vector.

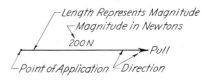

Fig. 9.24 Vector addition.

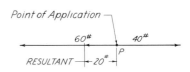

Fig. 9.25 Forces in opposite directions.

Fig. 9.26 Parallelogram of forces.

Fig. 9.27 Vector triangle.

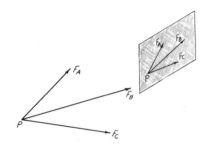

Fig. 9.28 Resultant of two forces.

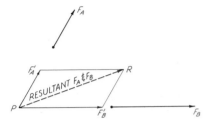

Fig. 9.29 Coplanar, concurrent forces.

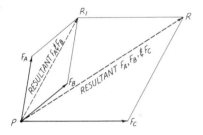

Fig. 9.30 Resultant of three or more forces with a common point of application (parallelogram method).

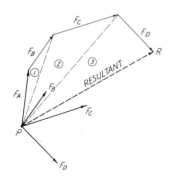

Fig. 9.31 Resultant of forces (polygon of forces).

P, the point of application. The resultant force on the body will not now be the sum of forces A and B, but instead will be the graphical addition of these forces as represented by the diagonal of a parallelogram having sides equal to the scaled length of the given forces. This single force R of 470 N would produce an effect upon the body that would be equivalent to the combined forces F_A and F_B. The single force which could replace any given force system, is known as the *resultant* (R) for the force system.

Figure 9.26 shows that the resultant R divides the parallelogram into two equal triangles. Therefore, R could have been found just as well by constructing a single triangle as shown in Fig. 9.27 provided that the vector F_B is drawn so that its tail-end touches the tip-end of F_A, and R is drawn with its arrow-end of to the tip-end of F_B. Since either of the triangles shown in Fig. 9.26 could have been drawn to determine R, it should be obvious the resultant is the same regardless of the order in which the vectors are added. However, it is important that they be added tip-end to tail-end and that the vector arrows show the true direction for the action of the concurrent forces in the given system.

To find the resultant of two forces, which are applied as shown in Fig. 9.28, it is first necessary to move the vector arrows along their lines of action to the intersection point P before one can apply the parallelogram method.

The forces of a system whose lines of action all lie in one plane are called *coplanar forces*. Should the lines of action pass through a common point, the point of application, the forces are said to be *concurrent*. Figure 9.29 shows a system of forces that are both concurrent and coplanar.

9.26

Addition of Vectors—Three or More Forces. The parallelogram method may be used to determine the resultant for a system of three or more forces that are concurrent and coplanar. In applying this method to three or more forces, it is necessary to draw a series of parallelograms, the number depending upon the number of vector quantities that are to be added graphically. For example, in Fig. 9.30 two parallelograms are required to determine the resultant R for the system. The resultant R_1 for forces F_A and F_B is determined by the first parallelogram to be drawn,

and then R_1 is combined in turn with F_C by forming the second and larger parallelogram. By combining the forces in this way R becomes the resultant for the complete system.

Where a considerable number of vectors form a system, a somewhat less complicated diagram results, and less work is required when the triangle method is extended and applied to the formation of a vector diagram such as the one shown in Fig. 9.31. In this case the diagram is formed by three vector triangles, one adjacent to the other, and the resultant of forces F_A and F_B is combined with F_C to form the second triangle. Finally, by combining the resultant of the three forces F_A, F_B, and F_C with F_D, the vector R is obtained, which represents the magnitude and direction of the resultant of the four forces. In the construction F_B, F_C, and F_D in the diagram must be drawn so as to be parallel respectively to their lines of action in the system. However, the order in which they are placed in the diagram is optional as long as one vector joins another tip to tail.

9.27

Vector Components. A component may be defined as one of two or more forces into which an original force may be resolved. The components, which together have the same action as the single force, are determined by a reversal of the process for vector addition, that is, the original force is resolved using the parallelogram method (see Fig. 9.32). The resolution of a plane vector into two components, horizontal and vertical, is illustrated in (*a*). In (*b*), the resolution of a force into components of specified direction is shown.

9.28

Forces in Equilibrium. A body is said to be in equilibrium when the opposing forces acting upon it are in balance. In such a state the resultant of the force system will be zero. The concurrent and coplanar force system shown in Fig. 9.33 is in a state of equilibrium, for the vector triangle closes and each vector follows the other tip to tail.

An *equilibrant* is the force that will balance two or more forces and produce equilibrium. It is a force that would equal the resultant of the system but would necessarily have to act in an opposite direction.

Figure 9.34 shows a weight supported by a short steel cable. The force to be determined is that needed to hold the weight in a state of equilibrium when it is swung from the position indicated by the broken lines into the position shown by solid lines.

This may be done by drawing a vector triangle with the forces in order from tip to tail. The 396 N force vector represents the equilibrant, the force that will balance the 686.7 N force and the 795 N tension force now in the cable. The reader may wonder at the increase in the tension force in the cable from a 686.7 N force when hanging straight down to a 795 N force when the cable is at an angle of 30° with the vertical. It might help to realize that as the weight is swung outward toward a position where the cable will be horizontal, both the tension force and the equilibrant will increase. Theoretically it would require forces infinitely large to hold the system in equilibrium with the cable in a horizontal position.

In solving a force system graphically it is possible to determine two unknowns in a coplanar system.

Now suppose that it is desired to determine the forces acting in the members of a simple truss as shown in Fig. 9.35(*a*). To determine these forces graphically, one should isolate the joint supporting the weight and draw a diagram, known as a free-body diagram to show the forces acting at the joint (*b*). Although the lines of this diagram may have any length, they must be parallel to the lines in the space diagram in (*a*). Since the boom will be in compression, a capital letter C has been placed along the line that represents the boom in the diagram. A letter T has been placed along the line for the cable because it will be in tension. Although the diagram may not have been essential in this particular case, such a diagram does play an important part in solving more complex systems.

In constructing the force polygon, it is necessary to start by drawing the vertical vector, for the load is the only force having a known magnitude and direction. After this vertical vector has been drawn to a length representing 4905 N, using a selected scale, the force polygon (triangle) may be completed by drawing the remaining lines representing the unknown forces parallel to their known lines of action as shown in (*a*). The force polygon will close since the force system is in equilibrium.

The magnitude of the unknown forces in the members of the truss can now be determined by measuring the lines of the

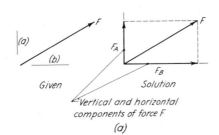
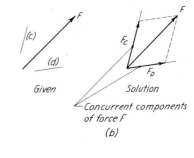

Fig. 9.32 Components.

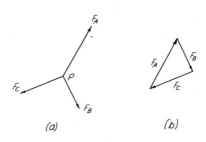
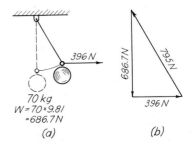

Fig. 9.33 Forces in equilibrium.

Fig. 9.34 Determination of forces (graphically).

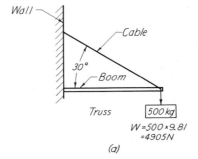
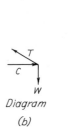
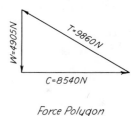

Fig. 9.35 Determination of the forces at the joint of a simple truss.

diagram using the same scale selected to lay out the length of the vertical vector. This method might be used to determine the forces acting in the members at any point in a truss.

9.29
Coplanar, Noncurrent Force Systems. Forces in one plane having lines of action that do not pass through a common point are said to be *coplanar, nonconcurrent forces* (Fig. 9.36).

9.30
Two Parallel Forces. When two forces are parallel and act in the same direction, their resultant will have a line of action that is parallel to the lines of action of the

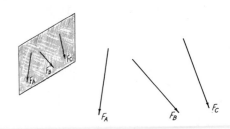

Fig. 9.36 Coplanar, nonconcurrent forces.

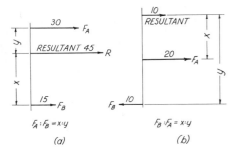

Fig. 9.37 Parallel forces.

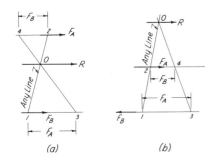

Fig. 9.38 Determination of the position of the resultant of parallel forces (graphical method).

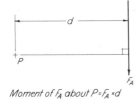

Moment of F_A about $P = F_A \times d$

Fig. 9.39 Moment of a force.

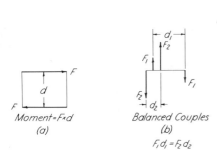

Fig. 9.40 Force couples.

given forces and it will be located between them. The magnitude of the resultant will be equal to the sum of the two forces [Fig. 9.37(a)], and it will act through a point that divides any perpendicular line joining the lines of action of the given forces inversely as the forces.

Should the two forces act in opposite directions, as shown in (b), the resultant will be located outside of them and will have the same direction as the greater force. Its magnitude will be equal to the difference between the two given forces. The proportion shown with the illustration in (b) may be used to determine the location of the point of application of the resultant. Those who prefer to determine graphically the location of the line of action for the resultant may use the method illustrated in Fig. 9.38. This method is based on well-known principles of geometry.

With the two forces F_A and F_B given, any line 1–2 is drawn joining their lines of action. From this line two distances must be laid off along the lines of action of the given forces. If the given forces act in the same direction, then the distances are laid off in opposite directions from the line 1–2, (a). If they act in opposite directions, the distances must be laid off on the same side of line 1–2, (b). In Fig. 9.38(a), a length equal by scale to F_A was laid off from point 1 on the line of action of F_B. Then from point 2 a length equal to F_B was marked off in an opposite direction. These measurements located points 3 and 4, the end points of the line intersecting line 1–2 at point O. Point O is on the line of action of the resultant R. In Fig. 9.38(b) this method has been applied to establish the location of the resultant for two forces acting in opposite directions.

9.31

Moment of a Force. The *moment of a force* with respect to a point is the product

of the force and the perpendicular distance from the given point to the line of action of the force. In the illustration, Fig. 9.39, the moment of the force F_A about point P is Mom. $= F_A \times d$. The perpendicular distance d is known as the lever arm of the force. Should the distance d be measured in inches and the force be given in pounds, the moment of the force will be in inch-pounds.

9.32

Force Couples (Fig. 9.40). Two equal forces that act in opposite directions are known as a *couple*. A couple does not have a resultant, and no single force can counteract the tendency to produce rotation. The measurement of this tendency is the moment of the couple that is the product of one of the forces and the perpendicular distance between them.

To prevent the rotation of a body that is acted upon by a couple, it is necessary to use two other forces that will form a second couple. The body acted upon by these couples will be in equilibrium if each couple tends to rotate the body in opposite directions and the moment of one couple is equal to the other.

9.33

String Polygon—Bow's Notation. A system for lettering space and force diagrams, known as *Bow's notation*, is widely used by technical authors. Its use in this chapter will tend to simplify the discussions which follow.

In the space diagram, shown in Fig. 9.41(a), each space from the line of action of one force to the line of action of the next one is given a lowercase letter such as *a*, *b*, *c*, and *d* in alphabetical order. Thus the line of action for any particular force can be designated by the letters of the areas on each side of it. For example, in Fig. 9.43 the line of action for the 1080-lb force, acting downward on the beam, would be designated as line of action *bc*. On the force diagram, corresponding capital letters are used at the ends of the vectors. In Fig. 9.41(b), AB represents the magnitude of *ab* in the space diagram and BC represents the magnitude of *bc*.

To find the resultant of three or more parallel forces graphically, the "funicular" or "string polygon" is used. The magnitude and direction of the required resultant for the system shown in Fig. 9.41 are known. The magnitude, representing the algebraic sum of the given forces, appears

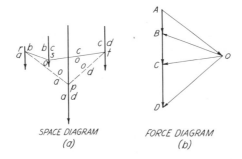

Fig. 9.41 Resultant of parallel forces—Bow's notation.

as the heavy line AD of the force polygon. It is required to determine the location of its line of action. With the forces located in the space diagram and the force polygon drawn, the steps for the solution are as follows:

1. Assume a pole point O and draw the rays OA, OB, OC, and OD. Each of the triangles formed is regarded as a vector triangle with one side representing the resultant of the forces represented by the other two sides. For example: If we consider AB to be a resultant force, then OA and OB are two component forces that could replace AB. For the second vector triangle, OB and BC have OC as their resultant. OC, when combined with CD, will have OD as the resultant. OA and OD combine with AD, the final resultant of the system.

2. Draw directly on the space diagram the corresponding strings of the funicular polygon. The funicular polygon may be started at any selected point r along the line of action ab. The string ob will then be parallel to OB of the force polygon. From point s, where ob intersects bc, draw oc parallel to OC. The line oc extended to cd establishes the location of point t. Line od drawn parallel to OD and line oa drawn parallel to OA intersect at point p. Point p is a point on the line of action of force ad, the resultant force AD for the given force system.

When one or more of a system of parallel forces are directed oppositely from the others, the magnitude and direction of the resultant will be equal to the algebraic sum of the original forces.

9.34

Coplanar, Nonconcurrent, Nonparallel Forces. In further study of coplanar and nonconcurrent forces it might be supposed that it is necessary to determine the magnitude, direction, and line of action of the one force that will establish a state of equilibrium when combined with the given forces AB, BC, and CD of the force system shown in Fig. 9.42. The direction and line of action of the original forces are given in both the space diagram in (a) and the force polygon in (b). The magnitude and direction of the force that will produce equilibrium is represented by DA, the force needed to close the force polygon. With the force polygon completed, the next step is to assume a pole point O and draw the rays OA, OB, OC, and OD. Now

OA and OB are component forces of AB, and AB might be replaced by these forces. To clarify this statement: each of the four triangles may be considered to be a vector triangle, and in the case of vector triangle OAB, AB can be regarded as the resultant for the other two forces OA and OB. It should be noted that component force OB of the vector triangle OBC must be equal and opposite in direction to component force OB of OAB.

All that remains to be done is to determine the line of action of the required force DA by drawing the string diagram as explained in Sec. 9.33, remembering that point r may be any point along the line of action of ab. The intersection point p for strings oa and od is a point along the line of action da of force DA. Although lines of action ab, bc, cd, and da were drawn to a length representing their exact magnitude in Fig. 9.42(a), they could have been drawn to a convenient length to allow for the construction of the string polygon, for these lines merely represent lines of action for the forces AB, BC, CD, and DA. The lines were presented in scaled length for illustrative purposes.

9.35

Equilibrium of Three or More Coplanar Parallel Forces. When a given system, consisting of three or more coplanar forces, is in equilibrium, both the force polygon and the funicular polygon must close. If the force polygon should close and the funicular polygon not close, the resultant of the given system will be found to be a force couple.

Two unknown forces of a parallel coplanar force system may be determined graphically by drawing the force and funicular polygons as shown in Fig. 9.43, since the forces are known to be in equilibrium, and all are vertical. Although one may be aware that the sum of the two reaction forces R_1 and R_2 is equal to the sum of forces AB, BC, CD, and DE, the magnitude of R_1 and R_2 as single forces is unknown. The location of point F in the force polygon, which is needed if one is to determine the magnitudes of R_1 and R_2, may be found by fulfilling the requirement that the funicular polygon be closed.

The funicular polygon is started at any convenient point along the known line of action of R_1, and successive strings are drawn parallel to corresponding rays of the force polygon. In area b the string will be parallel to OB, in area c the string will be parallel to OC, and so on, until the

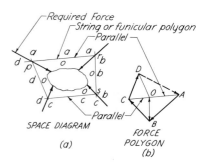

Fig. 9.42 Funicular or string polygon.

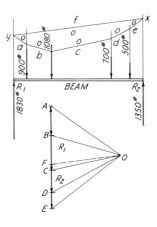

Fig. 9.43 To determine the reaction forces of a loaded beam.

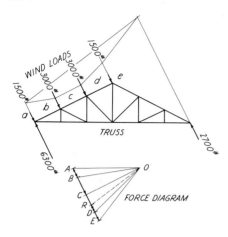

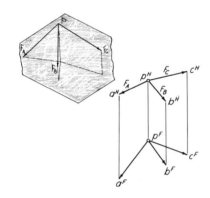

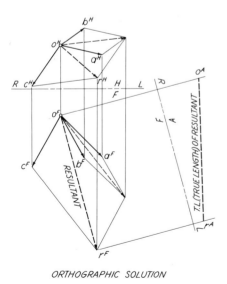

ORTHOGRAPHIC SOLUTION

Fig. 9.44 Determination of the reactions for wind loads.

Fig. 9.45 Concurrent, noncoplanar forces.

Fig. 9.46 Determination of the resultant of concurrent, noncoplanar forces.

string that is parallel to OE has been drawn. String of may then be added to close the funicular polygon. This closing line from point x to the starting point y determines the position of OF in the force polygon, for ray OF must be parallel to the string of. The magnitude of the reaction R_1 is represented to scale by vector FA, and R_2 is represented by the vector FE.

The graphical method for determining the values of wind load reactions for a roof truss having both ends fixed is shown in Fig. 9.44. The solution given is practically identical with the solution applied to the beam in Fig. 9.43.

9.36

Concurrent, Noncoplanar Force Systems. Up to this point in our study of force systems, the student's attention has been directed solely to systems lying in one plane in order that the graphical methods dealing with the composition and resolution of forces could be presented in a clear and simple manner, free from the thinking needed for understanding force systems involving the third dimension.

In dealing with noncoplanar forces it is necessary to use at least two views to represent a structure in space. Although the methods as applied to coplanar force systems for solving problems may be extended to noncoplanar systems, the vector diagram for noncoplanar forces must have two views instead of one view as in the case of coplanar forces. If the student is to understand the discussions that are to follow he must grasp the idea that for the composition and resolution of noncoplanar forces he will work with two distinct and separate space representations, the space diagram for the given structure and the related vector diagram (force polygon). Figure 9.45 shows two views of a concurrent, noncoplanar force system not in equilibrium.

There are a few basic relationships that exist between a space diagram and its related vector diagram, which must be kept in mind when solving noncoplanar force problems. These relationships are: (1) in corresponding views (H-view and H-view, or F-view and F-view), each vector in the vector diagram will be parallel to its corresponding representation in the space diagram; (2) if a system of concurrent, noncoplanar forces is in equilibrium, the force polygon in space closes, and the projection on each plane will close; and (3) the true magnitude of a force can be measured only in the vector diagram when

it appears in true length or is made to do so.

9.37

Determination of the Resultant of a Force System of Concurrent, Noncoplanar Forces. The parallelogram method for determining the resultant of concurrent forces as explained in Sec. 9.26 may be employed to find the resultant of the three forces OA, OB, and OC in Fig. 9.46. Any number of given concurrent, noncoplanar forces can be combined into their resultant by this method. In the illustration, forces OA and OB were combined into their resultant, which is the diagonal of the smaller parallelogram; then this resultant in turn was combined with the third force OC to obtain the final resultant R for the given system. Since the true magnitude of R can be scaled only in a view showing its true length, an auxiliary view was projected from the front view. The true length of R could also have been determined by revolution.

Since the single force needed to hold a force system in balance, known as the *equilibrant*, is equal to the resultant in magnitude but is opposite in direction, this method might be used to determine the equilibrant for a system of concurrent, noncoplanar forces.

In presenting this problem and the two problems that follow, it has been assumed that the student has read the previous sections of this chapter and that his knowledge of the principles of projection is sufficient for him to find the true length of a line, having two views given, and to draw the view of a plane so that it will appear as an edge.

9.38

To Find the Three Unknown Forces of a Simple Load-bearing Frame—Special Case. In dealing with the simple load-bearing frame in Fig. 9.47, it should be realized that this is a special case rather than a general one, for two of the truss members appear as a single line in the frontal view of the space diagram. This condition considerably simplifies the task of finding the unknown forces acting in the members, and it is this particular spatial situation that makes this a special case. However, it should be pointed out now that this condition must exist or be set up in a projected view when a vector solution is to be applied to any problem involving a system of concurrent, noncoplanar forces. More will be said about the necessity for having two unknown forces coincide in one of the views in the discussion in Sec. 9.40.

After the space diagram has been drawn to scale, the steps toward the final solution are as follows:

1. Draw a free-body diagram showing the joint at A, the joint at which the load is applied. The lines of this diagram may be of any length, but each line in it must be parallel respectively to a corresponding line in the view of the space diagram to which the free-body diagram is related. In this case, it is the horizontal (top) view. A modified form of Bow's notation was used for convenience in identifying the forces. The vertical load has been shown pulled to one side in order that this force can be made to fall within the range of the notation. This diagram is important to the solution of this problem, for it enables one to see and note the direction of all of the forces that the members exert upon the joint (note arrowheads). Capital letters were used on the free-body diagram to identify the spaces between the forces, rather than lowercase letters as is customary, so that lowercase letters could be used for the ends of the views of the vectors in the vector diagram.

2. Using a selected scale, start the two views of the vector diagram (b) by laying out vector RS representing the only known force, in this case the 1000-lb load. Since RS is a force acting in a vertical direction $r^F s^F$ will be in true length in the F-view and will appear as point $r^H s^H$ in the H-view.

3. Complete the H- and F-views of the vector diagram. Since each vector line in the top view must be parallel to a corresponding line in the top view of the space diagram, $s^H t^H$ must be drawn parallel to $a^H b^H$, $t^H u^H$ parallel to $a^H c^H$, and $u^H r^H$ parallel to $a^H d^H$. Since the forces acting at joint A are in equilibrium, the vector triangle will close and the vectors will appear tip to tail.

In the frontal view of the vector diagram, $s^F t^F$ will be parallel to $a^F b^F$, $t^F u^F$ will be parallel to $a^F c^F$, and $r^F u^F$ will be parallel to $d^F a^F$.

4. Determine the magnitude of the forces acting on joint A. Since vector RU shows its true length in the F-view, the true magnitude of the force represented may be determined by scaling $r^F u^F$ using the same scale used to lay

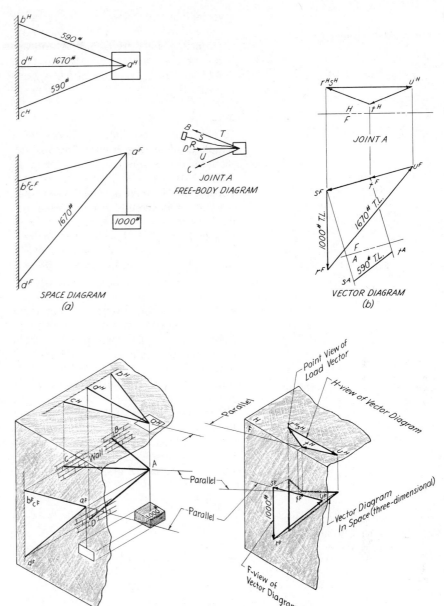

Fig. 9.47 Solution of a concurrent, noncoplanar force system—special case.

out the length of $r^F s^F$. Although it is known that vectors ST and TU are equal in magnitude, it is necessary to find the true length representation of one or the other of these vectors by some approved method before scaling to determine the true value of the force.

An arrowhead may now be added to the line of action of each force in the free-body diagram to indicate the direction of the action. Since the free-body diagram was related to the top view of the space diagram, the arrowhead for each force will point in the same direction as does the arrowhead on the corresponding vector in the H-view of the vector diagram. These arrowheads show that the forces in members AB and AC are acting away from joint A and are therefore *tension forces*. The force in AD acts toward A and thus is a *compression force*.

9.39
To Find the Three Unknown Forces of a Simple Load-bearing Frame—Composition Method (Fig. 9.48). Since the forces in AB and AC lie in an inclined plane that appears as an edge in the F-view, they may be composed into a single force that will have the same effect in the force system as the forces it replaces. This replacement force along with the force in AD and the load now become the forces that would be acting in a simple load-bearing truss (see Fig. 9.35). After the vertical vector of this concurrent coplanar force system has been drawn to a selected scale, the triangular force polygon in (b) may be completed by drawing the remaining lines representing the forces in AD and the resultant R. These lines are drawn parallel to their known lines of action as shown in the F-view of the space diagram in (a). Finally, the true length view of R must be transferred to the auxiliary view in (a) and resolved into its component forces, acting in AB and AC, that collec-

tively have the same action as the resultant R.

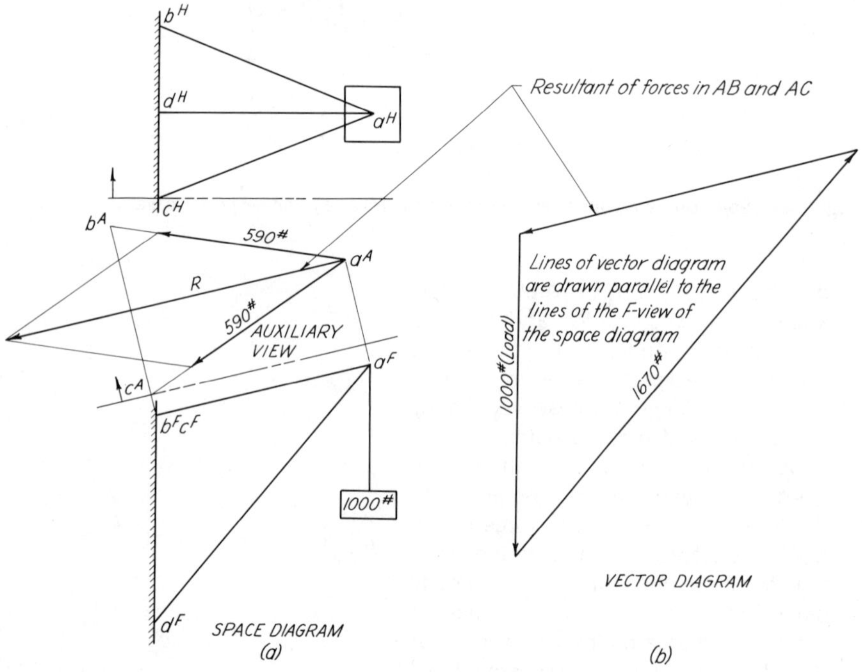

9.40
To Find the Three Unknown Forces of a Simple Load-bearing Truss—General Case. For the general case shown in Fig. 9.49, the known force is in a vertical position as in the previous problem, but no two of the three unknown forces appear coincident in either of the two given views. For this reason, it is necessary at the very start to add a complete auxiliary view to the space diagram that will combine with the existing top view to give a point view of one member and a line view of two of the three unknown forces. To obtain this desired situation, one should start with the following steps, which will transform the general case into the special case with which one should now be familiar.

> **Step 1.** Draw a true length line in the plane of two of the members. In Fig. 9.49(a) this line is DE, which appears in true length (TL) in the H-view.

> **Step 2.** Draw the needed auxiliary view, taken so the DE will appear as a point ($d^A e^A$) and OB and OC will be coincident (line $o^A b^A c^A$). This construction involves finding the edge view of a plane (see Sec. 9.13). In this particular case, the auxiliary view has height in common with the frontal view.

> **Step 3.** Draw the two views of the vector diagram by assuming the H-view and the A-view to be the given views of the special case. Proceed by the steps set forth for the special case in Sec. 9.38.

> **Step 4.** Determine the magnitude of the forces and add arrowheads to the freebody diagram to show the direction of action of the forces acting on point O.

9.41
In practice, engineers and technologists find wide use for methods that solve problems through the use of three-dimensional vector diagrams, for any quantity having both magnitude and direction may be represented by a vector. And, although the examples used in the chapter dealt with static structures, which are in the field of the structural engineer, vector diagram methods are used frequently by the electrical engineer for solving problems arising in his field and by the mechanical engineer for problems dealing with bodies in motion. The student will without a doubt encounter some of these methods in a textbook for a later course or will have them presented to him by his instructor.

Fig. 9.48 To determine the three unknown forces of a simple load-bearing frame—composition method.

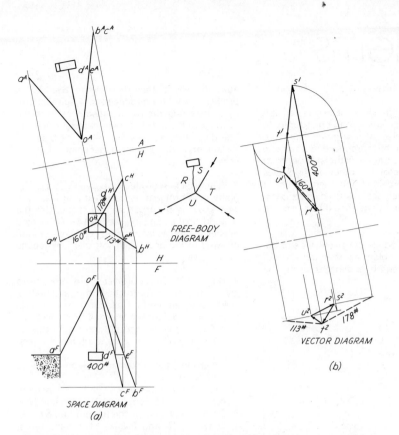

FREE-BODY
DIAGRAM

SPACE DIAGRAM
(a)

VECTOR DIAGRAM

(b)

Fig. 9.49 Solution of a concurrent, noncoplanar force system—general case.

Problems

Descriptive geometry

Problems 1 through 5 have been selected and arranged to offer the student an opportunity to apply basic principles of descriptive geometry.

The problems can be reproduced to a suitable size by transferring the needed distances from the drawing to one or the other of the scales that have been provided for each group of problems. If the inch scale is used, metric values in problem instructions should be converted to decimal inches.

1. (Fig. 9.50). Reproduce the given views of the line or lines of a problem as assigned and determine the true length or lengths using the auxiliary view method explained in Sec. 9.7. A problem may be reproduced to a suitable size by transferring the needed distances from the drawing to the given scale to determine values. The distances, as they are determined, should be laid off on the drawing paper using a full-size scale.

2. (Fig. 9.51). Reproduce the given views of the plane of a problem as assigned and draw the view showing the true size and shape. Use the method explained in Sec. 9.14. Determine needed distances by transferring them from the drawing to the accompanying scale, by means of the dividers.

3. (Fig. 9.52). These problems are intended to give some needed practice in manipulating views to obtain certain relationships of points and lines. Determine needed distances by transferring them from the drawing to the accompanying scale, by means of the dividers.

1. Determine the distance between points A and B.
2. Draw the H- and F-views of a 13 mm perpendicular erected from point N of the line MN.
3. Draw the H-view of the 95 mm line ST.
4. Draw the H- and F-views of a plane represented by an equilateral triangle and containing line AB as one of the edges. The added plane ABC is to be at an angle of 30° with plane $ABDE$.
5. If the figure $MNOP$ is a plane surface, an edge view of the surface would appear as a line. Draw such a view to determine whether or not $MNOP$ is a plane.
6. A vertical pole with top O is held in place by three guy wires. Determine the slope in tangent value of the angle for the guy wire that has a bearing of N 23° W.

4. (Fig. 9.53). In this group of problems it is required to determine the shortest distance between skew lines and the angle formed by intersecting lines.

1. Show proof that the plane $ABCD$ is an oblique plane.

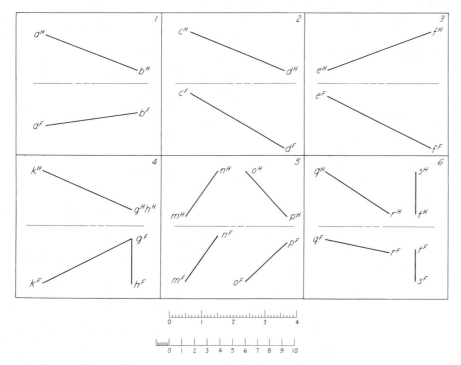

Fig. 9.50 True length of a line.

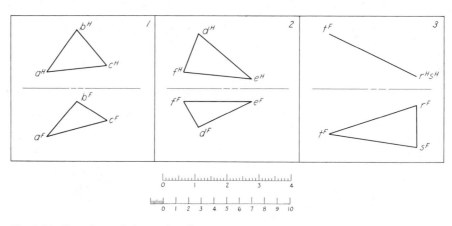

Fig. 9.51 True size and shape of a plane.

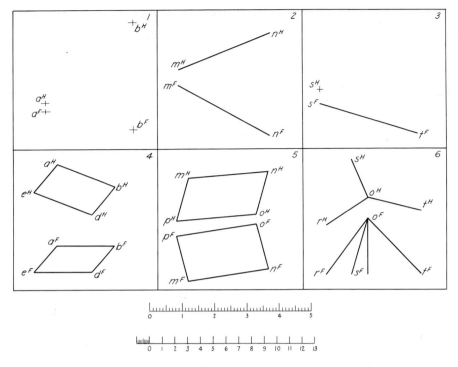

Fig. 9.52 Relationships of points and lines.

2. Determine the shortest distance between the lines *MN* and *ST*.
3. Through point *K* on line *GH* draw the *F*- and *H*-views of a line that will be perpendicular to line *EF*.
4. Determine the angle between line *AB* and a line intersecting *AB* and *CD* at the level of point *E*.

5. Through a point on line *MN* that is 32 mm from point *N*, draw the *F*- and *H*-views of a line that will be perpendicular to line *ST*.
6. Erect a 25 mm perpendicular at point *K* in the plane *EFGH*. Connect the outer end point *L* of the perpendicular with *F*. Determine the angle between *LF* and *KF*.

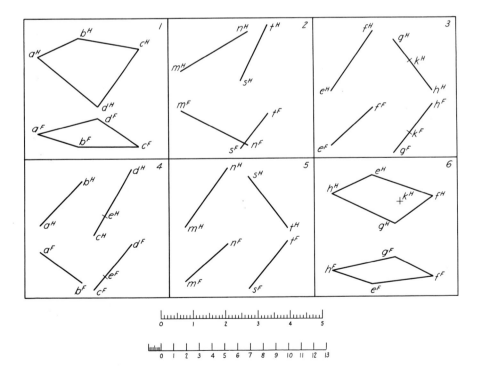

Fig. 9.53 Shortest distance between skew lines and the angle formed by intersecting lines.

5. (Fig. 9.54). These problems require that the student determine the angle between a line and a plane and the angle between two given planes.

1. Determine the angle between the planes *MNQP* and *RST*.
2. Determine the angle between the line *ST* and
   (a) The *H*-plane.
   (b) The *F*-plane.
3. The line *EF* has a bearing of N 53° E. What angle does this line make with the *P*-plane?
4. Draw the *F*- and *H*-views of a line through point *K* that forms an angle of 35° with plane *MNQR*.
5. The top and front views of planes *ABC* and *RST* are partially drawn.
   (a) Complete the views including the line of intersection.
   (b) Determine the angle between the line of intersection and the *H*-plane of projection.
6. Two views of a plane *RST* and the top view of an arrow are shown. The arrow, pointing downward and toward the left, is in a plane that forms an angle of 68° with plane *RST*.

The arrow point is 6 mm from the plane *RST*.
(a) Draw the front view of the arrow.
(b) Draw the top and front views of the line of intersection of the 68° plane and plane *RST*.

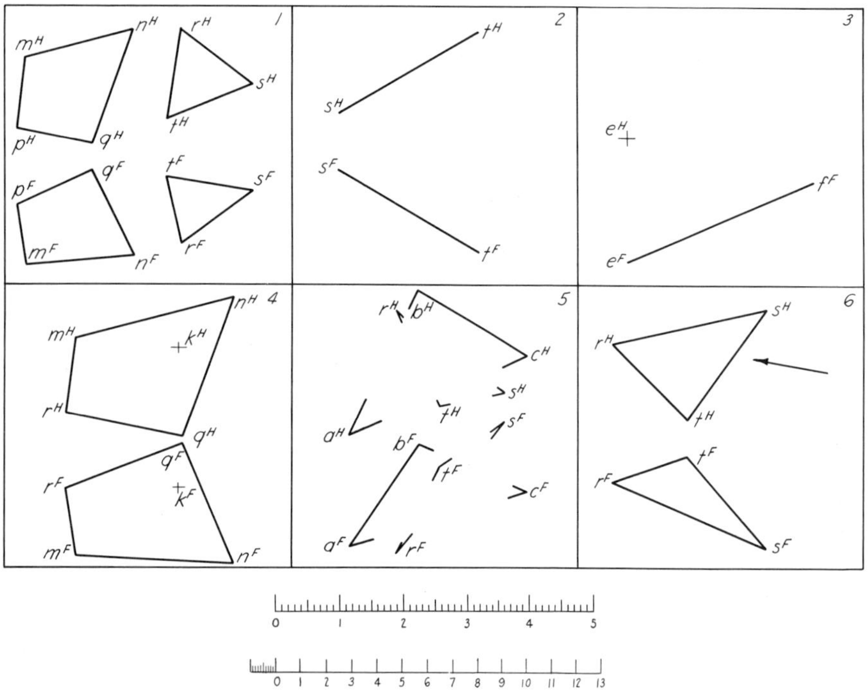

Vector geometry

The following problems have been selected to emphasize the basic principles underlying vector geometry. By solving a limited number of the problems presented, the student should find that he has a working knowledge of some vector methods that are useful for solving problems in design that involve the determination of the magnitude of forces as well as their composition and resolution. The student is to select his own scale remembering that a drawing made to a large scale usually assures more accurate results.

1. (Fig. 9.55). A force of 408 N acts downward at an angle of 60° with the horizontal. Determine the vertical and horizontal components of this force.

2. (Fig. 9.56). Determine the resultant force for the given coplanar, concurrent force system.

3. (Fig. 9.57). Determine the magnitude of the force F_C and the angle that the resultant force R makes with the horizontal for the given coplanar, concurrent force system.

4. (Fig. 9.58). Determine the magnitude and direction of the equilibrant for the given coplanar, concurrent force system.

5. (Fig. 9.59). A block weighing 45 lb is to be pulled up an inclined plane sloping at an angle of 30° with the horizontal. If the frictional resistance is 16 lb, what is the magnitude of the force F_M that is required to move the block uniformly up the plane?

Fig. 9.54 Angle between a line and a plane and the angle between two planes.

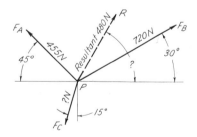

Fig. 9.57

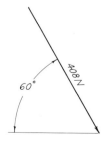

Fig. 9.55

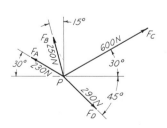

Fig. 9.56 Force system.

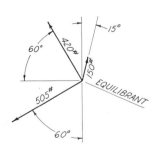

Fig. 9.58

6. (Fig. 9.60). A horizontal beam AB is hinged at B as shown. The end of the beam at A is connected by a cable to a hook in the wall at C. The load at A is 250 lb. Using the dimensions as given, determine the tension force in the cable and the reaction on the hinge at B. The weight of the beam is to be neglected.

7. (Fig. 9.61). A 600-lb (272.2-kg) load is supported by cables as shown. Determine the magnitude of the tension in the cables.

8. (Fig. 9.62). A ship that is being pulled through the entrance of a harbor is headed due east through a cross current moving at 4 knots as shown. If the ship is moving at 12 knots, what is the speed of the tugboat?

9. (Fig. 9.63). Determine the magnitude of the reactions R_1 and R_2 of the beam with loads as shown.

10. (Fig. 9.64). Determine the magnitude of the reactions R_1 and R_2 of the beam.

11. (Fig. 9.65). Determine the magnitude of the reactions R_1 and R_2 for the roof truss shown. Each of the six panels is of the same length.

12. (Fig. 9.66). Determine the magnitude of the reactions R_1 and R_2 to the wind loads acting on the roof truss as shown.

13. (Fig. 9.67). A tripod with a 40 kg load is set up on a level floor as shown. Determine the stresses in the three legs due to the vertical load on the top.

14–15. (Figs. 9.68 and 9.69). Determine the stresses in the members of the space frame shown.

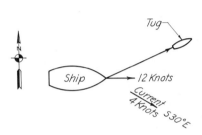

Fig. 9.61 Cable support system. (Values shown in [ ] are in the metric system.)

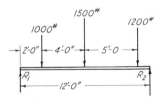

Fig. 9.62

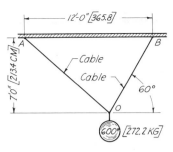

Fig. 9.63

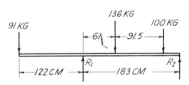

Fig. 9.64 Simple beam.

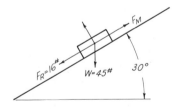

Fig. 9.59

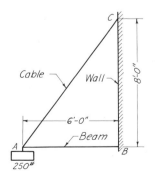

Fig. 9.60

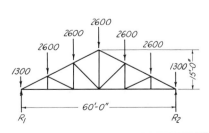

Fig. 9.65

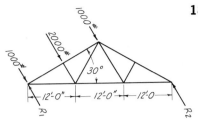

Fig. 9.66

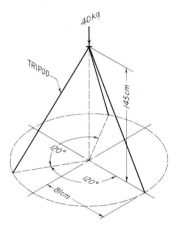

Fig. 9.67

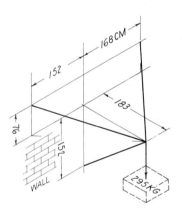

Fig. 9.68 Space frame.

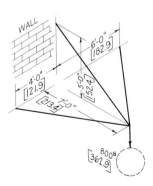

Fig. 9.69 Space frame. (Values shown in [ ] are in the metric system.)

This advanced concept train is now being built under contract. These advanced-technology cars feature an aluminum structure with exterior panels made of vandal-resistant acrylic plastic. Designed to conserve electricity, the electric propulsion system of ACT-1 includes a flywheel within the cars which store up energy that can be used to supply the electric motors when needed. In the development of the ACT-1 cars, many of the problems involving intersections and developments were solved by graphical methods. *(Courtesy U.S. Office of Transportation)*

10

Developments and intersections

10.1

Introduction. Intersections and developments are logically a part of the subject of descriptive geometry. A few of the many applications that can be handled without advanced study in projection, however, are presented in this chapter.

Desired lines of intersection between geometric surfaces may be obtained by applying the principles of projection with which the student is already familiar. Although developments are laid out and are not drawn by actual projection in the manner of exterior views, their construction nevertheless requires the application of orthographic projection in finding the true lengths of elements and edges.

10.2

Geometric Surfaces. A geometric surface is generated by the motion of a geometric line, either straight or curved. Surfaces that are generated by a moving straight line are known as *ruled surfaces*, and those generated by a curved line are known as *double-curved surfaces*. Any position of the generating line, known as a *generatrix*, is called an *element of the surface*.

Ruled surfaces include planes, single-curved surfaces, and warped surfaces.

A *plane* is generated by a straight line moving in such a manner that one point touches another straight line as it moves parallel to its original position.

A *single-curved surface* is generated by a straight line moving so that in any two of its near positions it is in the same plane.

A *warped surface* is generated by a straight line moving so that it does not lie in the same plane in any two near positions.

Double-curved surfaces include surfaces that are generated by a curved line moving in accordance with some mathematical law.

10.3

Geometric Objects. Geometric solids are bounded by geometric surfaces. They may be classified as follows:

1. Solids bounded by plane surfaces: Tetrahedron, cube, prism, pyramid, and others.

2. Solids bounded by single-curved surfaces: Cone and cylinder (generated by a moving straight line).

3. Solids bounded by warped surfaces: Conoid, cylindroid, hyperboloid of one nappe, and warped cone.

4. Solids bounded by double-curved surfaces: Sphere, spheroid, torus, paraboloid, hyperboloid, and so on (surfaces of revolution generated by curved lines).

A □ Developments

10.4

Introduction. A layout of the complete surface of an object is called a *development* or *pattern*. The development of an object bounded by plane surfaces may be thought of as being obtained by turning the object, as illustrated in Figs. 10.1 and 10.2, to unroll the imaginary enclosing surface on a plane. Practically, the drawing operation consists of drawing the successive surfaces in their true sizes with their common edges joined.

The surfaces of cones and cylinders also may be unrolled on a plane. The development of a right cylinder (Fig. 10.3) is a rectangle having a width equal to the altitude of the cylinder and a length equal to the cylinder's computed circumference (πd). The development of a right circular cone (Fig. 10.4) is a sector of a circle having a radius equal to the slant height of the cone and an arc length equal to the circumference of its base.

Warped and double-curved surfaces cannot be developed accurately, but they may be developed by some approximate method. Ordinarily, an approximate pattern will prove to be sufficiently accurate

for practical purposes if the material of which the piece is to be made is somewhat flexible.

Plane and single-curved surfaces (prisms, pyramids, cylinders, and cones), which can be accurately developed, are said to be developable. Warped and double-curved surfaces, which can be only approximately developed, are said to be nondevelopable.

10.5

Practical Developments. On many industrial drawings, a development must be shown to furnish the necessary information for making a pattern to facilitate the cutting of a desired shape from sheet metal. Because of the rapid advance of the art of manufacturing an ever-increasing number of pieces by folding, rolling, or pressing cut sheet-metal shapes, one must have a broad knowledge of the methods of constructing varied types of developments. Patterns also are used in stonecutting as guides for shaping irregular faces.

A development of a surface should be drawn with the inside face up, as it theoretically would be if the surface were unrolled or unfolded, as illustrated in Figs. 10.1–10.4. This practice is further justified because sheet-metal workers must make the necessary punch marks for folding on the inside surface.

Although in actual sheet-metal work extra metal must be allowed for lap at seams, no allowance will be shown on the developments in this chapter. Many other practical considerations have been purposely ignored, as well, in order to avoid confusing the beginner.

10.6

To Develop a Right Truncated Prism. Before the development of the lateral surface of a prism can be drawn, the true lengths of the edges and the true size of a right section must be determined. On the right truncated prism, shown in Fig. 10.5, the true lengths of the prism edges are shown in the front view and the true size of the right section is shown in the top view.

The lateral surface is "unfolded" by first drawing a "stretch-out line" and marking off the widths of the faces (distances 1–2, 2–3, 3–4, and so on, from the top view) along it in succession. Through these points light construction lines are then drawn perpendicular to the line 1_D1_D, and the length of the respective edge is set off on each by projecting from

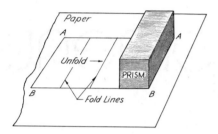

Fig. 10.1 Development of a prism.

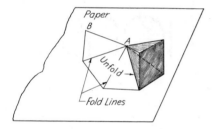

Fig. 10.2 Development of a pyramid.

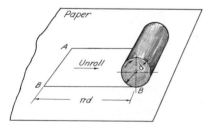

Fig. 10.3 Development of a cylinder.

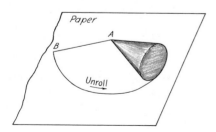

Fig. 10.4 Development of a cone.

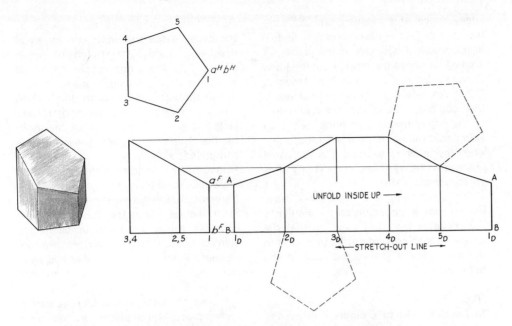

Fig. 10.5 Standard method of developing the lateral surface of a right prism.

the front view. When projecting edge lengths to the development, the points should be taken in a clockwise order around the perimeter, as indicated by the order of the numbers in the top view. The outline of the development is completed by joining these points. Thus far, nothing has been said about the lower base or the inclined upper face. These may be joined to the development of the lateral surface, if so desired.

In sheet-metal work, it is usual practice to make the seam on the shortest element in order to save time and conserve solder or rivets.

10.7
To Develop an Oblique Prism. The lateral surface of an oblique prism, such as the one shown in Fig. 10.6, is developed by the same general method used for a right prism. Similarly, the true lengths of the edges are shown in the front view, but it is necessary to find the true size of the right section by auxiliary plane construction. The widths of the faces, as taken from the auxiliary right section, are set off along the stretch-out line, and perpendicular construction lines representing the edges are drawn through the division points. The lengths of the portions of each respective

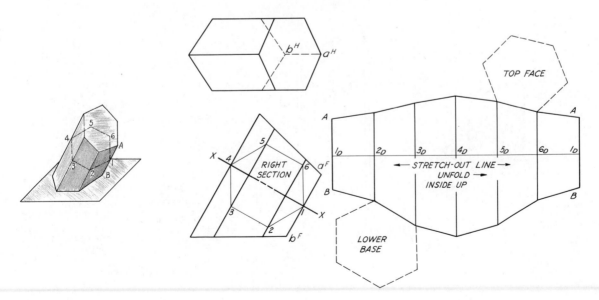

Fig. 10.6 Development of an oblique prism.

edge, above and below plane XX, are transferred to the corresponding line in development. Distances above plane XX are laid off above the stretch-out line, and distances below XX are laid off below it. The development of the lateral surface is then completed by joining the end points of the edges by straight lines. Since an actual fold will be made at each edge line when the prism is formed, it is the usual practice to heavy these edge (fold) lines on the development.

The stretch-out line might well have been drawn in a position perpendicular to the edges of the front view, so that the length of each edge might be projected to the development (as in the case of the right prism).

10.8
To Develop a Right Cylinder. When the lateral surface of a right cylinder is rolled out on a plane, the base develops into a straight line (Fig. 10.7). The length of this line, which is equal to the circumference of a right section ($\pi \times$ diam), may be calculated and laid off as the stretch-out line 1_D1_D.

Since the cylinder can be thought of as being a many-sided prism, the development may be constructed in a manner similar to the method illustrated in Fig. 10.5. The elements drawn on the surface of the cylinder serve as edges of the many-sided prism. Twelve or 24 of these elements ordinarily are used, the number depending on the size of the cylinder. Usually they are spaced by dividing the

circumference of the base, as shown by the circle in the top view, into an equal number of parts. The stretch-out line is divided into the same number of equal parts, and perpendicular elements are drawn through each division point. Then the true length of each element is projected to its respective representation on the development, and the development is completed by joining the points with a smooth curve. In joining the points, it is advisable to sketch the curve in lightly, freehand, before using the French curve. Since the surface of the finished cylindrical piece forms a continuous curve, the elements on the development are not heavied. When the development is symmetrical, as in this case, only one-half need be drawn.

A piece of this type might form a part of a two-piece, three-piece, or four-piece elbow. The pieces are usually developed as illustrated in Fig. 10.8. The stretch-out line of each section is equal in length to the computed perimeter of a right section.

10.9
To Develop an Oblique Cylinder. Since an oblique cylinder theoretically may be thought of as enclosing a regular oblique prism having an infinite number of sides, the development of the lateral surface of the cylinder shown in Fig. 10.9 may be constructed by using a method similar to the method illustrated in Fig. 10.6. The circumference of the right section becomes stretch-out line 1_D1_D for the development.

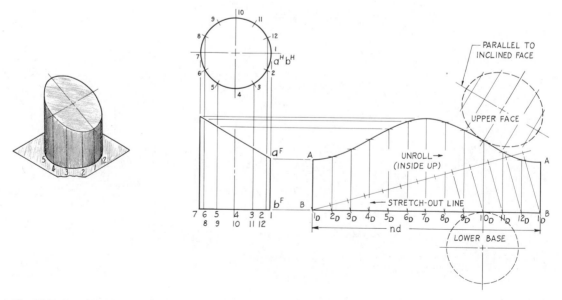

Fig. 10.7 Development of a right circular cylinder.

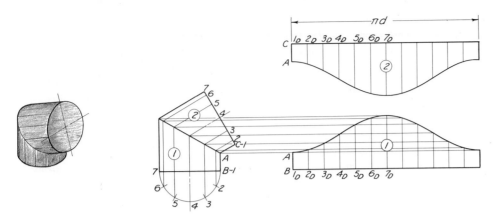

Fig. 10.8   Two-piece elbow.

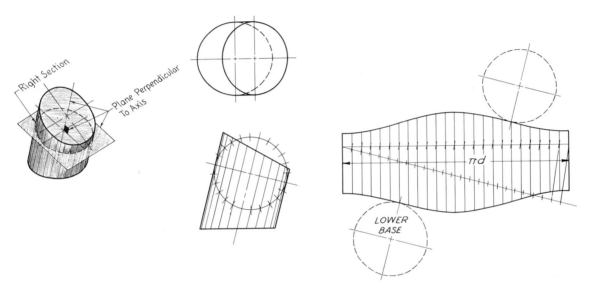

Fig. 10.9   Development of an oblique cylinder.

10.10

To Determine the True Length of a Line.
In order to construct the development of
the lateral surface of some objects, it frequently is necessary to determine the true
lengths of oblique lines that represent the
edges. The general method for determining the true lengths of lines inclined to all
of the coordinate planes of projection has
been explained in detail in Sec. 5.29.

10.11

True-length Diagrams.   When it is necessary in developing a surface to find the
true lengths of a number of edges or elements, some confusion may be avoided by
constructing a true-length diagram adjacent to the orthographic view, as shown in
Fig. 10.10. The elements were revolved
into a position parallel to the F (frontal)
plane so that their true lengths show in
the diagram. This practice prevents the

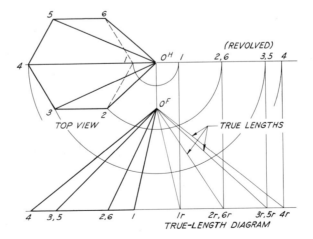

Fig. 10.10   True-length diagram (the revolution method).

front view in the illustration from being cluttered with lines, some of which would represent elements and others their true lengths.

Figure 10.12 shows a diagram that gives the true lengths of the edges of the pyramid. Each line representing the true length of an edge is the hypotenuse of a right triangle whose altitude is the altitude of the edge in the front view and whose base is equal to the length of the projection of the edge in the top view. The lengths of the top projections of the edges of the pyramid are laid off horizontally from the vertical line, which could have been drawn at any distance from the front view. Since all the edges have the same altitude, this line is a common vertical leg for all the right triangles in the diagram. The true-length diagram shown in Fig. 10.10 could very well have been constructed by this method.

10.12

To Develop a Right Pyramid. To develop (unfold) the lateral surface of a right pyramid, it is first necessary to determine the true lengths of the edges and the true size of the base. With this information, the development can be constructed by laying out the faces in successive order with their common edges joined. If the surface is imagined to be unfolded by turning the pyramid, as shown in Fig. 10.2, each triangular face is revolved into the plane of

the paper about the edge that is common to it and the preceding face.

Since the edges of the pyramid shown in Fig. 10.11 are all equal in length, it is necessary only to find the length of the one edge $A1$ by revolving it into the position a^F1r. The edges of the base, 1–2, 2–3, and so on, are parallel to the horizontal plane of projection and consequently show in their true length in the top view. With this information, the development is easily completed by constructing the four triangular surfaces.

10.13

To Develop the Surface of a Frustum of a Pyramid. To develop the lateral surface of the frustum of a pyramid (Fig. 10.12), it is necessary to determine the true lengths of edges of the complete pyramid as well as the true lengths of edges of the frustum. The desired development is obtained by first constructing the development of the complete pyramid and then laying off the true lengths of the edges of the frustum on the corresponding lines of the development.

It may be noted with interest that the true length of the edge $B3$ is equal to the length $b'3'$ on the true-length line a^F3' and that the location of point b' can be established by the shortcut method of projecting horizontally from point b^F. Point b' on a^F3' is the true revolved position of point B, because the path of point B is in a horizontal plane that projects as a line in the front view.

10.14

To Develop a Right Cone. As previously explained in Sec. 10.4, the development of a regular right circular cone is a sector of a circle. The development will have a radius equal to the slant height of the cone and an included angle at the center equal to $(r/s) \times 360°$ (Fig. 10.13). In this equation, r is the radius of the base and s is the slant height.

10.15

To Develop a Right Truncated Cone. The development of a right truncated cone must be constructed by a modified method of triangulation, in order to develop the outline of the elliptical inclined surface. This commonly used method is based on the theoretical assumption that a cone is a pyramid having an infinite number of sides. The development of the incomplete right cone shown in Fig. 10.14 is constructed on a layout of the whole

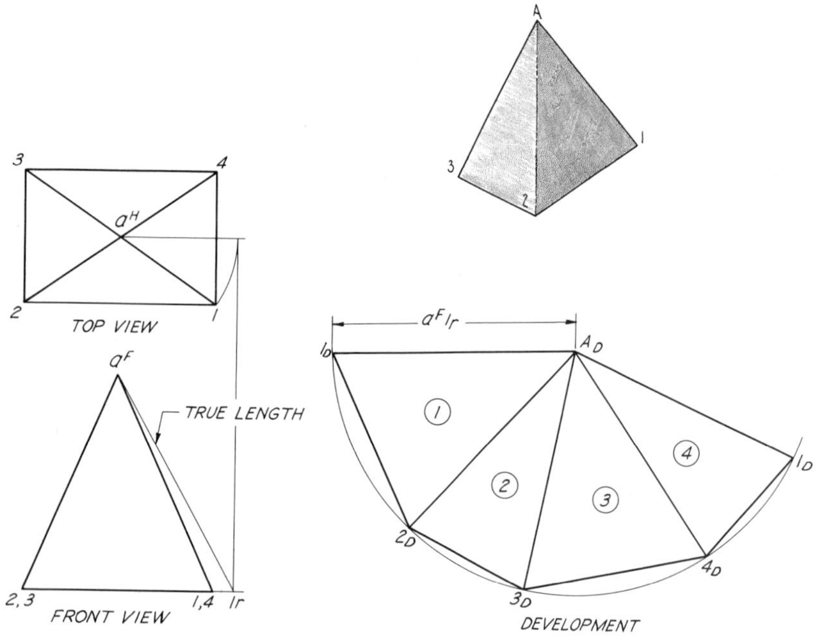

Fig. 10.11 Development of a rectangular right pyramid.

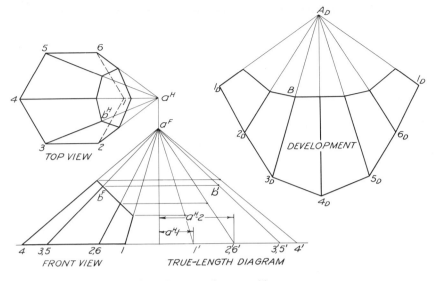

Fig. 10.12 Development of the frustum of a pyramid.

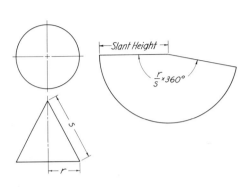

Fig. 10.13 Development of a right cone.

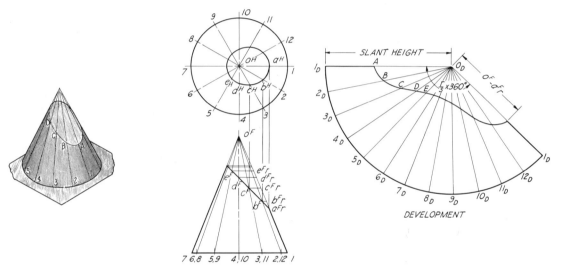

Fig. 10.14 Development of a truncated cone.

cone by a method similar to the standard method illustrated for the frustum of a pyramid in Fig. 10.12.

Elements are drawn on the surface of the cone to serve as edges of the many-sided pyramid. Either 12 or 24 are used, depending on the size of the cone. Their location is established on the developed sector by dividing the arc representing the unrolled base into the same number of equal divisions, into which the top view of the base has been divided. At this point in the procedure, it is necessary to determine the true lengths of the elements of the frustum in the same manner that the true lengths of the edges of the frustum of a pyramid were obtained in Fig. 10.12. With this information, the desired devel-

opment can be completed by setting off the true lengths on the corresponding lines of the development and joining the points thus obtained with a smooth curve.

10.16

Triangulation Method of Developing Approximately Developable Surfaces. A nondevelopable surface may be developed approximately if the surface is assumed to be composed of a number of small developable surfaces (Fig. 10.15). The particular method ordinarily used for warped surfaces and the surfaces of oblique cones is known as the *triangulation method*. The procedure consists of completely covering the lateral surface with numerous small triangles that will lie approximately on

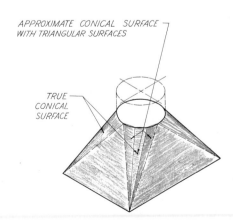

Fig. 10.15 Triangulation of a surface.

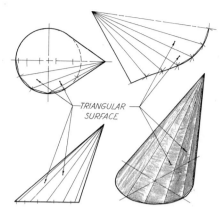

TRIANGULAR SURFACE

Fig. 10.16   Triangulation of an oblique cone.

the surface (Fig. 10.16). These triangles, when laid out in their true size with their common edges joined, produce an approximate development that is accurate enough for most practical purposes.

Although this method of triangulation is sometimes used to develop the lateral surface of a right circular cone, it is not recommended for such a purpose. The resulting development is not as accurate as it would be if constructed by one of the standard methods (Secs. 10.14 and 10.15).

10.17

To Develop an Oblique Cone Using the Triangulation Method.   A development of the lateral surface of an oblique cone is constructed by a method similar to that

used for an oblique pyramid. The surface is divided into a number of unequal triangles having sides that are elements on the cone and bases that are the chords of short arcs of the base.

The first step in developing an oblique cone (Fig. 10.17) is to divide the circle representing the base into a convenient number of equal parts and draw elements on the surface of the cone through the division points (1, 2, 3, 4, 5, and so on). To construct the triangles forming the development, it is necessary to know the true lengths of the elements (sides of the triangles) and chords. In the illustration, all the chords are equal. Their true lengths are shown in the top view. The true lengths of the oblique elements may be determined by one of the standard methods explained in Sec. 10.11.

Since the seam should be made along the shortest element, $A1$ will lie on the selected starting line for the development and $A7$ will be on the center line. To obtain the development, the triangles are constructed in order, starting with the triangle A–1–2 and proceeding around the cone in a clockwise direction (as shown by the arrow in the top view). The first step in constructing triangle A–1–2 is to set off the true length a^F1' along the starting line. With point A_D of the development as a center, and with a radius equal to a^F2', strike an arc; then, with point 1_D as a center, and with a radius equal to the chord 1–2, strike an arc across the first arc to locate point 2_D. The triangle $A_D2_D3_D$ and the remaining triangles are formed in exactly the same manner. When all the triangles have been laid out, the development of the whole conical surface is completed by drawing a smooth curve through the end points of the elements.

10.18

Transition Pieces.   A few of the many types of transition pieces used for connecting pipes and openings of different shapes and sizes are illustrated pictorially in Fig. 10.18.

10.19

To Develop a Transition Piece Connecting Rectangular Pipes.   The transition piece shown in Fig. 10.19 is designed to connect two rectangular pipes of different sizes on different axes. Since the piece is a frustum of a pyramid, it can be accurately developed by the method explained in Sec. 10.13.

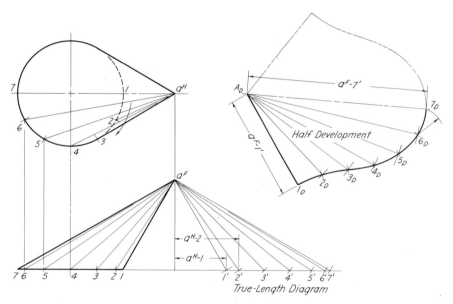

Half Development

True-Length Diagram

Fig. 10.17   Development of an oblique cone.

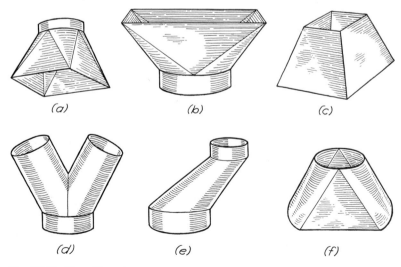

(a)   (b)   (c)

(d)   (e)   (f)

Fig. 10.18   Transition pieces.

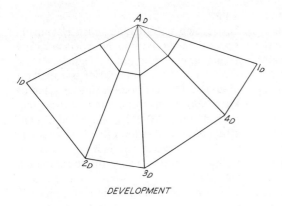

DEVELOPMENT

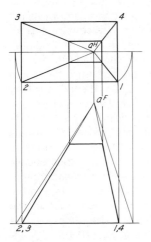

Fig. 10.19 Transition piece.

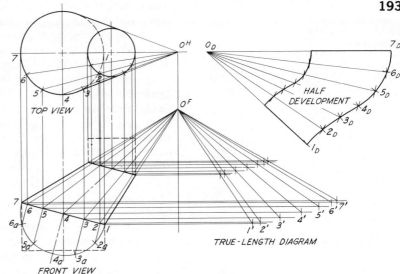

Fig. 10.20 Transition piece connecting two pipes.

10.20

To Develop a Transition Piece Connecting Two Circular Pipes. The transition piece shown in Fig. 10.20 connects two circular pipes on different axes. Since the piece is a frustum of an oblique cone, the surface must be triangulated, as explained in Sec. 10.17, and the development must be constructed by laying out the triangles in their true size in regular order. The general procedure is the same as that illustrated in Fig. 10.17. In this case, however, since the true size of the base is not shown in the top view, it is necessary to construct a partial auxiliary view to find the true lengths of chords between the end points of the elements.

10.21

To Develop a Transition Piece Connecting a Circular and a Square Pipe. A detailed analysis of the transition piece shown in Fig. 10.21 reveals that it is composed of four isosceles triangles whose bases form the square base of the piece and four conical surfaces that are parts of oblique cones. It is not difficult to develop this type of transition piece because, since the

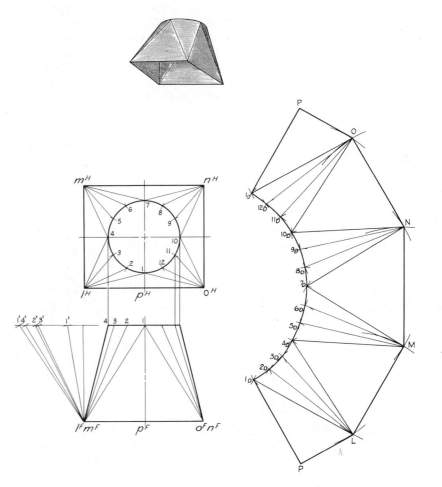

Fig. 10.21 Transition piece connecting a circular and square pipe.

whole surface may be "broken up" into component surfaces, the development may be constructed by developing the first and then each succeeding component surface separately (Fig. 10.15). The surfaces are developed around the piece in a clockwise direction, in such a manner that each successive surface is joined to the preceding surface at their common element. In the illustration, the triangles *1LO, 4LM, 7MN,* and *10NO* are clearly shown in top view. Two of these, *1LO* and *10NO,* are visible on the pictorial drawing. The apexes of the conical surfaces are located at the corners of the base.

Before starting the development, it is necessary to determine the true lengths of the elements by constructing a true-length diagram, as explained in Sec. 10.11. The true lengths of the edges of the lower base (*LM, MN, NO,* and *OL*) and the true lengths of the chords (*1–2, 2–3, 3–4,* and so on) of the short arcs of the upper base are shown in the top view. The development is constructed in the following manner: First, the triangle 1_DPL is constructed, using the length p^Hl^H taken from the top view and true lengths from the diagram. Next, using the method explained in Sec. 10.17, the conical surface, whose apex is at *L,* is developed in an attached position. Triangle 4_DLM is then added, and so on, until all component surfaces have been drawn.

10.22
To Develop a Transition Piece Having an Approximately Developable Surface by the Triangulation Method. Figure 10.22 shows a half development of a transition piece that has a warped surface instead of a partially conical one, such as that discussed in Sec. 10.21. The method of constructing the development is somewhat similar, however, in that it is formed by laying out, in true size, a number of small triangles that approximate the surface. The true size of the circular intersection is shown in the top view, and the true size of the elliptical intersection is shown in the auxiliary view, which was constructed for that purpose.

The front half of the circle in the top view should be divided into the same number of equal parts as the half-auxiliary view. By joining the division points, the lateral surface may be initially divided into narrow quadrilaterals. These in turn may be subdivided into triangles by drawing diagonals, which, though theoretically curved lines, are assumed to be straight. The true lengths of the elements and the diagonals are found by constructing two separate true-length diagrams by the method illustrated in Fig. 10.12.

10.23
To Develop a Sphere. The surface of a sphere is a double-curved surface that can

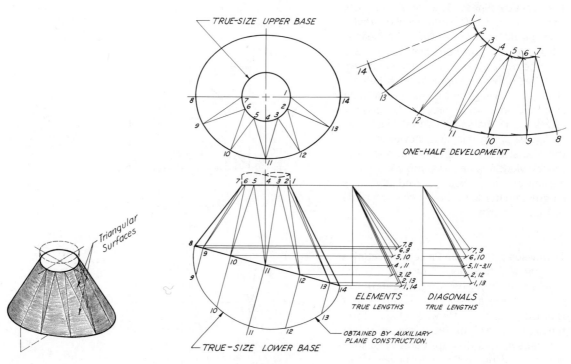

Fig. 10.22 Development of transition piece by triangulation.

be developed only by some approximate method. The standard methods commonly used are illustrated in Fig. 10.23.

In (*a*) the surface is divided into a number of equal meridian sections of cylinders. The developed surfaces of these form an approximate development of the sphere. In drawing the development it is necessary to develop the surface of only one section, for this can be used as a pattern for the developed surface of each of the others.

In (*b*) the sphere is cut by parallel planes, which divide it into a number of horizontal sections, the surfaces of which approximate the surface of the sphere. Each of these sections may be considered the frustum of a right cone whose apex is located at the intersection of the chords extended.

B □ Intersections

10.24

Lines of Intersection of Geometric Surfaces. The line of intersection of two surfaces is a line that is common to both. It may be considered the line that would contain the points in which the elements of one surface would pierce the other. Almost every line on a practical orthographic representation is a line of intersection; therefore, the following discussion may be deemed an extended study of the same subject. The methods presented in this chapter are the recognized easy procedures for finding the more complicated lines of intersection created by intersecting geometric surfaces.

In order to complete a view of a working drawing or a view necessary for developing the surfaces of intersecting geometric shapes, one frequently must find the line of intersection between surfaces. On an ordinary working drawing the line of intersection may be "faked in" through a few critical points. On a sheet-metal drawing, however, a sufficient number of points must be located to obtain an accurate line of intersection and an ultimately accurate development.

The line of intersection of two surfaces is found by determining a number of points common to both surfaces and drawing a line or lines through these points in correct order. The resulting line of intersection may be straight, curved, or straight and curved. The problem of finding such a line may be solved by one of

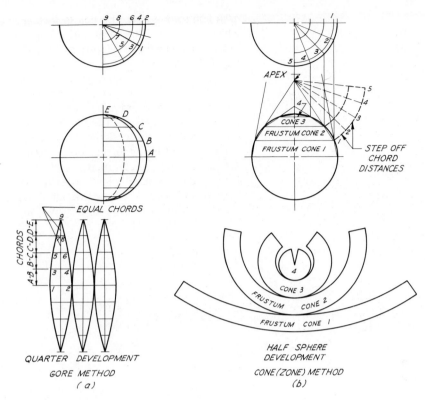

Fig. 10.23 Approximate development of a sphere.

two general methods, depending on the type of surfaces involved.

For the purpose of simplifying this discussion of intersections, it should be assumed that all problems are divided into these two general groups:

Group I. Problems involving two geometric figures, both of which are composed of plane surfaces.

Group II. Problems involving geometric figures which have either single-curved or double-curved surfaces.

For instance, the procedure for finding the line of intersection of two prisms is the same as that for finding the line of intersection of a prism and a pyramid; hence, both problems belong in the same group (Group I). Since the problem of finding the line of intersection of two cylinders and the problem of finding the line of intersection of a cylinder and a cone both involve single-curved surfaces, these two also belong in the same group (Group II).

Problems of the first group are solved by locating the points through which the edges of each of two geometric shapes pierce the other. These points are vertices of the line of intersection. Whenever one of two intersecting plane surfaces appears as a line in one view, the points through which the lines of the other surfaces penetrate it usually may be found by inspecting that view.

Problems of the second group may be solved by drawing elements on the lateral surface of one geometric shape in the region of the line of intersection. The points at which these elements intersect the surface of the other geometric shape are points that are common to both surfaces and consequently lie on their line of intersection. A curve, traced through these points with the aid of a French curve, will be a representation of the required intersection. To obtain accurate results, some of the elements must be drawn through certain critical points at which the curve changes sharply in direction. These points usually are located on contour elements. Hence, the usual practice is to space the elements equally around the surface, starting with a contour element.

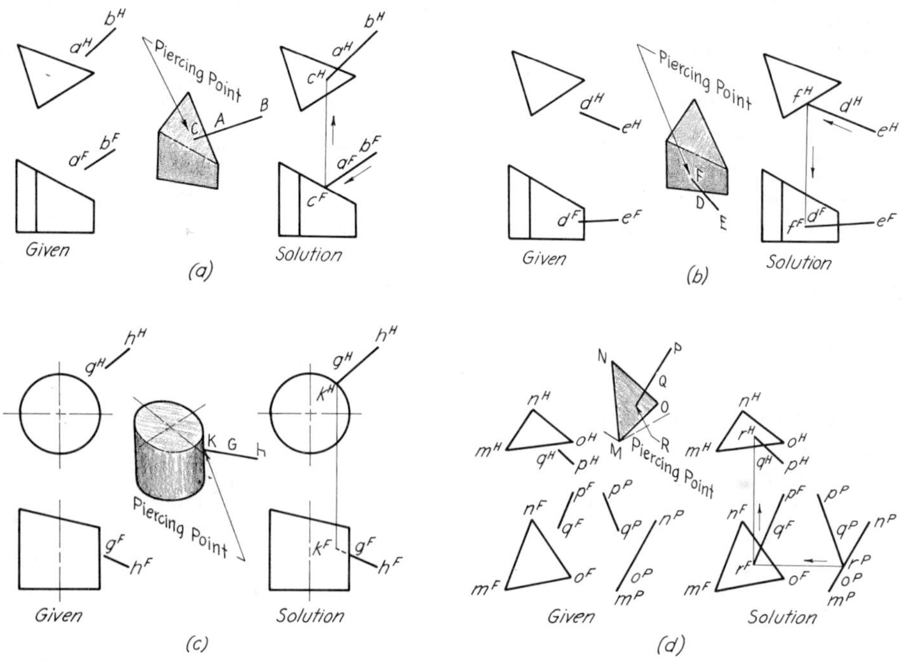

10.25
Determination of a Piercing Point by Inspection (Fig. 10.24). It is easy to determine where a given line pierces a surface when the surface appears as an edge view (line) in one of the given views. For example, when the given line AB is extended as shown in (a), the F-view of the piercing point C is observed to be at c^F, where the frontal view of the line AB extended intersects the line view of the surface. With the position of c^F known, the H-view of point C can be quickly found by projecting upward to the H-view of AB extended.

In (b) the H-view (f^H) of the piercing point F is found first by extending $d^H e^H$ to intersect the edge view of the surface pierced by the line. By projecting downward, f^F is located on $d^F e^F$ extended.

In (c) the views of the piercing point K are found in the same manner as in (b), the only difference being that the edge view of the surface pierced by the line appears as a circle arc in the H-view instead of a straight line. It should be noted that a part of the line is invisible in the F-view because the piercing point is on the rear side of the cylinder.

The F- and H-views of the piercing point R in (d) may be found easily by projection after the P-view (r^P) of R has been once established by extending $p^P q^P$ to intersect the line view of the surface.

10.26
Determination of a Piercing Point Using a Line-projecting Plane. When a line pierces a given oblique plane and an edge view is not given, as in Fig. 10.25, a line-projecting plane (cutting plane) may be used to establish a line of intersection that will contain the piercing point. In the illustration, a vertical projecting plane was selected that would contain the given line RS and intersect the given plane ABC along line DE, as illustrated by the pictorial drawing.

Solution: Draw the H-view of the projecting plane through $r^H s^H$ to establish $d^H e^H$, as shown in the H-view in (c). Locate $d^F e^F$ and draw the F-view of the

Fig. 10.24 Determination of a piercing point by inspection.

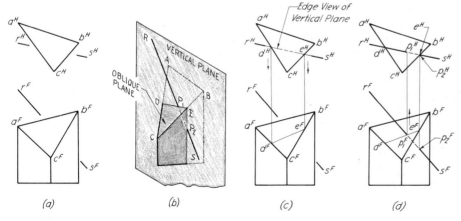

Fig. 10.25 Use of a line-projecting plane.

line of intersection. Then, complete the line view $r^F s^F$ to establish $p_1{}^F$ at the point of intersection of $r^F s^F$ and $d^F e^F$. Finally, locate $p_1{}^H$ on $r^H s^H$ by projecting upward from $p_1{}^F$, as shown in (d).

10.27

To Find Where a Line Pierces a Geometric Solid–cylinder–cone–sphere Using Projecting Planes (Fig. 10.26). The points where a line pierces a cylinder, cone, or sphere may be found easily through the use of a projecting plane (cutting plane) that contains the given line, as illustrated in (a), (b), and (c).

In (a) the intersections of the projecting plane and the cylinder are straight-line elements because the projecting plane used is parallel to the axis of the cylinder. The use of planes parallel to the axis permits the rapid solution of this type of problem. As shown by the pictorial drawing, the vertical projecting plane cuts elements on both the right and left sides of the cylinder. The line AB intersects the element RS at C and the other element at D. Points C and D are the piercing points.

The piercing points of a line and a cone are the points of intersection of the line and the two specific elements of the cone that lie in the projecting plane containing the line as shown in (b). The vertex of the cone and the given line fix the position of the projecting plane, the plane of the elements. In the illustration, the vertical projecting plane, taken through the line EF and the vertex of the cone T, cuts the base of the cone at U and V, the points needed to establish the F-views of the elements lying in the plane. The points of intersection of the given line EF and these elements are points G and H, the points where the line pierces the cone. If the given line had not been in a position to intersect the axis of the cone, it would have been necessary to use an oblique cutting plane through the apex.

A projecting plane that contains a line piercing a sphere will cut a circle on the surface of the sphere; therefore, points where the given line intersects the circle will be points where the line pierces the sphere. [See the pictorial drawing in (c).] In the illustrations, a vertical projecting plane was used containing the given line JK. The F-views of the piercing points M and N ($m^F n^F$) were found first at the points of intersection of the line and the circle. The H-views ($m^H n^H$) of the piercing points were found by projecting upward from m^F and n^F in the F-view.

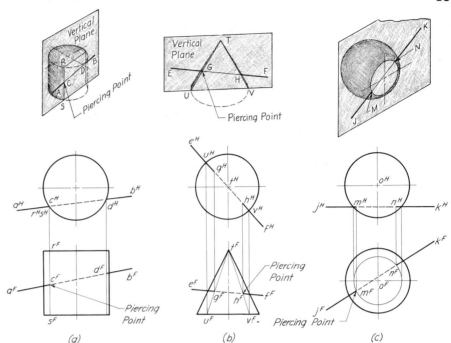

Fig. 10.26 To determine where a line pierces a geometric solid.

10.28

To Determine the Points Where a Line Pierces a Cone, General Case. In Sec. 10.27, the statement was made that the piercing points of a line and a cone are the points of intersection of the line and the two specific elements of the cone that lie in the projecting plane containing the line and the apex of the cone [Fig. 10.26(b)]. This pertained to a special condition for which a line-projecting plane could be used. For other cases, the following general statement applies: The piercing points of a line and any surface must lie on the lines of intersection of the given surface and a cutting plane that contains the line. Obviously, an infinite number of cutting planes could have been assumed that would have contained the line AB in Fig. 10.27, but all would have resulted in curved lines of intersection, except in the case of the one plane that was selected to pass through the apex O of the cone. As can be noted by observing the pictorial drawing, this choice gives a plane that intersects the cone along two straight-line elements.

Solution: (1) Form a cutting plane containing the line AB and the apex O by drawing a line from O to an assumed point M on line AB. Lines AB and OM define the cutting plane. (2) Extend the

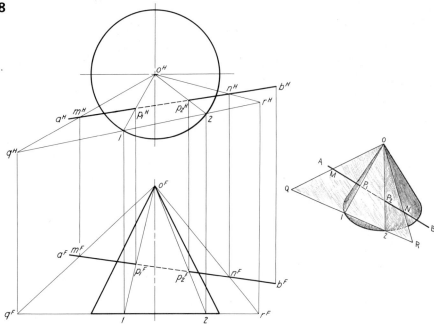

Fig. 10.27 To determine the points where a line pierces a cone—general case.

elements of intersection o^F1 and o^F2 in the F-view. Points $p_1{}^F$ and $p_2{}^F$, where the F-views of these elements intersect a^Fb^F, are the F-views of the piercing points of the line AB and the given cone. (4) Project the F-views of the piercing points to the H-view to locate $p_1{}^H$ and $p_2{}^H$.

When the line and cone are both oblique an added view (auxiliary view), showing the base as an edge, may be needed to obtain a quick and accurate solution employing the steps illustrated in Fig. 10.27.

10.29
To Find the Intersection of Two Planes, Line-projecting Plane Method. The intersection of two oblique planes may be determined by finding where two of the lines of one plane pierce the other plane, as illustrated by the pictorial drawing in Fig. 10.28. The procedure that is illustrated employs line-projecting planes to find the piercing points of the lines XY and XZ and the oblique plane RST. Therefore, it might be said that the solution requires the determination of the piercing point of a line and an oblique plane, as explained in Sec. 10.26.

> **Given:** The oblique planes RST and XYZ.
>
> **Solution:** Since the (vertical) line-projecting plane C_1P_1 is to contain the line XY of the plane XYZ, draw the line-view representation of this projecting plane to coincide with x^Hy^H. Next, project the line of intersection AB between the line-projecting plane C_1P_1 and plane RST from the top view, where it appears as a^Hb^H in the edge-view representation of C_1P_1 to the front view. Then, since it is evident that the line AB is not parallel to XY, which lies in the projecting plane (see F-view), the line XY intersects AB. The location of this intersection at E is established first in the F-view, where the line x^Fy^F intersects a^Fb^F at e^F. The H-view of E, that is, e^H, is found by projecting upward from e^F in the F-view to the line view x^Hy^H. The other end of the line of intersection between the two given planes at F is found by using the line-projecting plane C_2P_2 and following the same procedure as for determining the location of point E.

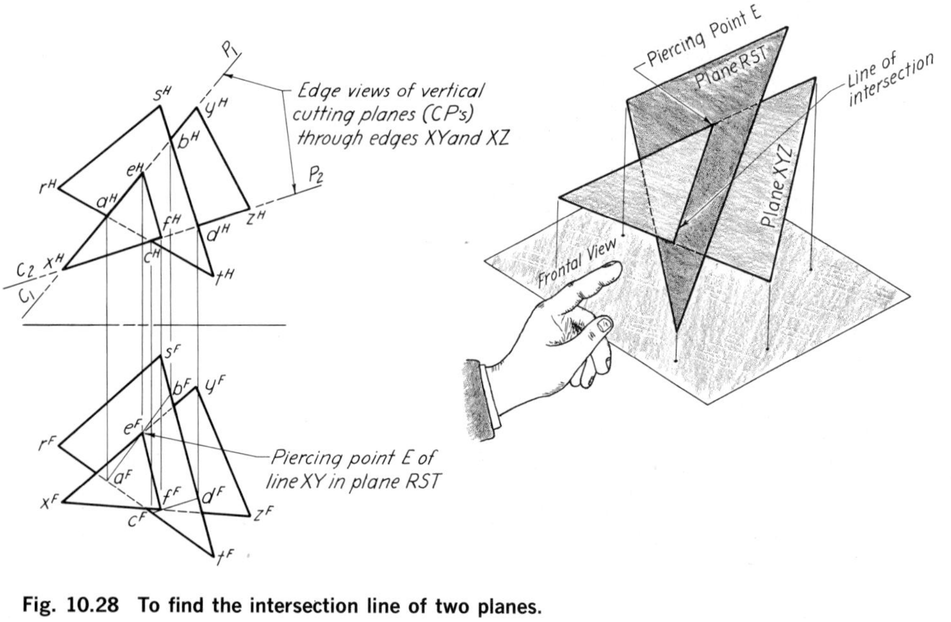

cutting plane until it intersects the plane of the base of the cone. In the pictorial illustration it should be noted that OM and ON extended establish the line of intersection QR between the cutting plane and the plane of the base of the cone. Line ON, drawn from O to an assumed point N, is an additional line in the plane. (3) Project points 1 and 2, where the line q^Hr^H intersects the curve of the base of the cone in the H-view, to the F-view, and draw the views of the

10.30
To Find the Intersection of a Cylinder and an Oblique Plane. There are two distinct and separate methods shown in Fig. 10.29 for finding the line of intersection of an oblique plane and a cylinder. Both meth-

Edge views of vertical cutting planes (CP's) through edges XY and XZ

Piercing point E of line XY in plane RST

Fig. 10.28 To find the intersection line of two planes.

ods appear together on the drawing at the left. The selected line method is illustrated pictorially in (*b*), while the cutting-plane method is shown in (*c*).

In the application of the selected line method, any line of the given plane, such as line BR, is drawn in the F- and H-views (*a*). It can be noted by observing, in the pictorial drawing in (*b*), that this particular line pierces the cylinder at points P_1 and P_2 to give two points on the line of intersection. On the multiview drawing, the locations of the H-views of points P_1 and P_2 can readily be recognized as being at the points labeled $p_1{}^H$ and $p_2{}^H$, where the line view $b^H r^H$ intersects the edge view of the surface of the cylinder. The F-views ($p_1{}^F$ and $p_2{}^F$) of points P_1 and P_2 were found by projecting downward from $p_1{}^H$ and $p_2{}^H$ to the line view $b^F r^F$. Additional points along the line of intersection, as needed, can be obtained by using other lines of the plane.

The line of intersection of this same plane and cylinder could have been almost as easily determined through the use of a series of line-projecting (cutting) planes passed parallel to the axis of the cylinder, as illustrated in (*c*). It should be noted that the vertical cutting plane shown cuts elements on the cylinder that intersect XY, the line of intersection of the cutting plane and the given plane, at points P_3 and P_4. Points P_3 [not visible in (*c*)] and P_4 are two points on the line of intersection, for they lie in both the cutting plane and the given plane and are on the surface of the cylinder. After the position of the cutting plane has been established in (*a*) by drawing the line representation in the H-view, x^H and y^H must be projected downward to the corresponding lines of the plane in the F-view. Line $x^F y^F$ as then drawn is the F-view of the line of intersection of the cutting plane and given plane. Finally, as the last step, the intersection elements that appear as points in the H-view at $p_3{}^H$ and $p_4{}^H$ must be drawn in the F-view. The F-views of points P_3 and $P_4(p_3{}^F$ and $p_4{}^F)$ are at the intersection of the F-views of the elements and the line $x^F y^F$.

A series of selected planes will give the points needed to complete the F-view of the intersection.

10.31

To Find the Intersection of a Cone and an Oblique Plane. When the intersecting plane is oblique, as is true in Fig. 10.30, it is usually desirable to employ the cut-

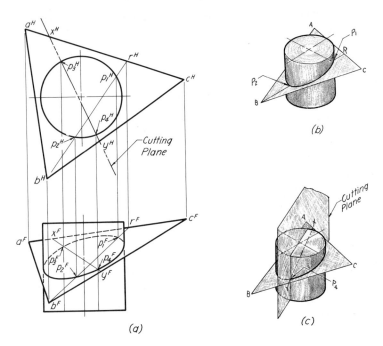

Fig. 10.29 To find the intersection of a cylinder and a plane.

ing-plane method shown in (*a*) rather than to resort to the use of an additional view, as in (*b*). Since the given cone is a right cone, any one vertical cutting plane, passing through the apex O, will simultaneously cut straight lines on the conical surface and across the given oblique plane. The pictorial illustration shows that the cutting plane XY intersects the cone along the two straight-line elements O–2 and O–8 and the plane along the line RS. Points G and H, where the elements intersect the line RS, are points along the required line of intersection because both points lie in the given plane $ABCD$ and are on the surface of the cone. In (*a*), the line representation of the cutting plane XY in the H-view establishes the position of the H-view ($r^H s^H$) of the line RS. The H-views of elements O–2 and O–8 also lie in the edge view of the cutting plane XY. With this much known, the F-views of the two elements and line RS may be drawn. Points g^F and h^F are at the intersection of $r^F s^F$ and o^F–2 and o^F–8, respectively. A series of cutting planes passed similarly furnishes the points needed to complete the solution.

At times one might prefer to determine the line of intersection through the use of a constructed auxiliary view that shows the given plane as an edge. In this case, when selected elements may be seen to intersect the line view of the plane in the auxiliary view, the solution becomes quite

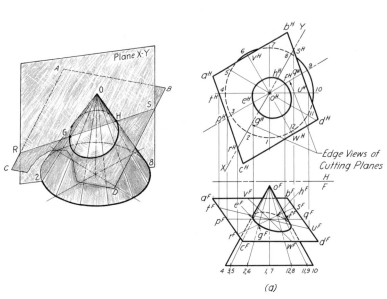

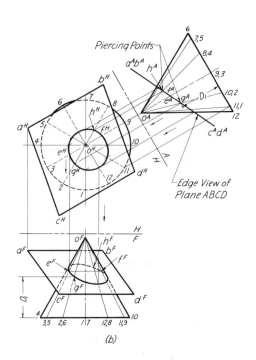

Fig. 10.30 To find the intersection of a cone and a plane.

simple, because all that is required is to project the point of intersection of an element and the line view of the plane to the F- and H-views of the same element. For example, the A-view (o^A–8) of the element O–8 can be seen to intersect the edge line $a^A b^A c^A d^A$ at h^A, the A-view of point H. By projecting to line o^H–8, the H-view of H (h^H) may be easily established. The F-view of H (h^F) lies directly below h^H on o^F–8. Through the projected

views of other points, located similarly, smooth curves may be drawn to form the F- and H-view representations, as shown.

10.32
To Find the Intersection of a Sphere and an Oblique Plane. Horizontal cutting planes have been used to find the line of intersection of the sphere and oblique plane shown in Fig. 10.31. Two approaches to the solution have been given on the line drawing. The horizontal cutting planes, as selected, cut circles from the sphere and straight lines from the given oblique plane. For example, the cutting plane CP_3 in the F-view cuts the horizontal line 3–3 from plane $ABCD$ and a circle from the sphere. This circle appears as an edge in the F-view and shows in its true diameter. In the H-view it will show in true shape. The line and circle intersect at two points that also have been identified by the number 3, the number assigned to the cutting plane in which these two points lie. These points now located in the H-view are projected to the CP_3 line in the F-view. The curved line through points, that have been determined by a series of planes, is a line common to both surfaces and is therefore the line of intersection.

This problem could also have been solved by using an auxiliary view showing the plane $ABCD$ as an edge. As before, the horizontal cutting planes will cut circles from the sphere and lines from the plane. However, in this case both the lines

Fig. 10.31 To find the intersection of a sphere and an oblique plane.

and the intersections show as points in the A-view. For each CP, these points must be projected from the auxiliary view to the corresponding circle in the H-view. The F-views of these points on the line of intersection may be found by projection and by using measurements taken from the A-view.

10.33

To Find the Intersection of Two Prisms. In Fig. 10.32, points A, C, and D, through which the edges of the horizontal prism pierce the vertical prism, are first found in the top view (a^H, c^H, and d^H) and are then projected downward to the corresponding edges in the front view. Point B, through which the edge of the vertical prism pierces the near face of the triangular prism, cannot be found in this manner because the side view from which it could be projected to the front view is not shown. Its location, however, can be established in the front view without even drawing a partial side view, if some scheme like the one illustrated in the pictorial drawing is used. In this scheme, the intersection line AB, whose direction is shown in the top view as line $a^H b^H$, is extended on the triangular face to point X on the top edge. Point x^H is projected to the corresponding edge in the front view and a light construction line is drawn between the points a^F and x^F. Since point B is located on line AX (see pictorial) at the point where the edge of the prism pierces the line, its location in the front view is at point b^F where the edge cuts the line $a^F x^F$.

10.34

To Find the Intersection of a Pyramid and a Prism (Fig. 10.33). The intersection of a right pyramid and a prism may be found by the same general method used for finding the intersection of two prisms (Sec. 10.33).

10.35

To Determine the Intersection of a Prism and a Pyramid Using Line-projecting Planes. Frequently, it becomes necessary to draw the line of intersection between two geometric shapes so positioned that the piercing points of edges cannot be found by inspection if only the principal views are to be used. In this case, one must resort to the method discussed in Sec. 10.26 to determine where a line, such as the edge line GD of the prism shown in Fig. 10.34, pierces a surface. As illustrated

by the pictorial drawing, a vertical plane passed through the edge DG of the prism, intersects the surface ABC of the pyramid along line MN that contains point D, the piercing point of DG. In (a), the H-view ($m^H n^H$) of the line MN lies along $d^H g^H$ extended to m^H on the edge of the pyra-

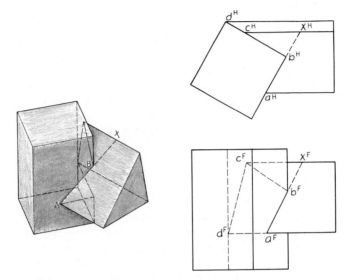

Fig. 10.32 Intersecting prisms.

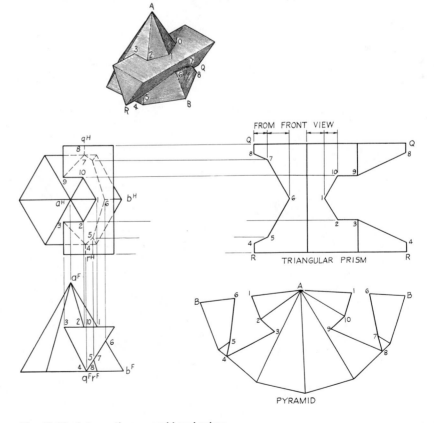

Fig. 10.33 Intersecting pyramid and prism.

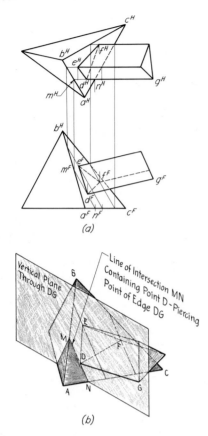

(a)

(b)

Fig. 10.34 Intersecting pyramid and prism.

mid, because the H-view of the cutting plane appears as an edge that coincides with $d^H g^H$. With the F-view of MN established by projecting downward from $m^H n^H$ in the H-view, the frontal view of the piercing point D is at d^F where the view of the edge line DG of the prism intersects $m^F n^F$. The H-view of D is found by projecting upward from d^F. The two other piercing points, at E and F, are found in the same manner using two other line-projecting planes.

10.36
To Construct a Development Using Auxiliary Views. When one of the components is oblique to the principal planes of projection, as is the prism in Fig. 10.35, the construction work needed for the development can be simplified somewhat through the use of an auxiliary view to find the true lengths of the edges and an oblique (secondary auxiliary) view to show a right section. Since the plane on which the auxiliary view is projected is a vertical one that is parallel to the edges of the prism, the distances perpendicular to the AH reference line are height distances. In making the construction, distances are taken from the F-view in the direction of the single-headed arrow for use in the auxiliary view in the direction indicated by a

similar arrow. For the oblique view, projected on an O-plane that is perpendicular to both the A-plane and the edges of the prism, distances are taken from the AH reference line in the direction of the arrow numbered 2 to be laid out from the reference line for the O-view. Since this is the second auxiliary, the arrows indicating the direction for equal distances have been given two heads.

If there is sufficient space available, the true-length measurements for the edges in the development may be projected directly from the auxiliary view showing the true lengths. The true distances between these edges, taken from the right section in the direction of the arrow, are laid off along the stretch-out line. The arrow on the development indicates the direction in which the successive faces are laid down when the prism is turned to unroll the lateral surface inside-up.

10.37
To Find the Intersection of Two Cylinders. If a series of elements are drawn on the surface of the small horizontal cylinder, as in Fig. 10.36, the points A, B, C, and D in which they intersect the vertical cylinder will be points on the line of intersection (see pictorial). These points, which are shown as a^H, b^H, c^H, and d^H in the top view, may be located in the front view by projecting them downward to the corresponding elements in the front view, where they are shown as points a^F, b^F, c^F, and d^F. The desired intersection is represented by a smooth curve drawn through these points.

10.38
To Find the Intersection of Two Cylinders Oblique to Each Other. The first step in finding the line of intersection of two cylinders that are oblique to each other (Fig. 10.37) is to draw a revolved right section of the oblique cylinder directly on the front view of that cylinder. If the circumference of the right section is then divided into a number of equal divisions and elements are drawn through the division points, the points A, B, C, and D in which the elements intersect the surface of the vertical cylinder will be points on the line of intersection (see pictorial). In the case of the illustration shown, these points are found first in the top view and then are projected downward to the corresponding elements in the front view. The line of intersection in the front view is represented by a smooth curve drawn through these points.

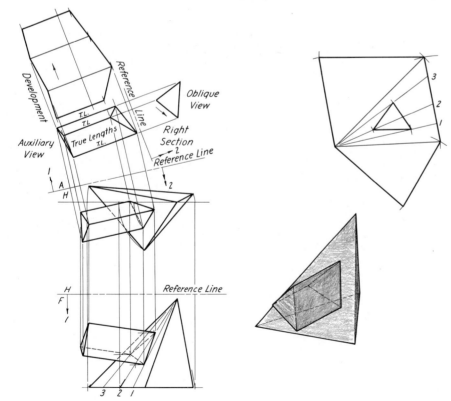

Fig. 10.35 To construct a development using auxiliary views.

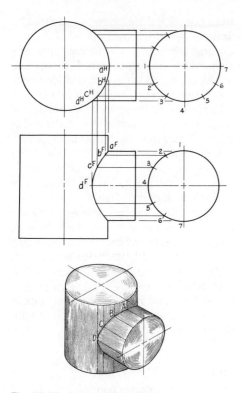

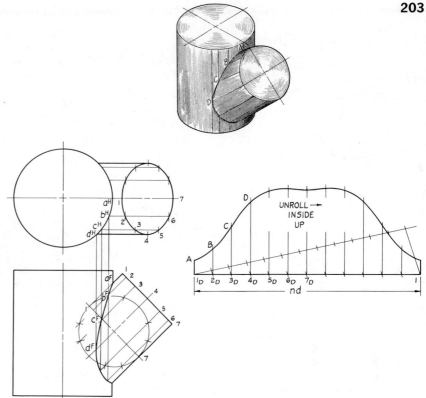

Fig. 10.36 Intersecting cylinders.

Fig. 10.37 Intersecting cylinders.

10.39

To Find the Intersection of Two Cylinders Using Line-projecting (Cutting) Planes. The line of intersection of the two cylinders shown in Fig. 10.37 could have been determined through the use of a series of parallel line-projecting (cutting) planes passed parallel to their axes (Fig. 10.38). The related straight-line elements cut on the cylinders by any one cutting plane, such as C, intersect on the line of intersection of the cylinders. As many line-projecting planes as are needed to obtain a smooth curve should be used and they should be placed rather close together where a curve changes sharply.

10.40

To Find the Intersection of a Cylinder and a Cone. The intersection of a cylinder and a cone may be found by assuming a number of elements on the surface of the cone. The points at which these elements cut the cylinder are on the line of intersection (see Figs. 10.39 and 10.40). In selecting the elements, it is the usual practice to divide the circumference of the base into a number of equal parts and draw elements through the division points. To obtain

needed points at locations where the intersection line will change suddenly in curvature, however, there should be additional elements.

In Fig. 10.39, the points at which the elements pierce the cylinder are first found in the top view and are then projected to the corresponding elements in the front view. A smooth curve through

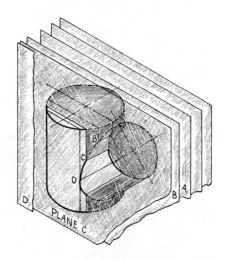

Fig. 10.38 To find the intersection of two cylinders using line-projecting planes.

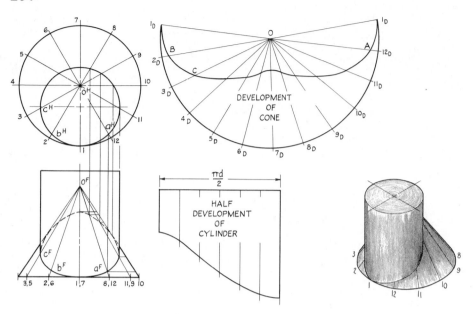

Fig. 10.39 Intersecting cylinder and cone.

from the conical surface and two straight-line elements from the cylindrical surface. Cutting plane 5, for example, cuts one element (numbered 5) from the near side of the cone and an upper and lower element from the surface of the cylinder. The intersection of these three elements, all numbered 5, establish the location of points A and B on the line of intersection.

An alternative method for finding the line of intersection of a cylinder and a right cone is illustrated in Fig. 10.41. Here horizontal cutting planes are passed through both geometric shapes in the region of their line of intersection. In each cutting plane, the circle cut on the surface of the cone will intersect elements cut on the cylinder at two points common to both surfaces (see pictorial). A curved line traced through a number of such points in different planes is a line common to both surfaces and is therefore the line of intersection.

these points forms the figure of the intersection.

To find the intersection of the cone and cylinder combination shown in Fig. 10.40, line-projecting cutting planes were passed through the vertex O parallel to the axis of the cylinder to cut intersecting elements on both geometric forms. The partial auxiliary is needed to establish these planes, because it is only in a view showing the axis of the cylinder as a point that these planes and the surface of the cylinder will show as edge views. Each cutting plane cuts one needed straight-line element

10.41

To Find the Intersection of an Oblique Cone and a Paraboloid. The method that is illustrated in Fig. 10.42 for finding the line of intersection of a cone and a paraboloid may be used effectively for the solution of many other intersection problems involving different geometric forms, such as two cones or a cone and a cylinder having parallel bases. For the arrangement as shown, a horizontal cutting plane will intersect each geometric form in a circle. When a number of horizontal cutting planes are passed through both geometric shapes in the region of their line of intersection, each plane cuts a circle on the cone that intersects a circle cut on the paraboloid at two points common to both surfaces. The curved line traced through

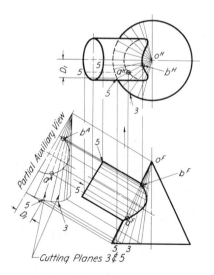

Fig. 10.40 To find the intersection of a cone and a cylinder.

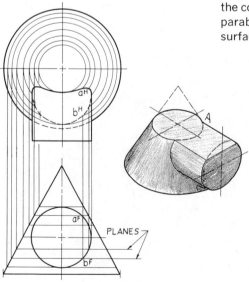

Fig. 10.41 Intersecting cylinder and cone.

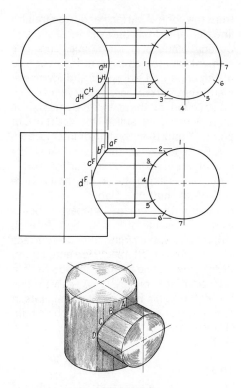

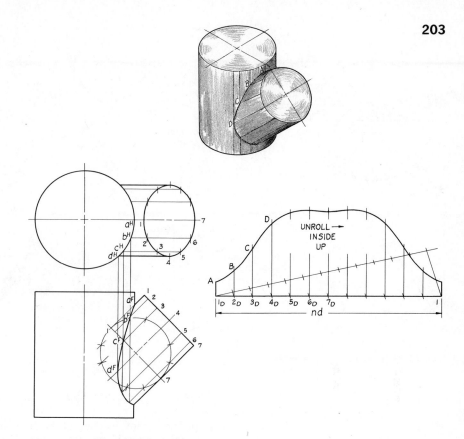

Fig. 10.36 Intersecting cylinders.

Fig. 10.37 Intersecting cylinders.

10.39

To Find the Intersection of Two Cylinders Using Line-projecting (Cutting) Planes. The line of intersection of the two cylinders shown in Fig. 10.37 could have been determined through the use of a series of parallel line-projecting (cutting) planes passed parallel to their axes (Fig. 10.38). The related straight-line elements cut on the cylinders by any one cutting plane, such as C, intersect on the line of intersection of the cylinders. As many line-projecting planes as are needed to obtain a smooth curve should be used and they should be placed rather close together where a curve changes sharply.

10.40

To Find the Intersection of a Cylinder and a Cone. The intersection of a cylinder and a cone may be found by assuming a number of elements on the surface of the cone. The points at which these elements cut the cylinder are on the line of intersection (see Figs. 10.39 and 10.40). In selecting the elements, it is the usual practice to divide the circumference of the base into a number of equal parts and draw elements through the division points. To obtain

needed points at locations where the intersection line will change suddenly in curvature, however, there should be additional elements.

In Fig. 10.39, the points at which the elements pierce the cylinder are first found in the top view and are then projected to the corresponding elements in the front view. A smooth curve through

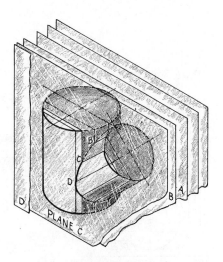

Fig. 10.38 To find the intersection of two cylinders using line-projecting planes.

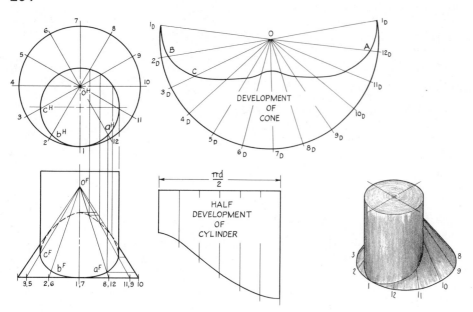

Fig. 10.39 Intersecting cylinder and cone.

from the conical surface and two straight-line elements from the cylindrical surface. Cutting plane 5, for example, cuts one element (numbered 5) from the near side of the cone and an upper and lower element from the surface of the cylinder. The intersection of these three elements, all numbered 5, establish the location of points A and B on the line of intersection.

An alternative method for finding the line of intersection of a cylinder and a right cone is illustrated in Fig. 10.41. Here horizontal cutting planes are passed through both geometric shapes in the region of their line of intersection. In each cutting plane, the circle cut on the surface of the cone will intersect elements cut on the cylinder at two points common to both surfaces (see pictorial). A curved line traced through a number of such points in different planes is a line common to both surfaces and is therefore the line of intersection.

these points forms the figure of the intersection.

To find the intersection of the cone and cylinder combination shown in Fig. 10.40, line-projecting cutting planes were passed through the vertex O parallel to the axis of the cylinder to cut intersecting elements on both geometric forms. The partial auxiliary is needed to establish these planes, because it is only in a view showing the axis of the cylinder as a point that these planes and the surface of the cylinder will show as edge views. Each cutting plane cuts one needed straight-line element

10.41
To Find the Intersection of an Oblique Cone and a Paraboloid. The method that is illustrated in Fig. 10.42 for finding the line of intersection of a cone and a paraboloid may be used effectively for the solution of many other intersection problems involving different geometric forms, such as two cones or a cone and a cylinder having parallel bases. For the arrangement as shown, a horizontal cutting plane will intersect each geometric form in a circle. When a number of horizontal cutting planes are passed through both geometric shapes in the region of their line of intersection, each plane cuts a circle on the cone that intersects a circle cut on the paraboloid at two points common to both surfaces. The curved line traced through

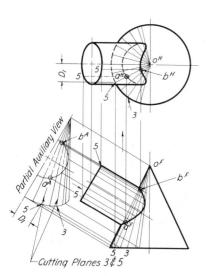

Fig. 10.40 To find the intersection of a cone and a cylinder.

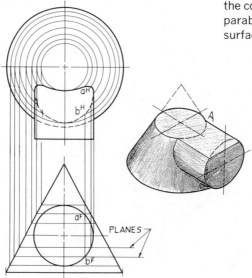

Fig. 10.41 Intersecting cylinder and cone.

the points that have been determined by the several planes is a line common to both surfaces and is therefore the line of intersection.

10.42

To Find the Intersection of a Prism and a Cylinder. In Fig. 10.43 it is required to find the intersection between the cylinder and two of the plane surfaces of the prism so that a pattern for the hole can be cut in the prism to match the cylinder. Although the intersection could have been secured merely by determining, through inspection, the piercing points of elements drawn arbitrarily on the cylinder, vertical line-projecting planes were used to illustrate an approach to this type of problem that is used by many people. The cutting planes are located in the view showing the right section of the cylinder. From this first step, the positions for the lines and elements cut on the prism and cylinder, respectively, can be determined by projection. Each line-projecting (cutting) plane cuts a line on a plane surface of the prism and related elements on the cylinder that intersect on the line of intersection. A sufficient number of line-projecting planes should be used to enable one to draw a smooth curved-line representation of the intersection. If more points are desired, at a location where a curve changes sharply, additional cutting planes may be added.

10.43

To Find the Intersection of a Prism and a Cone. The complete line of intersection may be found by drawing elements on the surface of the cone (Fig. 10.44) to locate points on the intersection as explained in Sec. 10.40. To obtain an accurate curve, however, some thought must be given to the placing of these elements. For instance, although most of the elements may be equally spaced on the cone to facilitate the construction of its development, additional ones should be drawn through the critical points and in regions where the line of intersection changes sharply in curvature. The elements are drawn on the view that will reveal points on the intersection, then the determined points are projected to the corresponding elements in the other view or views. In this particular illustration a part of the line of intersection in the top view is a portion of the arc of a circle that would be cut by a horizontal plane containing the bottom surface of the prism.

If the surfaces of the prism are parallel to the axis of the cone, as in Fig. 10.45, the line of intersection will be made up of the tips of a series of hyperbolas. The intersection may be found by passing planes that will cut circles on the surface of the cone. The points at which these cutting circles pierce the faces of the prism are points common to the lateral surfaces of both shapes and are therefore points on the required line of intersection. It should be noted that the resulting solution represents a chamfered bolthead.

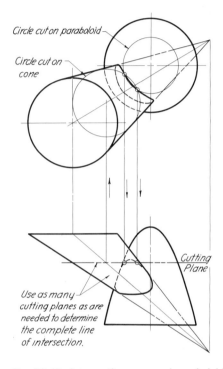

Fig. 10.42 Intersecting cone and paraboloid.

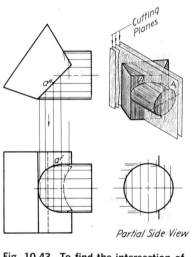

Fig. 10.43 **To find the intersection of a prism and a cylinder.**

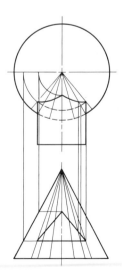

Fig. 10.44 **Intersecting cone and prism.**

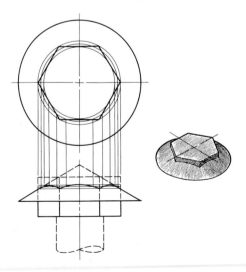

Fig. 10.45 **Intersecting cone and hexagonal prism.**

Problems

Note: It is common practice under the metric system to give all dimensions, that are not critical, in full millimeters. It is recommended that the student attempt to do likewise in laying out and dimensioning a problem that has been dual-dimensioned.

1. (Fig. 10.46). Develop the lateral surface of one or more of the prisms as assigned.

2. (Fig. 10.47). Develop the lateral surface of one or more of the pyramids as assigned. Make construction lines light. Show construction for finding the true lengths of the lines.

3. (Fig. 10.48). Develop the lateral surface of one or more of the cylinders as assigned. Use a hard pencil for construction lines and make them light.

4. (Fig. 10.49). Develop the lateral surface of one or more of the cones as assigned. Show all construction. Use a hard pencil for construction lines and make them light. In each case start with the shortest element and unroll, inside-up. It is suggested that 12 elements be used, in order to secure a reasonably accurate development.

5. (Fig. 10.50). Develop the lateral surface of one or more of the transition pieces as assigned. Show all construction lines in light, sharp pencil lines. Use a sufficient number of elements on the curved surfaces to ensure an accurate development.

6. (Fig. 10.51). Develop the sheet-metal connections. On pieces 3 and 4, use a sufficient number of elements to obtain a smooth curve and an accurate development.

7–8. (Figs. 10.52 and 10.53). Draw the line of intersection of the intersecting geometric shapes as assigned. Show the invisible portions of the lines of intersection as well as the visible. Consider that the interior is open.

9–10. (Figs. 10.54 and 10.55). Draw the line of intersection of the intersecting geometric shapes as assigned. It is suggested that the elements used to find points along the intersection be spaced 15° apart. Do not erase the construction lines. One shape does not pass through the other.

11. (Fig. 10.56). Draw the line of intersection of the intersecting geometric shapes as assigned. Show the invisible portions of the line of intersection as well as the visible. The interior of the combination is hollow. One shape does not pass through the other. Show construction with light, sharp lines drawn with a hard pencil.

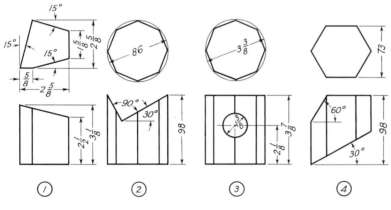

Fig. 10.46   Prisms.

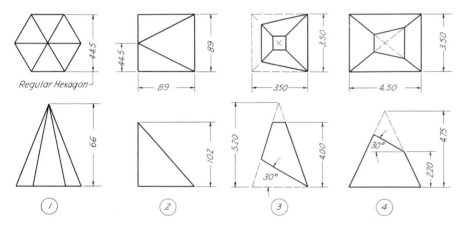

Fig. 10.47   Pyramids.

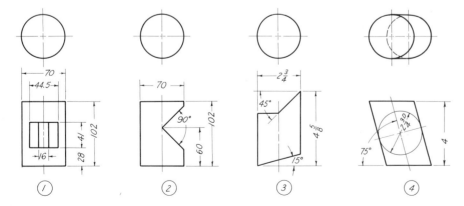

Fig. 10.48   Cylinders.

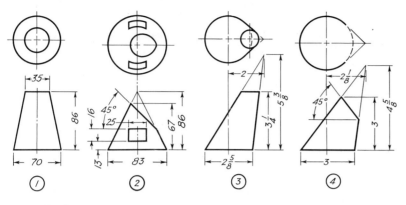

Fig. 10.49   Cones.

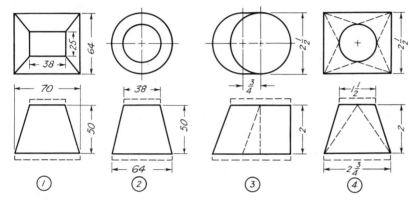

Fig. 10.50   Transition pieces.

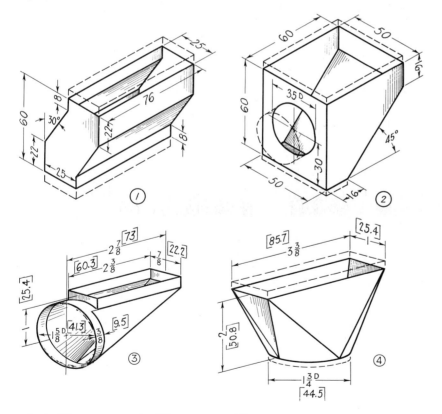

3-4 Dimensions in[ ] are in millimeters

Fig. 10.51   Sheet-metal connections (transitions).

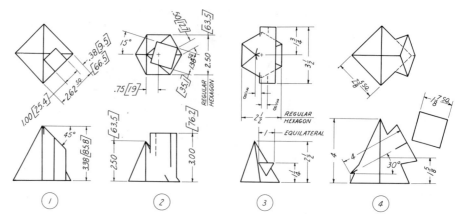

Fig. 10.52 Intersecting surfaces.

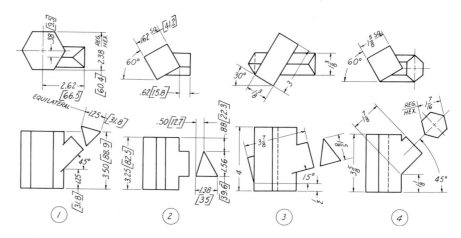

Fig. 10.53 Intersecting surfaces.

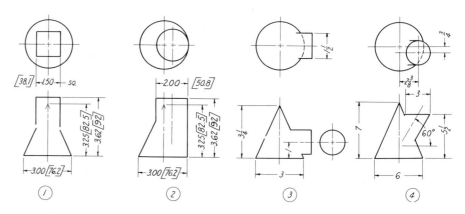

Fig. 10.54 Intersecting surfaces.

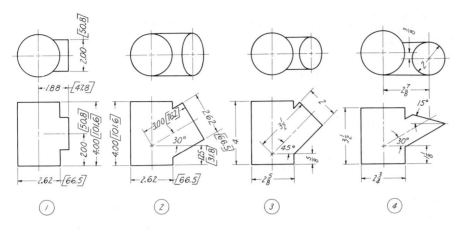

Fig. 10.55 Intersecting surfaces.

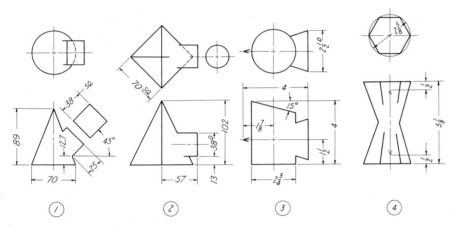

Fig. 10.56 Intersecting surfaces.

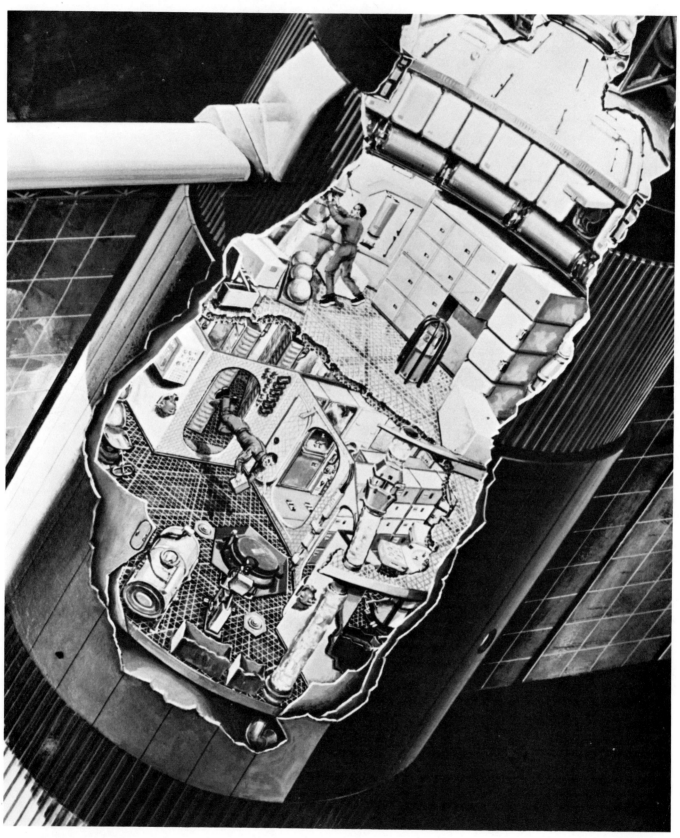

Shown is a pictorial illustration that was prepared by a NASA artist to promote the Apollo Applications Program. Illustrations of this type are used to sell an idea to administrators and political leaders. (*Courtesy National Aeronautics and Space Administration*)

11

Pictorial presentation

11.1

Introduction. An orthographic drawing of two or more views describes an object accurately in form and size, but, since each of the views shows only two dimensions without any suggestion of depth, such a drawing can convey information only to those who are familiar with graphic representation. For this reason, multiview drawings are used mainly by engineers, draftsmen, contractors, and shopmen.

Frequently, however, engineers and draftsmen find they must use conventional picture drawings to convey specific information to persons who do not possess the trained imagination necessary to construct mentally an object from views. To make such drawings, several special schemes of one-plane pictorial drawing have been devised that combine the pictorial effect of perspective with the advantage of having the principal dimensions to scale. But pictorial drawings, in spite of certain advantages, have disadvantages that limit their use. A few of these are as follows:

1. Some drawings frequently have a distorted, unreal appearance that is disagreeable.

2. The time required for execution is, in many cases, greater than for an orthographic drawing.

3. They are difficult to dimension.

4. Some of the lines cannot be measured.

Even with these limitations, pictorial drawings are used extensively for technical publications, Patent Office records, piping diagrams, and furniture designs. Occasionally they are used, in one form or another, to supplement and clarify machine and structural details that would be difficult to visualize (Fig. 11.1).

11.2
Divisions of Pictorial Drawing. Single-plane pictorial drawings are classified in three general divisions: (1) axonometric projection, (2) oblique projection, and (3) perspective projection (Fig. 11.2).

Perspective methods produce the most realistic drawings, but the necessary construction is more difficult and tedious than the construction required for the conventional methods classified under the other two divisions. For this reason, engineers customarily use some form of either axonometric or oblique projection. Modified methods, which are not theoretically correct, are often used to produce desired effects.

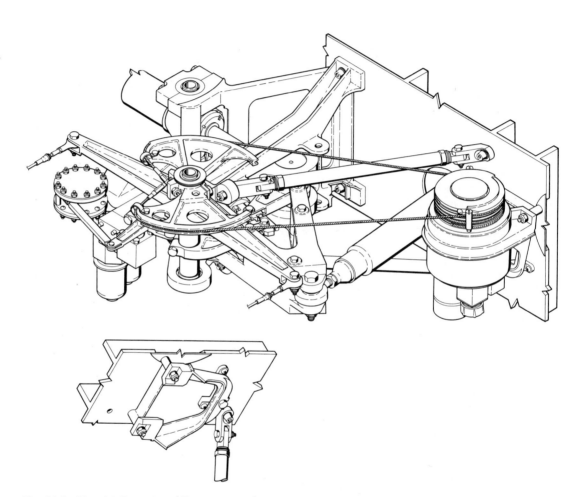

Fig. 11.1 Pictorial illustration. (*Courtesy Lockheed Aircraft Corp.*)

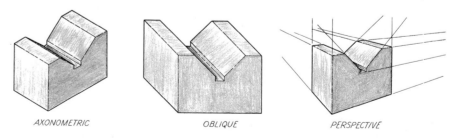

AXONOMETRIC   OBLIQUE   PERSPECTIVE

Fig. 11.2 Axonometric, oblique, and perspective projection.

A □ Axonometric projection

11.3

Divisions of Axonometric Projection. Theoretically, axonometric projection is a form of orthographic projection. The distinguishing difference is that only one plane is used instead of two or more, and the object is turned from its customary position so that three faces are displayed (Fig. 11.3). Since an object may be placed in a countless number of positions relative to the picture plane, an infinite number of views may be drawn, which will vary in general proportions, lengths of edges, and sizes of angles. For practical reasons, a few of these possible positions have been classified in such a manner as to give the recognized divisions of axonometric projection: (1) isometric, (2) dimetric, and (3) trimetric.

Isometric projection is the simplest of these, because the principal axes make equal angles with the plane of projection and the edges are therefore foreshortened equally.

11.4

Isometric Projection. If the cube in Fig. 11.3 is revolved through an angle of 45° about an imaginary vertical axis, as shown in II, and then tilted forward until its body diagonal is perpendicular to the vertical plane, the edges will be foreshortened equally and the cube will be in the correct position to produce an isometric projection.

The three front edges, called *isometric axes*, make angles of approximately 35° 16' with the vertical plane of projection, or picture plane. In this form of pictorial, the angles between the projections of these axes are 120°, and the projected lengths of the edges of an object, along and parallel to these axes, are approximately 81% of their true lengths. It should be observed that the 90° angles of the cube appear in the isometric projection as either 120° or 60°.

Now, if instead of turning and tilting the object in relation to a principal plane of projection, an auxiliary plane is used that will be perpendicular to the body diagonal, the view projected on the plane will be an axonometric projection. Since the auxiliary plane will be inclined to the principal planes on which the front, top, and side views would be projected, the auxiliary view, taken in a position perpendicular to the body diagonal, will be a secondary auxiliary view, as shown in Fig. 11.4.

11.5

Isometric Drawing. Objects seldom are drawn in true isometric projection, the use of an isometric scale being inconvenient and impractical. Instead, a conventional method is used in which all foreshortening is ignored, and actual true lengths are laid off along isometric axes and isometric lines. To avoid confusion and to set this method apart from true isometric projection, it is called isometric drawing.

The isometric drawing of a figure is slightly larger (approximately 22½%) than

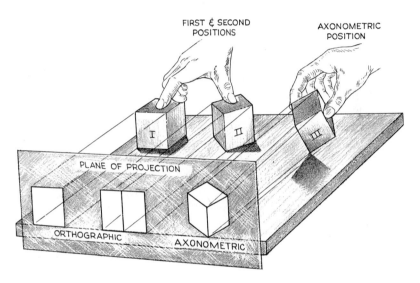

Fig. 11.3 Theory of axonometric projection.

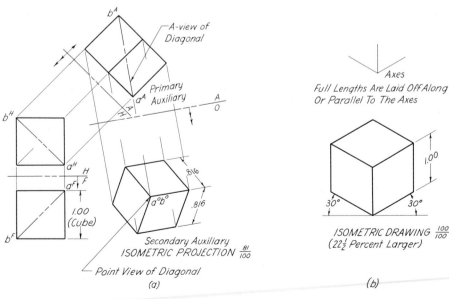

Fig. 11.4 Comparison of isometric projection and isometric drawing.

the isometric projection, but, since the proportions are the same, the increased size does not affect the pictorial value of the representation (see Fig. 11.4). The use of a regular scale makes it possible for a draftsman to produce a satisfactory drawing with a minimum expenditure of time and effort.

In isometric drawing, lines that are parallel to the isometric axes are called *isometric lines.*

11.6

To Make an Isometric Drawing of a Rectangular Object. The procedure followed in making an isometric drawing of a rectangular block is illustrated in Fig. 11.5. The three axes that establish the front

edges, as shown in (b), should be drawn through point A so that one extends vertically downward and the other two upward to the right and left at an angle of 30° from the horizontal. Then the actual lengths of the edges may be set off, as shown in (c) and (d), and the remainder of the view completed by drawing lines parallel to the axes through the corners thus located, as in (e) and (f).

Hidden lines, unless absolutely necessary for clearness, always should be omitted on a pictorial representation.

11.7

Nonisometric Lines. Those lines that are inclined and are not parallel to the isometric axes are called *nonisometric lines.* Since a line of this type does not appear in its true length and cannot be measured directly, its position and projected length must be established by locating its extremities. In Fig. 11.6, AB and CD, which represent the edges of the block, are nonisometric lines. The location of AB is established in the pictorial view by locating points A and B. Point A is on the top edge, X distance from the left-side surface. Point B is on the upper edge of the base, Y distance from the right-side surface. All other lines coincide with or are parallel to the axes and, therefore, may be measured off with the scale.

The pictorial representation of an irregular solid containing a number of nonisometric lines may be conveniently constructed by the box method; that is, the object may be enclosed in a rectangular box so that both isometric and nonisometric lines may be located by points of contact with its surfaces and edges (see Fig. 11.7).

A study of Figs. 11.6 and 11.7 reveals the important fact that lines that are parallel on an object are parallel in the pictorial view, and, conversely, lines that are not parallel on the object are not parallel on the view. It is often possible to eliminate much tedious construction work by the practical application of the principle of parallel lines.

11.8

Coordinate Construction Method. When an object contains a number of inclined surfaces, such as the one shown in Fig. 11.8, the use of the coordinate construction method is desirable. In this method, the end points of the edges are located in relation to an assumed isometric base line located on an isometric reference plane.

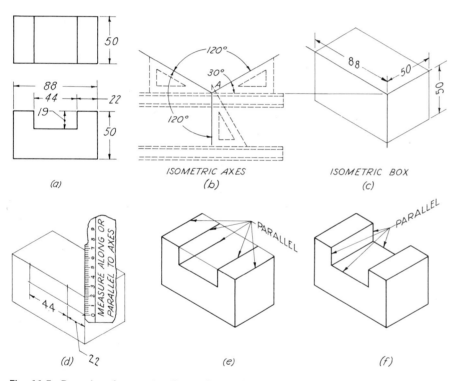

Fig. 11.5 Procedure for constructing an isometric drawing.

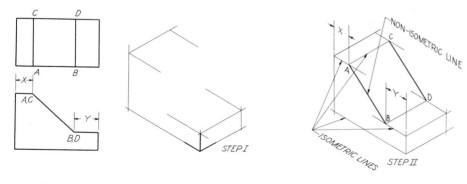

Fig. 11.6 Nonisometric lines.

For example, the line RL is used as a base line from which measurements are made along isometric lines, as shown. The distances required to locate point A are taken directly from the orthographic views.

Irregular curved edges are most easily drawn in isometric by the offset method, which is a modification of the coordinate construction method (Fig. 11.9). The position of the curve can be readily established by plotted points located by measuring along isometric lines.

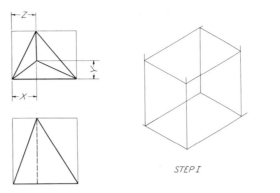
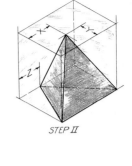

Fig. 11.7 Box construction.

11.9

Angles in Isometric Drawing. Since angles specified in degrees do not appear in true size on an isometric drawing, angular measurements must be converted in some manner to linear measurements that can be laid off along isometric lines. Usually, one or two measurements taken from an orthographic view may be laid off along isometric lines on the pictorial drawing to locate an inclined edge that has been specified by an angular dimension. The scale used for the orthographic view must be the same as the one being used in preparing the pictorial drawing.

In Fig. 11.10(a), the position of the inclined line AB was established on the isometric drawing by using the distance X taken from the front view of the orthographic drawing. When an orthographic drawing has already been prepared to a different scale than the scale being used for the pictorial representation, one can draw a partial orthographic view and take off the needed dimensions. A practical application of this idea is shown in (b). By making the construction of a partial view at the place where the angle is to appear on the isometric drawing, the position of the required line can be obtained graphically.

If desired, the tangent method, as explained in Sec. 3.11, may be used, as shown in (c). In using this method, a length equal to 10 units (any scale) is laid off along an isometric line that is to form one side of the angle. Then, a distance equal to 10 times the tangent of the angle is set off along a second isometric line that represents the second leg of the right triangle in pictorial. A line drawn through the end points of these lines will be the required line at the specified angle.

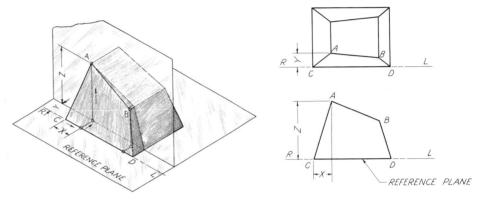

Fig. 11.8 Coordinate construction.

pears as an ellipse. The tedious construction required for plotting an ellipse accurately (Figs. 11.11 and 11.12) often is avoided by using some approximate method of drawing. The representation thus obtained is accurate enough for most work, although the true ellipse, which is slightly narrower and longer, is more pleasing in shape (Fig. 11.11). For an approximate construction, a four-center method is generally used.

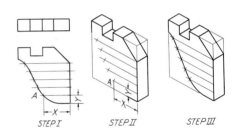

Fig. 11.9 Offset construction.

11.10

Circle and Circle Arcs in Isometric Drawing. In isometric drawing, a circle ap-

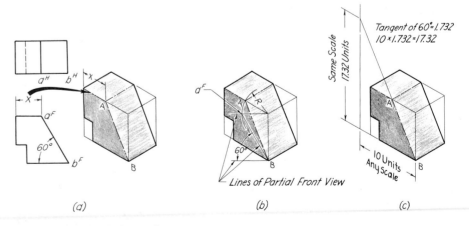

Fig. 11.10 Angles in isometric.

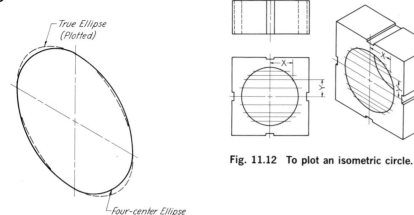

Fig. 11.11 Pictorial ellipses.

Fig. 11.12 To plot an isometric circle.

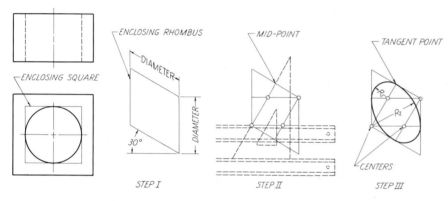

Fig. 11.13 Four-center approximation.

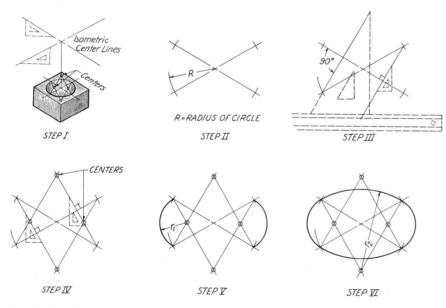

Fig. 11.14 Steps in drawing a four-center isometric circle (ellipse).

To draw an ellipse representing a pictorial circle, a square is conceived to be circumscribed about the circle in the orthographic projection. When transferred to the isometric plane in the pictorial view, the square becomes a rhombus (isometric square) and the circle an ellipse tangent to the rhombus at the midpoints of its sides. If the ellipse is to be drawn by the four-center method (Fig. 11.13), the points of intersection of the perpendicular bisectors of the sides of the rhombus will be centers for the four arcs forming the approximate ellipse. The two intersections that lie on the corners of the rhombus are centers for the two large arcs, while the remaining intersections are centers for the two small arcs. Furthermore, the length along the perpendicular from the center of each arc to the point at which the arc is tangent to the rhombus (midpoint) will be the radius. All construction lines required by this method may be made with a T-square and a 30° × 60° triangle.

The amount of work may be still further shortened and the accuracy of the construction improved by following the procedure shown in Fig. 11.14. The steps in this method are as follows:

Step I. Draw the isometric center lines of the required circle.

Step II. Using a radius equal to the radius of the circle, strike arcs across the isometric center lines.

Steps III and IV. Through each of these points of intersection erect a perpendicular to the other isometric center line.

Steps V and VI. Using the intersection points of the perpendiculars as centers and lengths along the perpendiculars as radii, draw the four arcs that form the ellipse (Fig. 11.15).

A circle arc will appear in pictorial representation as a segment of an ellipse. Therefore, it may be drawn by using as much of the four-center method as is required to locate the needed centers (Fig. 11.16). For example, to draw a quarter circle, it is only necessary to lay off the true radius of the arc along isometric lines drawn through the center and to draw intersecting perpendiculars through these points.

To draw isometric concentric circles by the four-center method, a set of centers must be located for each circle (Fig. 11.17).

When several circles of the same diameter occur in parallel planes, the construc-

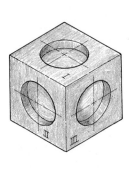

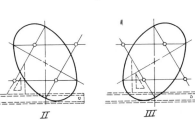

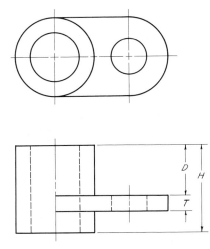

Fig. 11.15 Isometric circles.

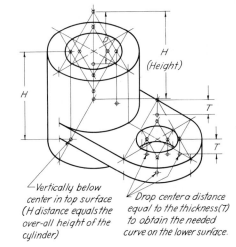

Fig. 11.16 Isometric circle arcs.

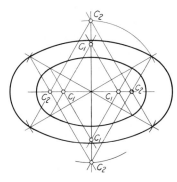

Fig. 11.17 Isometric concentric circles.

tion may be simplified. Figure 11.18 shows two views of an object and its corresponding isometric drawing. In Fig. 11.18, the centers for the ellipse representing the upper base of the large cylinder are found in the usual way, while the centers for the lower base are located by moving the centers for the upper base downward a distance equal to the height of the cylinder. By observing that portion of the object projecting to the right, it can be noted that corresponding centers lie along an isometric line parallel to the axis of the cylinder.

Circles and circle arcs in nonisometric planes may be plotted by using the offset or coordinate method. Sufficient points for establishing a curve must be located by transferring measurements from the orthographic views to isometric lines in the pictorial view. There is a rapid and easy way for drawing the cylindrical portion of the object shown in Fig. 11.19. The semicircular arc must be plotted on the rear surface as the first step. Then, after this has been done, each point is brought forward to the inclined face. The offset

Fig. 11.18 Isometric parallel circles.

Vertically below center in top surface (H distance equals the over-all height of the cylinder)

Drop center a distance equal to the thickness (T) to obtain the needed curve on the lower surface.

Fig. 11.19 Circles in nonisometric planes.

(a)

(b)

Surface A (Draw and use for constructing inclined surface B)

Surface B

Bring points forward from surface A

Surface A

Surface B

218

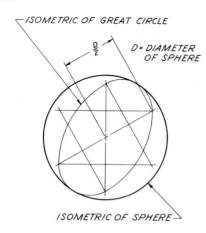

Fig. 11.20 Isometric drawing of a sphere.

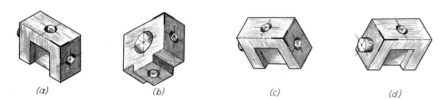

Fig. 11.21 Convenient positions of axes.

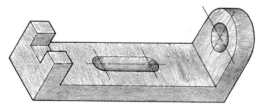

Fig. 11.22 Main axis horizontal—long objects.

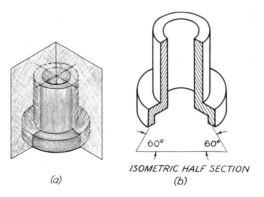

Fig. 11.23 Isometric half section.

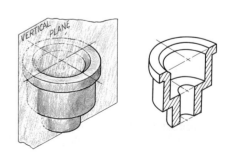

Fig. 11.24 Isometric full section.

distances (D_1, D_2, D_3, etc.) at each level are taken from the side view in (a).

The pictorial representation of a sphere is the envelope of all of the great circles that could be drawn on the surface. In isometric drawing, the great circles appear as ellipses and a circle is their envelope. In practice it is necessary to draw only one ellipse, using the true radius of the sphere and the four-center method of construction. The diameter of the circle is the long diameter of the ellipse (Fig. 11.20).

11.11
Positions of Isometric Axes. It is sometimes desirable to place the principal isometric axes so that an object will be in position to reveal certain faces to a better advantage (Fig. 11.21).

The difference in direction should cause no confusion, since the angle between the axes and the procedure followed in constructing the view are the same for any position. The choice of the direction may depend on the construction of the object,

but usually this is determined by the position from which the object is ordinarily viewed.

Reversed axes (b) are used in architectural work to show a feature as it would be seen from a natural position below.

Sometimes long objects are drawn with the long axis horizontal, as shown in Fig. 11.22.

11.12
Isometric Sectional Views (Fig. 11.23). Generally, an isometric sectional view is used for showing the inner construction of an object when there is a complicated interior to be explained or when it is desirable to emphasize features that would not appear in a usual outside view. Sectioning in isometric drawing is based on the same principles as sectioning in orthographic drawing. Isometric planes are used for cutting an object, and the general procedure followed in constructing the representation is the same as for an exterior view.

Figure 11.23 shows an isometric half section. It is easier, in this case, to outline the outside view of the object in full and then remove a front quarter with isometric planes.

Figure 11.24 illustrates a full section in isometric. The accepted procedure for constructing this form of sectional view is to draw the cut face and then add the portion that lies behind.

Section lines should be sloped at an angle that produces the best effect, but they should never be drawn parallel to object lines. In Fig. 11.25, (a) illustrates the slope that is correct for most drawings, while (b), (c), and (d) show the poor effect produced when this phase of section lining is ignored. Ordinarily, isometric section lines are drawn at 60°.

11.13
Dimetric Projection. The view of an object that has been so placed that two of its

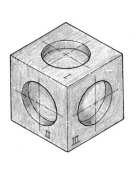
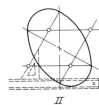
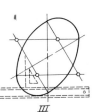
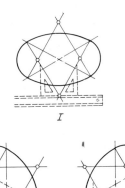

Fig. 11.15 Isometric circles.

Fig. 11.16 Isometric circle arcs.

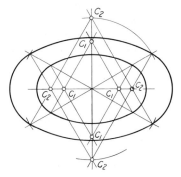

Fig. 11.17 Isometric concentric circles.

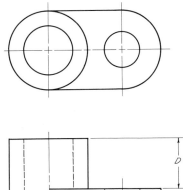
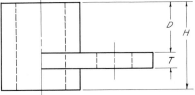

Fig. 11.18 Isometric parallel circles.

tion may be simplified. Figure 11.18 shows two views of an object and its corresponding isometric drawing. In Fig. 11.18, the centers for the ellipse representing the upper base of the large cylinder are found in the usual way, while the centers for the lower base are located by moving the centers for the upper base downward a distance equal to the height of the cylinder. By observing that portion of the object projecting to the right, it can be noted that corresponding centers lie along an isometric line parallel to the axis of the cylinder.

Circles and circle arcs in nonisometric planes may be plotted by using the offset or coordinate method. Sufficient points for establishing a curve must be located by transferring measurements from the orthographic views to isometric lines in the pictorial view. There is a rapid and easy way for drawing the cylindrical portion of the object shown in Fig. 11.19. The semicircular arc must be plotted on the rear surface as the first step. Then, after this has been done, each point is brought forward to the inclined face. The offset

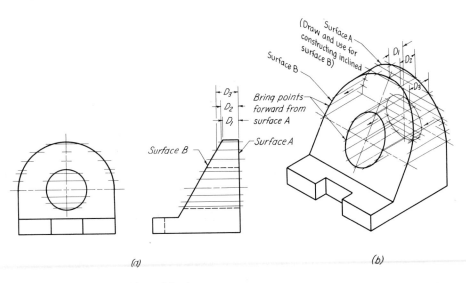

Fig. 11.19 Circles in nonisometric planes.

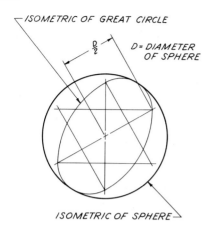

Fig. 11.20 Isometric drawing of a sphere.

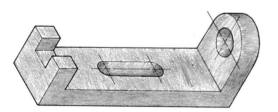

Fig. 11.21 Convenient positions of axes.

(a)   (b)   (c)   (d)

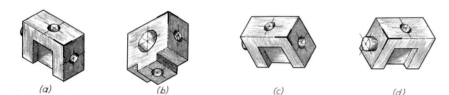

Fig. 11.22 Main axis horizontal—long objects.

distances (D_1, D_2, D_3, etc.) at each level are taken from the side view in (a).

The pictorial representation of a sphere is the envelope of all of the great circles that could be drawn on the surface. In isometric drawing, the great circles appear as ellipses and a circle is their envelope. In practice it is necessary to draw only one ellipse, using the true radius of the sphere and the four-center method of construction. The diameter of the circle is the long diameter of the ellipse (Fig. 11.20).

11.11

Positions of Isometric Axes. It is sometimes desirable to place the principal isometric axes so that an object will be in position to reveal certain faces to a better advantage (Fig. 11.21).

The difference in direction should cause no confusion, since the angle between the axes and the procedure followed in constructing the view are the same for any position. The choice of the direction may depend on the construction of the object, but usually this is determined by the position from which the object is ordinarily viewed.

Reversed axes (b) are used in architectural work to show a feature as it would be seen from a natural position below.

Sometimes long objects are drawn with the long axis horizontal, as shown in Fig. 11.22.

11.12

Isometric Sectional Views (Fig. 11.23). Generally, an isometric sectional view is used for showing the inner construction of an object when there is a complicated interior to be explained or when it is desirable to emphasize features that would not appear in a usual outside view. Sectioning in isometric drawing is based on the same principles as sectioning in orthographic drawing. Isometric planes are used for cutting an object, and the general procedure followed in constructing the representation is the same as for an exterior view.

Figure 11.23 shows an isometric half section. It is easier, in this case, to outline the outside view of the object in full and then remove a front quarter with isometric planes.

Figure 11.24 illustrates a full section in isometric. The accepted procedure for constructing this form of sectional view is to draw the cut face and then add the portion that lies behind.

Section lines should be sloped at an angle that produces the best effect, but they should never be drawn parallel to object lines. In Fig. 11.25, (a) illustrates the slope that is correct for most drawings, while (b), (c), and (d) show the poor effect produced when this phase of section lining is ignored. Ordinarily, isometric section lines are drawn at 60°.

11.13

Dimetric Projection. The view of an object that has been so placed that two of its

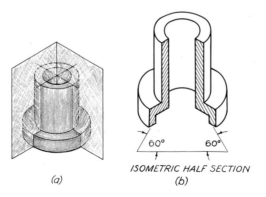

Fig. 11.23 Isometric half section.

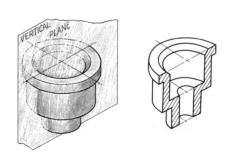

Fig. 11.24 Isometric full section.

axes make equal angles with the plane of projection is called a *dimetric projection.* The third axis may make either a smaller or larger angle. All of the edges along or parallel to the first two axes are foreshortened equally, while those parallel to the third axis are foreshortened a different amount. It might be said that dimetric projection, a division of axonometric projection, is like isometric projection in that the object must be placed to satisfy specific conditions. Similarly, a dimetric projection may be drawn by using the auxiliary view method. The secondary auxiliary view is the dimetric projection. The procedure is the same as for an isometric projection (Fig. 11.4), except that the line of sight is taken in the direction necessary to obtain the desired dimetric projection. Obviously, an infinite number of dimetric projections is possible.

In practical application, dimetric projection is sometimes modified so that regular scales can be used to lay off measurements to assumed ratios. This is called *dimetric drawing* [Fig. 11.26(a)].

The angles and scales may be worked out* for any ratios, such as $1:1:\frac{1}{2}$ (full size:full size:half size); $1:1:\frac{3}{4}$ (full size: full size:three-fourths size). For example, the angles for the ratios $1:1:\frac{1}{2}$ are $7°\ 11'$ and $41°\ 25'$. After the scales have been assumed and the angles computed, an enclosing box may be drawn in conformity to the angles and the view completed by following the general procedure used in isometric drawing, except that two scales must be used. The positions commonly used, along with the scale ratios and corresponding angles, are shown in Fig. 11.26(b). The first scale given in each ratio is for the vertical axis. Since two of the axes are foreshortened equally, while the third is foreshortened in different ratio, obviously two scales must be used. This is an effective method of representation.

11.14

Trimetric Projection (Fig. 11.28). A trimetric projection of an object is the view obtained when each of the three axes makes a different angle with the plane of projection. As might be expected, a trimetric projection may be constructed by drawing successive auxiliary views. How-

*Formula: $\cos \alpha = -\sqrt{2s_1^2 s_2^2 - s_2^4 / 2s_1 s_2}$. In this formula, α is one of the equal angles, s_1 is one of the equal scales, and s_2 is the third scale.

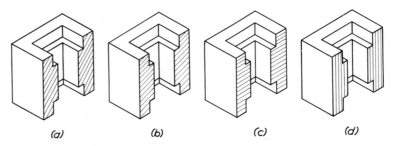

Fig. 11.25 Section lining.

ever, since there are an unlimited number of possible lines of sight that will produce unequal foreshortening in the directions of the three axes, considerable thought must be given to the selection of a position that will show the object pictorially to the best advantage in the second auxiliary view. Making this decision is not easy. This form of pictorial representation has been used to some extent by certain aircraft companies for the preparation of production illustrations.

11.15

To Construct an Axonometric Projection (Isometric, Dimetric, or Trimetric) Directly from the Orthographic Views. A true axonometric projection can be prepared by projecting directly from two or more of the orthographic views of an object. The method, commonly called the method of intersections, was developed some years prior to 1937, the year that it was first published by L. Eckhart of the Vienna College of Engineering. Since then it has been rather widely used in the United States by those whose principal task is the preparation of pictorial drawings of complicated parts for which the orthographic views may be available. The underlying principles and basic procedure for the preparation of an isometric projection have been illustrated in Fig. 11.27.

In (a) the relationship of the axonometric plane to the three principal planes has been shown pictorially, and it must be understood at this point in the discussion that, since the picture plane and the axonometric plane are coincident, the axonometric triangle will be in true size. Also two other facts must be noted. First, since axonometric projection is true orthographic projection, the projection lines are perpendicular to the plane of projection. Second, when a point is projected on adjacent planes [see (b)] and one of the planes is revolved into the plane of the other, the two views of the point will lie on a single line that is perpendicular to the

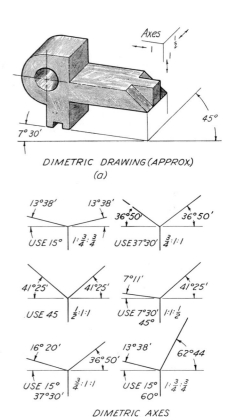

DIMETRIC DRAWING (APPROX.)
(a)

DIMETRIC AXES
(b)

Fig. 11.26 Approximate dimetric drawing.

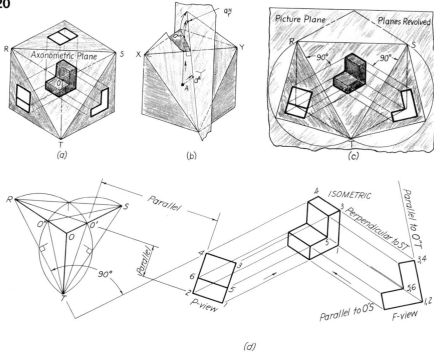

Fig. 11.27 To construct an axonometric projection from the orthographic views.

fold line—that is, the line of intersection XY of the two planes.

In (c) the principal planes have been shown revolved into the plane of the axonometric triangle, and it should now become evident to the reader how orthographic views might be used to obtain an axonometric projection once their revolved positions have been determined. By observing the drawing in (c) one can see that the projecting lines are perpendicular to the axis of rotation and that the principal edges of each view are parallel to the corresponding edges of the revolved plane.

To start the graphical diagram in (d) three mutually perpendicular lines were drawn to represent the three intersecting edges OR, OS, and OT of the cube [see (a)]. By geometry, OR will be perpendicular to ST, OS will be perpendicular to RT, and OT will be perpendicular to RS. With this diagram as now laid out, the right triangle OST may readily be revolved into the axonometric plane RST. One of the advantages of starting the diagram with the axes, as was done in the illustration, is that the forward triangle may then be revolved into the axonometric plane to determine the position of the orthographic views, instead of the rear triangle as in (c), which would require the construction of the entire box. In revolving the right tri-

angle OST, which must show in true size as $O''ST$ in the axonometric plane, it is necessary to apply the geometric principle that any two lines will form a right angle that are drawn from a point on the circumference of a circle to the end points of its diameter. Hence, point O moves along OR, which is perpendicular to ST, to a position on the semicircle that has been constructed on ST as a diameter. Similarly, O moves to O' on OS when the right triangle ORT has been revolved into the axonometric plane.

With the diagram constructed as explained, the step-by-step procedure for preparing the axonometric drawing shown is as follows:

1. Draw the F-view in some convenient location with the principal edges of the view parallel to the revolved positions of the two axes for the view, that is, with the base edge parallel to $O''S$ and the edge that is perpendicular to the base parallel to $O''T$.

2. Draw the P-view at another location with the edges of the view parallel to the revolved position of the axes for the P-view.

3. From the corners of the F-view draw projecting lines perpendicular to ST and parallel to OR.

4. From the corners of the P-view draw projecting lines perpendicular to RT and parallel to OS.

The intersection of these projecting lines with those from the F-view establishes the desired isometric projection.

When none of the angles formed by the coordinate axes of the diagram are equal, the resulting pictorial representation will be a trimetric projection. When two of the angles are equal a dimetric projection is obtained. In Fig. 11.27(d) all three of the angles were made equal because an isometric projection was deemed desirable (since engineers and draftsmen are accustomed to seeing isometric drawings and sketches in their daily work).

In preparing the trimetric projection shown in Fig. 11.28, a preliminary pictorial sketch was made to show the general appearance desired for the finished pictorial. Then the axes in the diagram were drawn very nearly parallel to the principal lines of the sketch. Otherwise, the procedure followed in constructing the pictorial representation was the same as has been explained for an isometric projection. To draw the elliptical projection of the circular arc, two points, numbered 1 and 2 in both orthographic views, were assumed along the arc. More points may be used when needed.

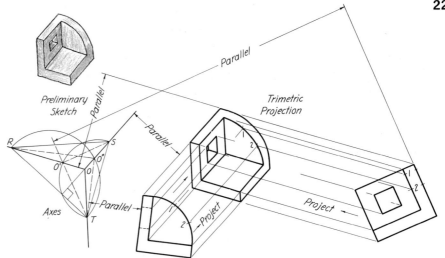

Fig. 11.28 To construct a trimetric projection.

B □ Oblique projection

11.16

Oblique Projection. In oblique projection, the view is produced by using parallel projectors that make some angle other than 90° with the plane of projection. Generally, one face is placed parallel to the picture plane and the projection lines are taken at 45°. This gives a view that is pictorial in appearance, as it shows the front and one or more additional faces of an object. In Fig. 11.29, the orthographic and oblique projections of a cube are shown. When the angle is 45°, as in this illustration, the representation is sometimes called *cavalier projection*. It is generally known, however, as an *oblique projection* or an *oblique drawing*.

11.17

Principle of Oblique Projection. The theory of oblique projection can be explained by imagining a vertical plane of projection in front of a cube parallel to one of its faces (Fig. 11.29). When the projectors make an angle of 45° in any direction with the picture plane, the length of any oblique projection $A'B'$ of the edge AB is equal to the true length of AB. Note that the projectors could be parallel to any element of a 45° cone having its base in the plane of projection. With projectors at this particular angle (45°), the face parallel to the plane is projected in its true size and shape and the edges perpendicular to the picture plane are projected in their true length. If the projectors make a greater angle, the oblique projection will be shorter, while if the angle is less, the projection will be longer.

11.18

Oblique Drawing. This form of drawing is based on three mutually perpendicular axes along which, or parallel to which, the necessary measurements are made for constructing the representation. Oblique drawing differs from isometric drawing principally in that two axes are always perpendicular to each other, while the third (receding axis) is at some convenient angle, such as 30°, 45°, or 60° with the horizontal (Fig. 11.31). It is somewhat

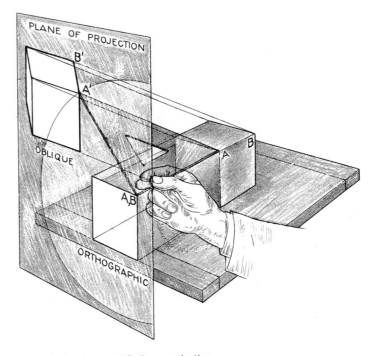

Fig. 11.29 Theory of Oblique projection.

more flexible and has the following advantages over isometric drawing: (1) circular or irregular outlines on the front face show in their true shape; (2) distortion can be reduced by foreshortening along the receding axis; and (3) a greater choice is permitted in the selection of the positions of the axes. A few of the various views that can be obtained by varying the inclination of the receding axis are illustrated in Fig. 11.30. Usually, the selection of the position is governed by the character of the object.

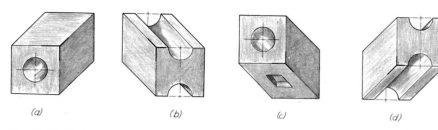

Fig. 11.30 Various positions of the receding axis.

and semicircle are shown parallel to the picture plane in order to avoid distortion and because, from the draftsman's standpoint, it is easier to draw a circle than to construct an ellipse.

In general, the procedure for constructing an oblique drawing is the same as for an isometric drawing.

11.20

Use of a Basic Plane. If the front face of an object is in one plane, it will appear in the oblique drawing exactly the same as in orthographic drawing. Note this fact in Fig. 11.31. But the front of many objects is composed of two or more parallel planes whose relationship must be carefully established. The convenient way to accomplish this is to use one of the planes as a basic (starting) plane and work from it in the direction of the receding axis, as illustrated in Fig. 11.32. Since the front surface A presents the contour shape, it should be selected as the basic plane and drawn first as the front of the oblique projection, as shown in (c). The center P of the circles for surface B may be located easily by measuring along the axis of the hole from O, in plane A, a distance X equal to the distance between the planes. The measurement must be made forward along the axis from O, because surface B is in front of A. The centers of the arcs on the surface C are located in a similar manner, except that the direction of making the measurements is from the basic plane toward the back.

When an object has one or more inclined surfaces with a curved outline, it may be drawn by constructing a right section and making offset measurements, as shown in Fig. 11.36.

11.19

To Make an Oblique Drawing. The procedure to be followed in constructing an oblique drawing of an adjustable guide is illustrated in Fig. 11.31. The three axes that establish the perpendicular edges in (b) are drawn through point O representing the front corner. OA and OB are perpendicular to each other and OC is at any desired angle (say, 30°) with the horizontal. After the width, height, and depth have been set off, the front face may be laid out in its true size and shape, as in (c), and the view can be completed by drawing lines parallel to the receding axis through the established corners. The circle

11.21

Rules for Placing an Object. Generally, the most irregular face, or the one containing the most circular outlines, should be placed parallel to the picture plane, in order to minimize distortion and simplify construction. By following this practice, all or most of the circles and circle arcs can be drawn with a compass, and the tedious construction that would be required to draw their elliptical representations in a receding plane is eliminated. In selecting the position of an object, two rules should be followed. The first is to place the face having the most irregular contour, or the most circular outlines, parallel to the picture plane. Note in Fig. 11.33 the advantage of following this rule.

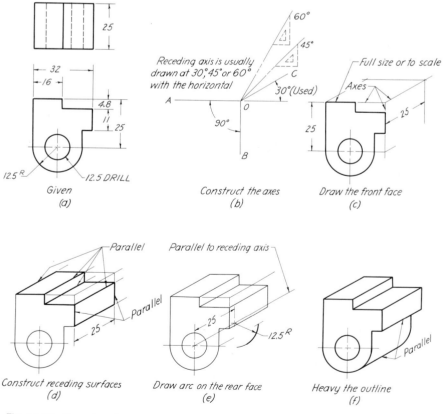

Fig. 11.31 Procedure for constructing an oblique drawing.

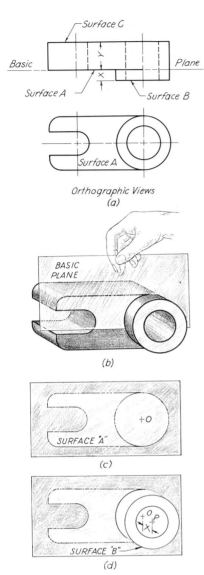

Surface C

Basic · · · Plane

Surface A — · · Surface B

Surface A

Orthographic Views
(a)

BASIC PLANE

(b)

SURFACE "A"   +O

(c)

SURFACE "B"

(d)

SURFACE C   Y — X
                Y
                Y
SURFACE A   SURFACE B

(e)

Fig. 11.32 Basic plane theory of construction.

When the longest face of an object is used as the front face, the pictorial view will be distorted to a lesser degree and, therefore, will have a more realistic and pleasing appearance. Hence, the second rule is to place the longest face parallel to the picture plane. Compare the views shown in Fig. 11.34 and note the greater distortion in (*a*) over (*b*).

If these two rules clash, the first should

govern. It is more desirable to have the irregular face show its true shape than it is to lessen the distortion in the direction of the receding axis.

11.22
Angles, Circles, and Circle Arcs in Oblique. As previously stated, angles, circles, and irregular outlines on surfaces parallel to the plane of projection show in true size and shape. When located on receding faces, the construction methods used in isometric drawing may usually be applied. Figure 11.35 shows the method of drawing the elliptical representation of a circle on an oblique face. Note that the method is identical with that used for constructing isometric circles, except for the slight change in the position of the axes.

Circle arcs and circles on inclined planes must be plotted by using the offset or coordinate method (Fig. 11.36).

11.23
Reduction of Measurements in the Direction of the Receding Axis. An oblique drawing often presents a distorted appearance that is unnatural and disagreeable to the eye. In some cases the view constructed by this scheme is so misleading in appearance that it is unsatisfactory

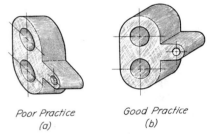

Poor Practice
(a)

Good Practice
(b)

Fig. 11.33 Irregular contour parallel to picture plane.

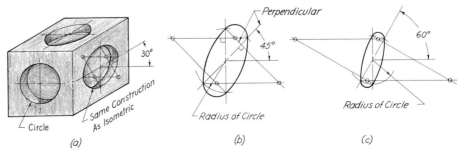

(a)   (b)

Fig. 11.34 Long axis parallel to picture plane.

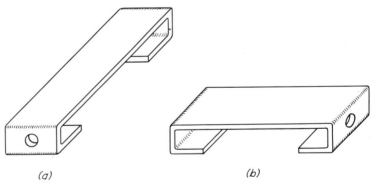

Circle   Same Construction As Isometric   30°

(a)

Perpendicular   45°

Radius of Circle

(b)

60°

Radius of Circle

(c)

Fig. 11.35 Oblique circles.

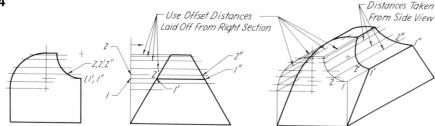

Fig. 11.36 Curved outlines on an inclined plane.

for any practical purpose. As a matter of interest, the effect of distortion is due to the fact that the receding lines are parallel and do not appear to converge as the eye is accustomed to anticipating (Fig. 11.37).

The appearance of excessive thickness can be overcome somewhat by reducing the length of the receding lines. For practical purposes, measurements are usually reduced one-half, but any scale of reduction may be arbitrarily adopted if the view obtained will be more realistic in appearance. When the receding lines are drawn one-half their actual length, the resulting pictorial view is called a *cabinet drawing*. Figure 11.38 shows an oblique drawing (*a*) and a cabinet drawing (*c*) of the same object, for the purpose of comparison.

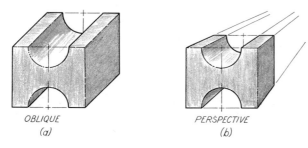

Fig. 11.37 Comparison of oblique and perspective.

11.24
Oblique Sectional Views. Oblique sectional views are drawn to show the interior construction of objects. The construction procedure is the same as for an isometric sectional view, except that oblique planes are used for cutting the object. An oblique half section is illustrated in Fig. 11.39.

11.25
Pictorial Dimensioning. The dimensioning of isometric and other forms of pictorial working drawings is done in accordance with the following rules:

1. Draw extension and dimension lines (except those dimension lines applying to cylindrical features) parallel to the pictorial axes in the plane of the surface to which they apply (Fig. 11.40).

2. If possible, apply dimensions to visible surfaces.

3. Place dimensions on the object, if, by so doing, better appearance, added clearness, and easy readings result.

4. Notes may be lettered either in pictorial or as on ordinary drawings. When lettered as on ordinary drawings the difficulties encountered in forming pictorial letters are avoided (Fig. 11.40).

5. Make the figures of a dimension appear to be lying in the plane of the surface whose dimension it indicates, by using vertical figures drawn in pictorial (Fig. 11.40). (*Note:* Guide lines and slope lines are drawn parallel to the pictorial axes.)

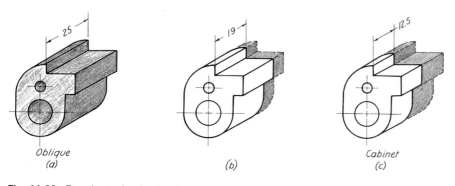

Fig. 11.38 Foreshortening in the direction of the receding axis.

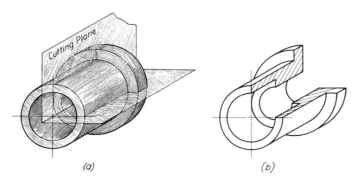

Fig. 11.39 Oblique half section.

11.26
Conventional Treatment of Pictorial Drawings. When it is desirable for an isometric

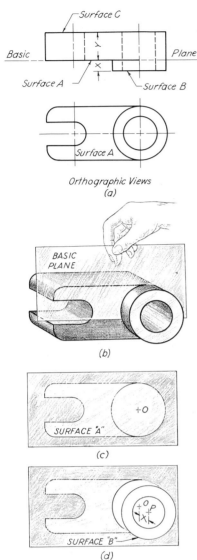

Orthographic Views
(a)

(b)

(c)

(d)

(e)

Fig. 11.32 Basic plane theory of construction.

When the longest face of an object is used as the front face, the pictorial view will be distorted to a lesser degree and, therefore, will have a more realistic and pleasing appearance. Hence, the second rule is to place the longest face parallel to the picture plane. Compare the views shown in Fig. 11.34 and note the greater distortion in (*a*) over (*b*).

If these two rules clash, the first should

govern. It is more desirable to have the irregular face show its true shape than it is to lessen the distortion in the direction of the receding axis.

11.22
Angles, Circles, and Circle Arcs in Oblique. As previously stated, angles, circles, and irregular outlines on surfaces parallel to the plane of projection show in true size and shape. When located on receding faces, the construction methods used in isometric drawing may usually be applied. Figure 11.35 shows the method of drawing the elliptical representation of a circle on an oblique face. Note that the method is identical with that used for constructing isometric circles, except for the slight change in the position of the axes.

Circle arcs and circles on inclined planes must be plotted by using the offset or coordinate method (Fig. 11.36).

11.23
Reduction of Measurements in the Direction of the Receding Axis. An oblique drawing often presents a distorted appearance that is unnatural and disagreeable to the eye. In some cases the view constructed by this scheme is so misleading in appearance that it is unsatisfactory

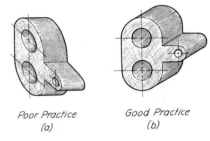

Poor Practice
(a)

Good Practice
(b)

Fig. 11.33 Irregular contour parallel to picture plane.

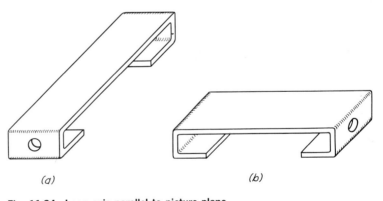

(a)

(b)

Fig. 11.34 Long axis parallel to picture plane.

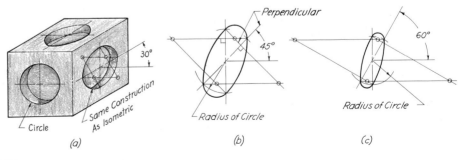

(a)

(b)

(c)

Fig. 11.35 Oblique circles.

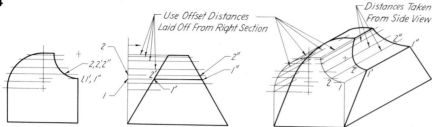

Fig. 11.36 Curved outlines on an inclined plane.

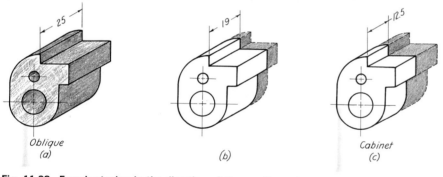

OBLIQUE
(a)

PERSPECTIVE
(b)

Fig. 11.37 Comparison of oblique and perspective.

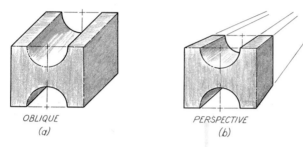

Oblique
(a)

(b)

Cabinet
(c)

Fig. 11.38 Foreshortening in the direction of the receding axis.

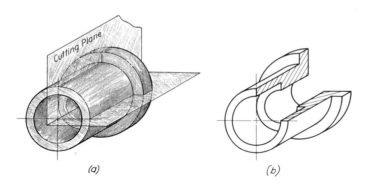

(a)

(b)

Fig. 11.39 Oblique half section.

for any practical purpose. As a matter of interest, the effect of distortion is due to the fact that the receding lines are parallel and do not appear to converge as the eye is accustomed to anticipating (Fig. 11.37).

The appearance of excessive thickness can be overcome somewhat by reducing the length of the receding lines. For practical purposes, measurements are usually reduced one-half, but any scale of reduction may be arbitrarily adopted if the view obtained will be more realistic in appearance. When the receding lines are drawn one-half their actual length, the resulting pictorial view is called a *cabinet drawing*. Figure 11.38 shows an oblique drawing (*a*) and a cabinet drawing (*c*) of the same object, for the purpose of comparison.

11.24
Oblique Sectional Views. Oblique sectional views are drawn to show the interior construction of objects. The construction procedure is the same as for an isometric sectional view, except that oblique planes are used for cutting the object. An oblique half section is illustrated in Fig. 11.39.

11.25
Pictorial Dimensioning. The dimensioning of isometric and other forms of pictorial working drawings is done in accordance with the following rules:

1. Draw extension and dimension lines (except those dimension lines applying to cylindrical features) parallel to the pictorial axes in the plane of the surface to which they apply (Fig. 11.40).

2. If possible, apply dimensions to visible surfaces.

3. Place dimensions on the object, if, by so doing, better appearance, added clearness, and easy readings result.

4. Notes may be lettered either in pictorial or as on ordinary drawings. When lettered as on ordinary drawings the difficulties encountered in forming pictorial letters are avoided (Fig. 11.40).

5. Make the figures of a dimension appear to be lying in the plane of the surface whose dimension it indicates, by using vertical figures drawn in pictorial (Fig. 11.40). (*Note:* Guide lines and slope lines are drawn parallel to the pictorial axes.)

11.26
Conventional Treatment of Pictorial Drawings. When it is desirable for an isometric

or an oblique drawing of a casting to present a somewhat more or less realistic appearance, it becomes necessary to represent the fillets and rounds on the unfinished surfaces. One method commonly used by draftsmen is shown in Figure 11.41(*a*). On the drawing in (*b*) all of the edges have been treated as if they were sharp. The conventional treatment for threads in pictorial is illustrated in (*b*) and (*c*).

C □ Perspective projection

11.27

Perspective. In perspective projection an object is shown much as the human eye or camera would see it at a particular point. Actually, it is a geometric method by which a picture can be projected on a picture plane in much the same way as in photography. Perspective drawing differs from the methods previously discussed in that the projectors or visual rays intersect at a common point known as the *station point* (Fig. 11.44).

Since the perspective shows an object as it appears instead of showing its true shape and size, it is rarely used by engineers. It is more extensively employed by architects to show the appearance of proposed buildings, by artist–draftsmen for production illustrations, and by illustrators in preparing advertising drawings.

Figure 11.1 shows a type of production illustration that has been widely used in assembly departments as an aid to those persons who find it difficult to read an orthographic assembly. This form of presentation, which may show a mechanism both exploded and assembled, has made it possible for industrial concerns to employ semitrained personnel. Figure 11.56 shows a type of industrial drawing made in perspective that has proved useful in aircraft plants. Because of the growing importance of this type of drawing, and also because engineers frequently will find perspective desirable for other purposes, its elementary principles should be discussed logically in this text. Other books on the subject, some of which are listed in the bibliography, should be studied by architectural students and those interested in a more thorough discussion of the various methods.

The fundamental concepts of perspective can be explained best if the reader will imagine himself looking through a picture

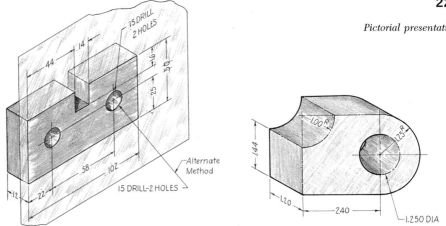

Fig. 11.40 Extension and dimension lines in isometric (left); numerals, fractions, and notes in oblique (right).

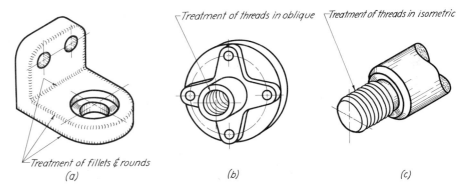

Fig. 11.41 Conventional treatment of fillets, rounds, and threads in pictorial.

plane at a formal garden with a small pool flanked by lampposts, as shown in Fig. 11.42. The point of observation, at which the rays from the eye to the objects in the scene meet, is called the *station point*, and the plane on which the view is formed by the piercing points of the visual rays is known as the *picture plane* (*PP*). The piercing points reproduce the scene, the size which depends on the location of the picture plane.

It should be noted that objects of the same height intercept a greater distance on the picture plane when close to it than when farther away. For example, rays from the lamppost at 2 intercept a distance 1–2 on the picture plane, while the rays from the pole at 4, which actually is the same height, intercept the lesser distance 3–4. From this fact it should be observed that the farther away an object is, the smaller it will appear, until a point is reached at which there will be no distance intercepted at all. This happens at the horizon.

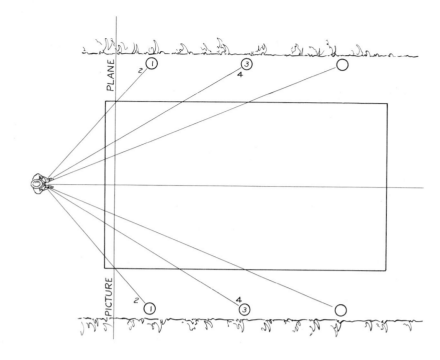

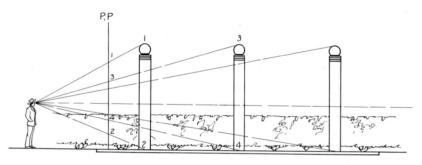

Fig. 11.42   Picture plane.

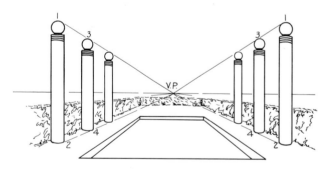

Fig. 11.43   The picture (perspective).

Figure 11.43 shows the scene observed by the man in Fig. 11.42 as it would be formed on the picture plane. The posts farther from the picture plane diminish in height, as each one has a height on the picture plane equal to the distance it intercepts, (Fig. 11.42). The lines of the pool

and hedge converge to the center of vision or vanishing point, which is located directly in front of the observer on the horizon.

11.28
Perspective Nomenclature.   Figure 11.44 illustrates pictorially the accepted nomenclature of perspective drawing. The *horizon line* is the line of intersection of the horizontal plane through the observation point (eye of the observer) and the picture plane. The horizontal plane is known as the *plane of the horizon*. The *ground line* is the line of intersection of the ground plane and the picture plane. The *CV point* is the center of vision of the observer. It is located directly in front of the eye in the plane of the horizon on the horizon line.

11.29
Location of Picture Plane.   The picture plane is usually placed between the object and the *SP* (station point). In parallel perspective (Sec. 11.35) it may be passed through a face of the object in order to show the true size and shape of the face.

11.30
Location of the Station Point.   Care must be exercised in selecting the location for the station point, for its position has much to do with the appearance of the finished perspective drawing. A poor choice of position may result in a distorted perspective that will be decidedly displeasing to the eye.

In general, the station point should be offset slightly to one side and should be located above or below the exact center of the object. However, it must be remembered that the center of vision must be near the center of interest for the viewer.

One should always think of the station point as the viewing point, and its location should be where the object can be viewed to the best advantage. It is desirable that it be at a distance from the picture plane equal to at least twice the maximum dimension (width, height, or depth) of the object, for at such a distance, or greater, the entire object can be viewed naturally, as a whole, without turning the head.

A wide angle of view is to be avoided in the interest of good picturization. It has been determined that best results are obtained when the visual rays from the station point (*SP*) to the object are kept within a cone having an angle of not more than 30° between diametrically opposing elements (see Fig. 11.45).

In locating an object in relation to the picture plane, it is advisable to place it so that both of the side faces do not make the same angle with the picture plane and thus will not be equally visible. It is common practice to choose angles of 30° and 60° for rectangular objects.

11.31

Position of the Object in Relation to the Horizon. When making a perspective of a tall object, such as a building, the horizon usually is assumed to be at a height above the ground plane equal to the standing height of a man's eye, normally about 5 ft 6 in.

A small object may be placed either above or below the horizon (eye level), depending on the view desired. If an object is above the horizon, it will be seen looking up from below, as shown in Fig. 11.46. Should the object be below the horizon line, it will be seen from above.

11.32

Lines. The following facts should be recognized concerning the perspective of lines:

1. Parallel horizontal lines vanish at a single *VP* (vanishing point). Usually the *VP* is at the point where a line parallel to the system through the *SP* pierces the *PP* (picture plane).

2. A system of horizontal lines has its *VP* on the horizon.

3. Vertical lines, since they pierce the picture plane at infinity, will appear vertical in perspective.

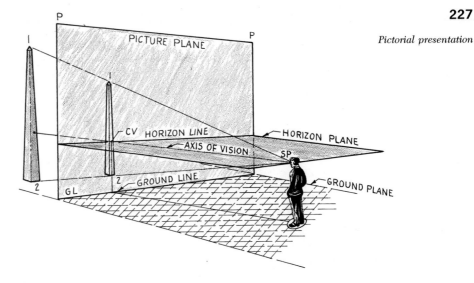

Fig. 11.44 Nomenclature.

4. When a line lies in the picture plane, it will show its true length because it will be its own perspective.

5. When a line lies behind the picture plane, its perspective will be shorter than the line.

11.33

Perspective by Multiview Projection. Adhering to the theory that a perspective drawing is formed on a picture plane by visual rays from the eye to the object, as illustrated in Fig. 11.42, a perspective can be drawn by using multiview projection (see Fig. 11.47). The multiview method may be the easiest for a student to understand but it is not often used by an expe-

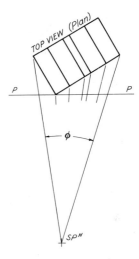

Fig. 11.45 Angle of vision.

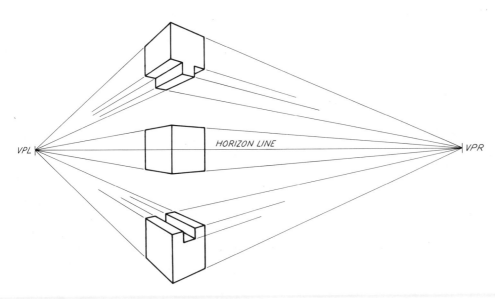

Fig. 11.46 Objects on, above, or below the horizon.

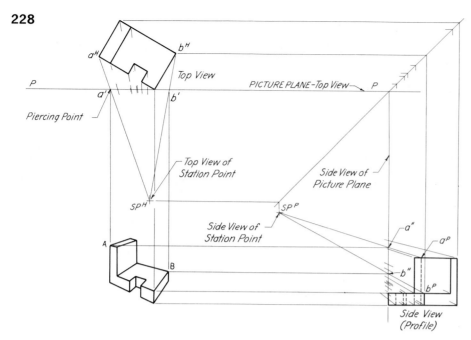

Fig. 11.47 Perspective drawing—orthographic method.

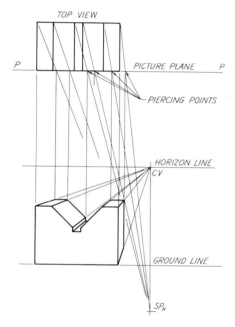

Fig. 11.48 Parallel perspective.

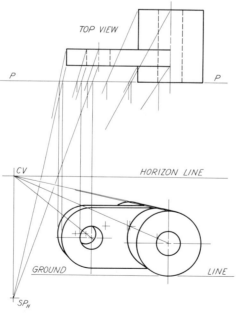

Fig. 11.49 Circles in parallel perspective.

After the preliminary layout has been completed point A may be located in the perspective by the following procedure:

Step I. Draw the top view ($SP^H a^H$) and the side view ($SP^P a^P$) of the visual ray from the eye to point A.

Step II. From a' (the top view of the piercing point of the ray) draw a projection line downward.

Step III. From a'' (the side view of the piercing point) draw a horizontal projection line to intersect the one drawn from a'.

Point A of the perspective is at this intersection.

Point B and the other points that are needed for the perspective representation are found in the same manner as point A.

11.34

Types of Perspective. In general, there are two types of perspective: *parallel perspective* and *angular perspective*. In parallel perspective, one of the principal faces is parallel to the picture plane and is its own perspective. All vertical lines are vertical, and the receding horizontal lines converge to a single vanishing point. In angular perspective, the object is placed so that the principal faces are at an angle with the picture plane. The horizontal lines converge at two vanishing points.

11.35

Parallel Perspective. Figure 11.48 shows the parallel perspective of a rectangular block. The PP line is the top view of the picture plane, SP_H is the top view of the station point, and CV is the center of vision. The receding horizontal lines vanish at CV. The front face, since it lies in the picture plane, is its own perspective and shows in its true size. The lines representing the edges back of the picture plane are found by projecting downward from the points at which the visual rays pierce the picture plane, as shown by the top views of the rays. Figure 11.49 shows a parallel perspective of a cylindrical machine part.

11.36

Angular Perspective. Figure 11.50 shows pictorially the graphical method for the preparation of a two-point perspective drawing of a cube. To visualize the true layout on the surface of a sheet of drawing paper, it is necessary to revolve mentally the horizontal plane downward into the vertical or picture plane. On completion of

rienced person because considerably more line work is required if the scene or object to be represented is at all complicated. The top and side views are drawn in multiview projection and the picture plane (as an edge view) and the station point are shown in each case. SP^H and its related views of the rays are in the top view, while SP^P and its projections of the rays belong to the side view. The front view of the picture plane is in the plane of the paper.

Sec. 11.36, it is suggested that the reader turn back and endeavor to associate the development of the perspective in Fig. 11.51 with the pictorial presentation in Fig. 11.50. For a full understanding of the construction in Fig. 11.51, it is necessary to differentiate between the lines that belong to the horizontal plane and those that are on the vertical or picture plane. In addition, it must be fully realized that there is a top view for the perspective that is a line and that in this line view lie the points that must be projected downward to the perspective representation (front view).

Figure 11.51 shows an angular perspective of a block. The block has been placed so that one vertical edge lies in the picture plane. The other vertical edges are parallel to the plane, while all of the horizontal lines are inclined to it so that they vanish at the two vanishing points, *VPL* and *VPR*, respectively.

In constructing the perspective shown in this illustration, an orthographic top view was drawn in such a position that the visible vertical faces made angles of 30° and 60° with the picture plane. Next, the location of the observer was assumed and the horizon line was established. The vanishing points *VPL* and *VPR* were found by drawing a 30° line and a 60° line through the *SP*. Since these lines are parallel to the two systems of receding horizontal lines, each will establish a required vanishing point at its intersection with the picture plane. The vertical line located in the picture plane, which is its own perspective, was selected as a measuring line on which to project vertical measurements from the orthographic front view. The lines shown from these division points along this line to the vanishing points (*VPL–VPR*) established the direction of the receding horizontal edge lines in the perspective. The positions of the back edges were determined by projecting downward from the points at which the projectors from the station point (*SP*) to the corners of the object pierced the picture plane, as shown by the top view of the object and projectors.

11.37

Use of Measuring Lines. Whenever the vertical front edge of an object lies in the picture plane, it can be laid off full length in the perspective, because theoretically, it will be in true length in the picture formed on the plane by the visual rays (see Fig. 11.51). Should the near vertical edge lie

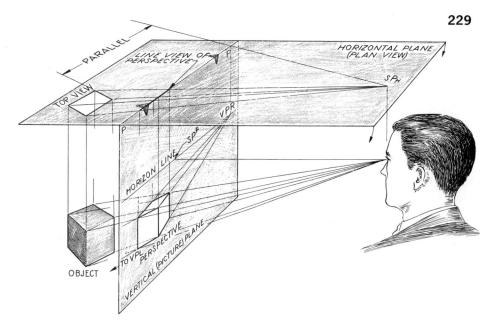

Fig. 11.50 Angular perspective.

behind the picture plane, as is the case with the edge line *AB* in Fig. 11.52, the use of a measuring line becomes desirable. The measuring line *a'b'* is the vertical edge *AB* moved forward to the picture plane, where it will appear in its true height. Some prefer to think of the vertical side as being extended to the picture

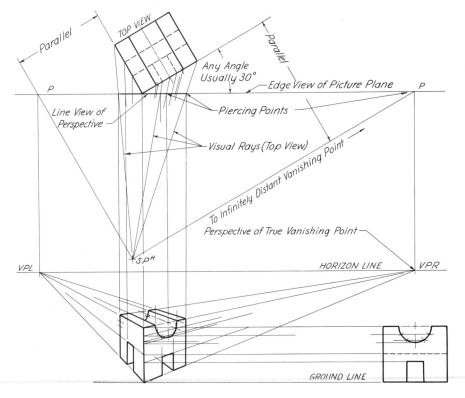

Fig. 11.51 Angular perspective.

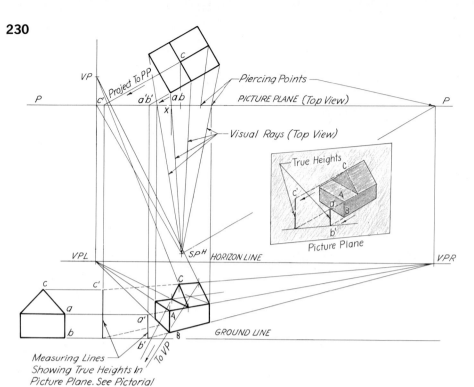

Fig. 11.52 Use of measuring lines.

the picture plane (point X) is projected downward to the front-view picturization. Points A and B must fall on the vanishing lines from a' and b' to VPR respectively.

A measuring line may be used to establish the "picture height" of any feature of an object. For example, in Fig. 11.52, the vertical measuring line through c' was used as needed to locate point C and the top line of the object in the perspective.

11.38

Vanishing Points for Inclined Lines. In general, the perspective of any inclined straight line can be found by locating the end points of the line perspectively. By this method, an end point may be located by drawing the perspective representations of any two horizontal lines intersecting at the end point. Where several parallel inclined lines vanish to the same vanishing point, it may prove to be worthwhile to locate the VP for the system in order to conserve time and achieve a higher degree of accuracy.

Just as in the case of other lines, the vanishing point for any inclined line may be established by finding the piercing point in the picture plane (PP) of a line through SP parallel to the given inclined line.

In Fig. 11.53 the inclined lines are AB and CD. If one is to understand the construction shown for locating vanishing point VPI, he must recognize that vertical planes vanish in vertical lines and that a line in a vertical plane will vanish at a point on the vanishing line of the plane. The preceding statements being true, the vanishing point VPI for line CD must lie at some point on a vertical line through VPR, since CD and CE lie in the same vertical plane. VPI is the point at which a line drawn parallel (in space) to CD through SP pierces the picture plane.

On the drawing, the distance D that VPI is above VPR may be found easily through the construction of a right triangle with SP^HQ as a base. Angle β is the slope angle for line CD (see side view). The needed distance D is the line QR, or the short leg of the triangle.

11.39

Use of Measuring Points. Frequently, it is desirable to use measuring points in preparing a perspective drawing, because their use permits the laying off of a series of direct measurements that can be transferred quickly and accurately to the perspective. Measuring points are, in real-

plane so that the true height of the side is revealed. The length and position of AB is established in the perspective picture by first drawing vanishing lines from a' and b' to VPR; then the top view of the edge in

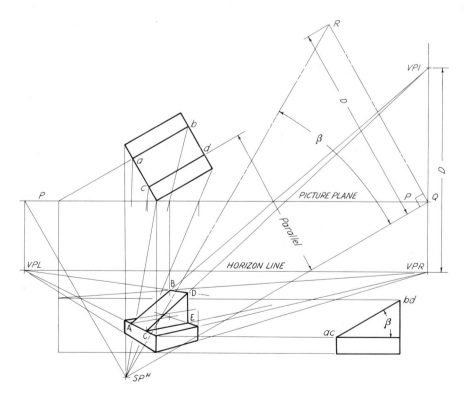

Fig. 11.53 Vanishing point for inclined lines.

ity, special vanishing points that are used to establish distances in perspective along perspective lines.

To understand the measuring point method, one must realize that, in theory, a vertical face is rotated into the picture plane so that direct measurements may be made along the horizontal ground line to the same scale as the top view. In Fig. 11.54 the vertical face containing line AB was revolved into the picture plane to position ab_r. The vanishing point of the line bb_r is MPR. The position of MPR was established by first drawing a line from SP parallel to bb_r to a piercing point in the picture plane and then projecting to the horizon line.

Another method for determining the location of measuring points is included in this same illustration. By this second method, the location of a measuring point can be readily found by swinging an arc with the plan view of the vanishing point as a center and using a radius equal to the distance from this center to the SP. For example, the location of MPR on the horizon was determined by swinging an arc from Y using the distance from Y to SP as a radius. The point of intersection of this arc and picture plane (in plan) projected to the horizon established the position of MPR for use in preparing the perspective drawing shown. MPL was similarly located.

The scaled lengths (D_1, D_2, and D_3) laid off along ab_r, when vanished to MPR, established the desired lengths along the line AB, which represents the perspective of the lower edge of the object. Height distances were laid off along the measuring line through the vertical edge lying in the picture plane at a. With width, height, and depth distances established, the perspective drawing was completed by following the procedure given in Sec. 11.36.

11.40

Circles in Perspective. If a circle is on a surface that is inclined to the picture plane (PP), its perspective will resemble an ellipse. It is the usual practice to construct the representation within an enclosing square by finding selected points along the curve in the perspective, as shown in Fig. 11.55(a). Any points might be used, but it is recommended that points be located on 30° and 60° lines.

In (a), the perspective representation of the circle was found by using visual rays and parallel horizontal lines in combination. In starting the construction, the po-

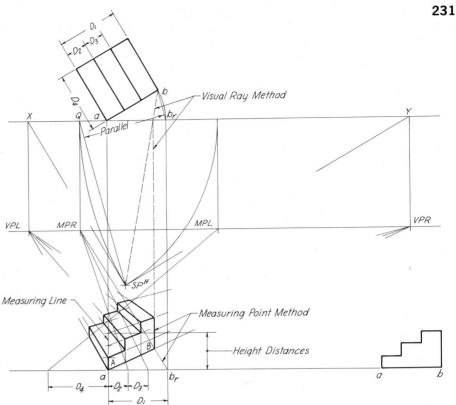

Fig. 11.54 Use of measuring points.

sitions of several selected points, located on the circumference and lying on horizontal lines in the plane of the circle, were established in both views. After these lines had been drawn in the perspective in the usual manner, the locations of the points along them were determined through the use of visual rays, as shown. Specifically,

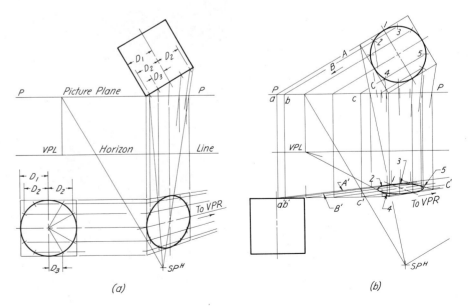

(a)   (b)

Fig. 11.55 Circles in perspective.

the position of a point in the perspective was found by projecting downward from the piercing point of the ray from the point and the picture plane (see line view) to the perspective view of the line on which the point must lie. In (*b*), the same method was applied to construct the perspective view of a circle in a horizontal plane. It should be noted in this case that the horizontal lines, as established in the top view, were extended to cut the picture plane so that the true height of these lines at the plane could be used in the perspective view for locating the end points of the perspective representations (at the left).

D □ Industrial illustrations

11.41

Technical Illustration Drawings. The design and manufacturing procedures of present-day mass production require various types of pictorial illustrations to communicate ideas and concepts to large numbers of persons who are all working toward a common objective. These illustration drawings are of value at all stages of a project from the design phase, where they may be only pencil-shaded freehand sketches, through all of the stages of production that we may consider to include not only the assembly but final installa-

tion of the systems as well. They are used in operation and maintenance manuals to make complex and difficult tasks understandable to those persons who may be unable to interpret conventional drawings (Fig. 11.56). Pictorial illustrations range from simple types of line drawings (Figs. 11.57 and 11.58) that have already been discussed to artists' renderings that have the realism of photographs. An artist's rendering, such as the one shown in Fig. 11.60, is usually prepared to reinforce an oral or written report that is to be presented before a decision-making group that consists to some extent of persons who would otherwise be unable to understand construction details. Drawings of this type are depended on to sell a project (Fig. 11.61).

Illustration drawings are used for many purposes in every field of engineering technology (Fig. 11.59). They appear in advertising literature, operation and service manuals, patent applications (Fig. 11.64), and textbooks (Fig. 11.71). Illustration drawings may be working drawings, assembly drawings, piping and wiring diagrams, and architectural and engineering renderings that are almost true-to-life. Typical examples of pictorial drawings that were prepared to facilitate assembly are shown in Figs. 11.1 and 16.11. Figure 22.1 shows an electronic diagram that appeared in a service manual.

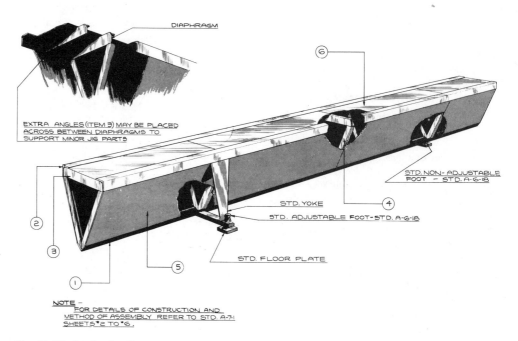

Fig. 11.56 Production illustration prepared in perspective. (*Courtesy Craftint Mfg. Co.*)

**EXPERIMENTAL
CATALYTIC CONVERTER**

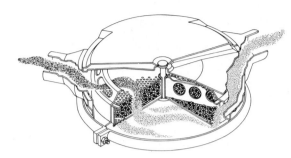

Fig. 11.57 This pictorial representation of a catalytic converter appeared in a technical report prepared by General Motors. Pictorial drawings such as this one are often needed in final design reports. Some engineers and technologists think that the utilization of a complex combination of components along with a catalytic converter in the exhaust line will prove to be the most effective method of treating hydrocarbons and carbon monoxide. (*Courtesy General Motors Corporation*)

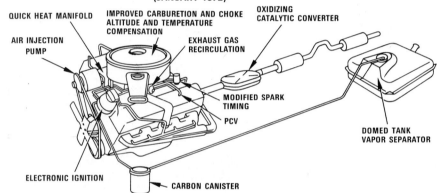

**ADVANCED EMISSION CONTROL SYSTEM
(JANUARY 1972)**

QUICK HEAT MANIFOLD

IMPROVED CARBURETION AND CHOKE ALTITUDE AND TEMPERATURE COMPENSATION

OXIDIZING CATALYTIC CONVERTER

AIR INJECTION PUMP

EXHAUST GAS RECIRCULATION

MODIFIED SPARK TIMING

PCV

DOMED TANK VAPOR SEPARATOR

ELECTRONIC IGNITION

CARBON CANISTER

Fig. 11.58 Items other than the catalytic converter that are a part of the total package of emission control are shown in the pictorial representation of the advanced emission control system under development at General Motors. This presentation was extracted from a technical report prepared by GM engineers. (*Courtesy General Motors Corporation*)

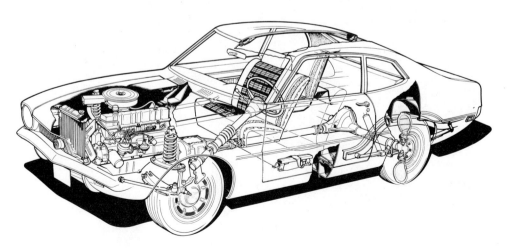

Fig. 11.59 Industrial illustration. (*Courtesy Ford Motor Company*)

234

Pictorial presentation

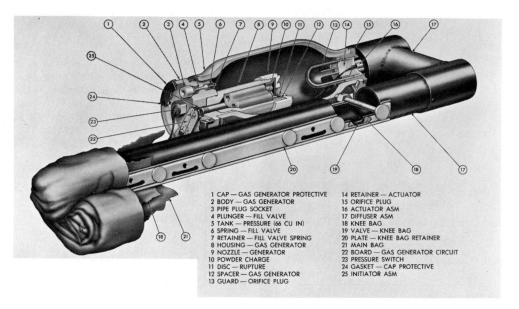

1 CAP — GAS GENERATOR PROTECTIVE	14 RETAINER — ACTUATOR
2 BODY — GAS GENERATOR	15 ORIFICE PLUG
3 PIPE PLUG SOCKET	16 ACTUATOR ASM
4 PLUNGER — FILL VALVE	17 DIFFUSER ASM
5 TANK — PRESSURE (66 CU IN)	18 KNEE BAG
6 SPRING — FILL VALVE	19 VALVE — KNEE BAG
7 RETAINER — FILL VALVE SPRING	20 PLATE — KNEE BAG RETAINER
8 HOUSING — GAS GENERATOR	21 MAIN BAG
9 NOZZLE — GENERATOR	22 BOARD — GAS GENERATOR CIRCUIT
10 POWDER CHARGE	23 PRESSURE SWITCH
11 DISC — RUPTURE	24 GASKET — CAP PROTECTIVE
12 SPACER — GAS GENERATOR	25 INITIATOR ASM
13 GUARD — ORIFICE PLUG	

Fig. 11.60 Components of the Allied Chemical Corporation air-bag modules to be used in the air-bag/seat-belt restraint system now under development by the Ford Motor Company. The system has a main bag (cushion) and a small knee bag. (*Courtesy Ford Motor Company*)

11.42
Design Illustrations. Design illustrations are prepared to clarify conventional engineering design and production drawings and written specifications. These are used for the communication of ideas and concepts concerning the details of complicated designs. Properly prepared, drawings of this type reveal the relationship of the components of a system so clearly that the principles of operation of a unit can be understood by almost everyone, even by persons who may be relatively unfamiliar with graphic methods. It is common practice to prepare a series of such drawings to clarify complex details of construction, to indicate the function of closely related parts, and reveal structural features. A pictorial of this type is shown in Fig. 11.62. It was used to supplement a technical paper presented at a recent Society of Automotive Engineers Congress held in Detroit.

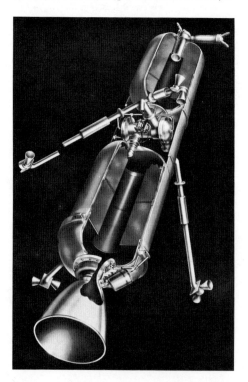

Fig. 11.61 Artist's concept of a rocket control system. (*Courtesy Aeronutronic Division, Philco-Ford*)

11.43
Shading Methods. A pictorial illustration can be improved and given a more realistic appearance by shading to produce the effect of surface texture. To this shades and shadows may be added to give additional realism. The use of pencil shading is most common. Ink shading (see Fig. 11.71) produces clean illustrations of high quality that are well suited for reproductions in texts and brochures. The techniques of surface shading by means of ink lines will be discussed in Sec. 11.44.

Some of the other more basic methods of representing surface textures under light and shade involve the use of Rossboard, Craftint paper, Zip-a-tone overlay film, and the airbrush.

Most of the shaded pictorial drawings in

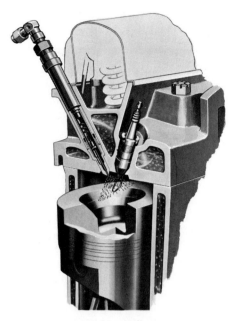

Fig. 11.62 New combustion process developed by the Ford Motor Company, cutaway view. (*Courtesy Ford Motor Company*)

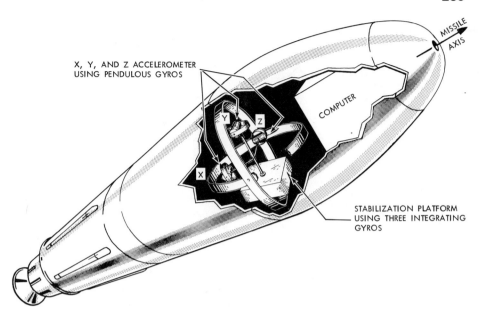

X, Y, AND Z ACCELEROMETER USING PENDULOUS GYROS

MISSILE AXIS

COMPUTER

STABILIZATION PLATFORM USING THREE INTEGRATING GYROS

Fig. 11.63 Pictorial representation of the principal elements of the inertial guidance system in a missile vehicle. (*Courtesy General Motors Corporation*)

this text were drawn on Rossboard. This popular drawing paper, with its rough plaster-type surface, is available in many textured patterns. Surface shading is done with a very soft pencil.

Craftint papers, single-tone or double-tone, have the pattern in the paper. Drawings are prepared on these papers in the usual manner and regular black waterproof drawing ink is used for the finished lines. Solid black areas are filled in with ink before the areas where shading is desired are brushed with a developer that brings out the surface pattern. These papers are available in many shading patterns. The drawing shown in Fig. 11.56 was prepared on Craftint paper.

Zip-a-tone clear cellulose overlay screens with printed shading patterns of dots and lines provide an easy method of surface shading suitable for high-quality printed reproductions (Fig. 11.63). The screen, backed with a clear adhesive, is applied as a sheet or partial sheet to the areas to be shaded. Unwanted portions are removed with a special cutting needle or razor blade, before the screen is rubbed down firmly to complete the bond between screen and paper.

A high degree of realism can be achieved using a small spray gun that in the language of the artist is known as an *airbrush*. This delicate instrument sprays a fine mist of diluted ink over the surface of the drawing to produce variations of

tone. A capable artist can produce a representation that has the realism of a photograph. The pictorial illustration in Fig. 11.61 is an excellent example of airbrush rendering by a commercial illustrator. The airbrush is sometimes used to retouch photographs to improve their appearance for reproduction.

11.44
Surface Shading by Means of Lines (Fig. 11.64). Line shading is a conventional method of representing, by ruled lines, the varying degrees of illumination on the surfaces of an object. It is a means of giving clearer definition to the shapes of objects and a finished appearance to certain types of drawings. In practice, line shading is used on Patent Office drawings, display drawings, and on some illustrations prepared for publications. It is never used on ordinary drawings, and for this reason few draftsmen ever gain the experience necessary to enable them to employ it effectively.

In shading surfaces, the bright areas are left white and the dark areas are represented by parallel shade lines (Figs. 11.65 and 11.66). Varying degrees of shade may be represented in one of the following ways:

1. By varying the weight of the lines while keeping the spacing uniform, as in Fig. 11.65.

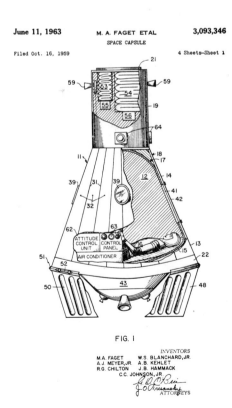

June 11, 1963   M. A. FAGET ETAL   3,093,346
                        SPACE CAPSULE
Filed Oct. 16, 1959   4 Sheets—Sheet 1

FIG. I

INVENTORS
M.A. FAGET   W.S. BLANCHARD, JR.
A.J. MEYER, JR.   A.B. KEHLET
R.G. CHILTON   J.B. HAMMACK
C.C. JOHNSON, JR.
ATTORNEYS

Fig. 11.64 Patent drawing with surface shading.

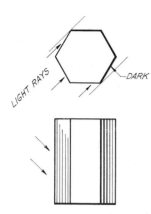

Fig. 11.65 Surface shading on a prism.

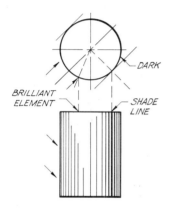

Fig. 11.66 Surface shading on a cylinder.

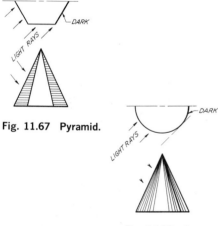

Fig. 11.67 Pyramid.

Fig. 11.68 Cone.

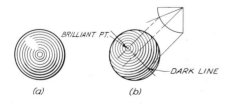

Fig. 11.69 Surface shading on a sphere.

2. By using uniform straight lines and varying the spacing.

3. By varying both the weight of the lines and the spacing, as in Fig. 11.66.

The rays of light are assumed to be parallel and coming from the left, over the shoulder of the draftsman (Fig. 11.65). In accordance with this, two of the visible faces of the hexagonal prism, shown in Fig. 11.65, would be illuminated, while the remaining visible inclined face would be dark. It should be noted that the general principle of shading is modified in the case of flat inclined surfaces, which, theoretically, would be uniformly lighted. Such surfaces are shaded in accordance with a conventional scheme, the governing rule of which may be stated as follows: *The portion of an illuminated inclined surface nearest the eye is the lightest, while the portion of a shaded inclined surface nearest the eye is the darkest.* In the application of this rule, the shading on an inclined illuminated face will increase in density as the face recedes, while on an unilluminated inclined surface it will decrease in density as the face recedes, as shown in Fig. 11.65.

A cylinder would be shaded as shown in Fig. 11.66. The lightest area on the surface is at the *brilliant line*, where light strikes it and is reflected directly to the eye; the darkest is along the *shade line*, where the light rays are tangent to the surface. The brilliant line passes through the point at which the bisector for the angle formed by a light ray to the center and the visual ray from the center line to the eye would pierce the external surface. In shading a cylinder, the density of the shading is increased in both directions, from the bright area along the brilliant line

to the contour element on the left and the shade line on the right. The density of the shading is slightly decreased on the shaded portion beyond the shade line. Since one would expect this area to be even darker, some draftsmen extend the dark portion to the contour element. On very small cylinders, the bright side (left side) usually is not shaded.

A pyramid may be shaded as shown in Fig. 11.67.

The surface of a cone will be the darkest at the element where the light rays are tangent to the surface (Fig. 11.68).

A sphere is shaded by drawing concentric circles having either the geometric center of the view or the "brilliant point" as a center (Fig. 11.69). The darkest portion of the surface is along the shade line, where the parallel rays of light are tangent to the sphere and the lightest portion is around the *brilliant point*. The *brilliant point* is where the bisector of the angle between a light ray to the center and the visual ray from the center to the eye pierces the external surface. The construction necessary to determine the dark line and the brilliant point is shown in Fig. 11.69(b).

Although good line shading requires much practice and some artistic sense, a skillful draftsman should not avoid shading the surfaces of an object simply because he never before has attempted to do so. After careful study, he should be able to produce fairly satisfactory results. Often the shading of a view makes it possible to eliminate another view that otherwise would be necessary.

Figures 11.70 and 11.71 show applications of line shading on technical illustrations.

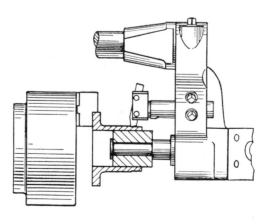

Fig. 11.70 Line-shaded drawing. (*From Doyle,* **Tool Engineering,** *Prentice-Hall, Inc.*)

Fig. 11.71 Line-shaded pictorial drawing. (*Courtesy Socony-Vacuum Oil Co.*)

Problems

The student will find that a preliminary sketch will facilitate the preparation of isometric and oblique drawings of the problems of this chapter. On such a sketch he may plan the procedure of construction. Since many technologists and designers frequently find it necessary to prepare pictorial sketches during discussions with untrained persons who cannot read orthographic views, it is recommended that some problems be sketched freehand on either plain or pictorial grid paper (see Fig. 6.34). Additional problems may be found at the end of Chapter 6.

Note: It is common practice under the metric system to give all dimensions, that are not critical, in full millimeters. It is recommended that the student attempt to do likewise in laying out and dimensioning a problem that has been dual-dimensioned.

1–9. (Figs. 11.72–11.80). Prepare instrumental isometric drawings or freehand sketches of the objects as assigned.

10–14. (Figs. 11.76–11.80). Prepare instrumental oblique drawings or freehand sketches of the objects as assigned.

15. (Fig. 11.81). Make an isometric drawing of the differential spider.

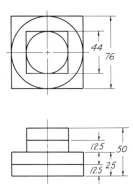

Fig. 11.72

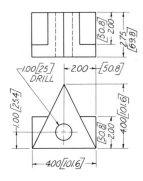

Fig. 11.73

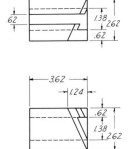

Fig. 11.74

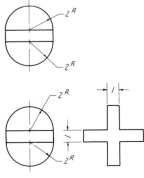

Fig. 11.75

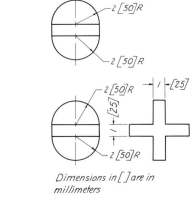

Dimensions in [ ] are in millimeters

Fig. 11.76

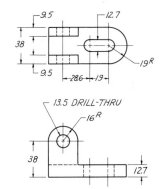

Fig. 11.77

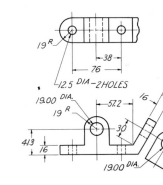

Fig. 11.78

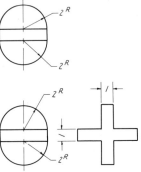

Fig. 11.79

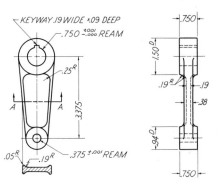

Fig. 11.80

238

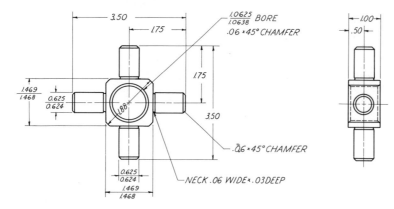

Fig. 11.81 Differential spider.

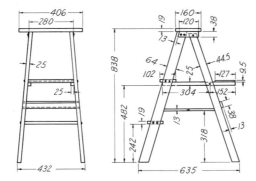

Fig. 11.82 Stepladder.

16. (Fig. 11.82). Make an isometric drawing of the stepladder. Select a suitable scale.

17. (Fig. 11.83). Make an isometric drawing of the sawhorse. Select a suitable scale.

18. (Fig. 11.84). Make an oblique drawing of the locomotive driver nut.

19. (Fig. 11.85). Make an oblique drawing of the adjustment cone.

20. (Fig. 11.86). Make an oblique drawing of the fork.

21. (Fig. 11.87). Make an oblique drawing of the feeder guide.

22. (Fig. 11.88). Make an isometric drawing of the hinge bracket.

23. (Fig. 11.89). Make an isometric drawing of the alignment bracket.

24. (Fig. 11.90). Make an isometric drawing of the stop block.

25. (Figs. 11.85–11.87). Make a parallel perspective drawing as assigned.

26. (Figs. 11.88–11.90). Make an angular perspective drawing as assigned.

27. (Fig. 11.91). Make a pictorial drawing (oblique, isometric, or perspective) of the slotted bell crank.

28. (Fig. 11.92). Make a pictorial drawing (oblique, isometric, or perspective) of the control guide.

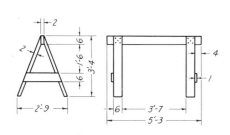

Fig. 11.83 Sawhorse.

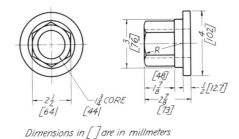

Dimensions in [ ] are in millimeters

Fig. 11.84 Locomotive driver nut.

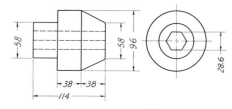

Dimensions are in millimeters

Fig. 11.85 Adjustment cone.

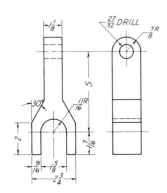

Fig. 11.86 Fork.

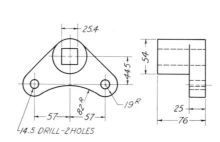

Fig. 11.87 Feeder guide.

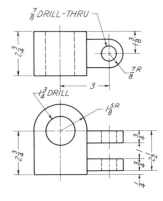

Fig. 11.88 Hinge bracket.

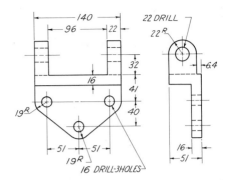

Fig. 11.89 Alignment bracket.

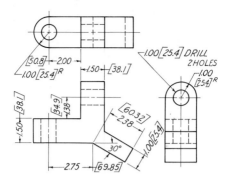

Fig. 11.90 Stop block.

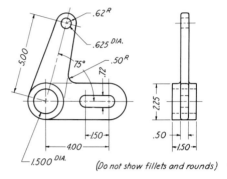

Fig. 11.91 Slotted bell crank.

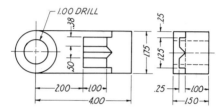

Fig. 11.92 Control guide.

Part  Design

"Four major U.S. corporations are now working together to investigate the feasibility of large scale satellite systems that would harness the abundant energy of the sun for use on earth. . . .

"A satellite solar power station of the size presently under consideration would be capable of producing 10,000 megawatts which is sufficient energy to meet the base load demands of a city such as New York in the year of 2000. A network of satellites such as these would have the potential to meet a significant portion of future energy needs. More importantly, it would be virtually pollution-free and would not consume irreplaceable energy sources on earth." (*Courtesy Grumman Corporation—one of four companies taking part in the investigation*)

12

The design process and graphics

A □ The Design process

12.1

Design. In the dictionary, design is defined as follows: (1) to form or conceive in the mind, (2) to contrive a plan, (3) to plan and fashion the form of a system (structure), and (4) to prepare the preliminary sketches and/or plans for a system that is to be produced. Engineering design is a decision-making process used for the development of engineering systems for which there is human need (Fig. 12.1). To design is to conceive, to innovate, to create. One may design an entirely new system or modify and rearrange existing things in a new way for improved usefulness or performance. Engineering design begins with the recognition of a social or economic need (Fig. 12.2). The need must first be translated into an acceptable idea by conceptualization and decision making. Then the idea must be tested against the physical laws of nature before one can be certain that it is workable. This requires that the designer have a full knowledge of the fundamental physical laws of the basic sciences, a working knowledge of the engineering sciences, and the ability to communicate ideas both graphically (Fig. 12.21) and orally. The designer should be well grounded in economics, have some knowledge of engineering materials and be familiar to a limited extent with manufacturing methods. In addition, some knowledge of both marketing and advertising will prove worthwhile, since usually what is produced must be distributed at a profit. Proficiency in designing can be attained only through total involvement,

244

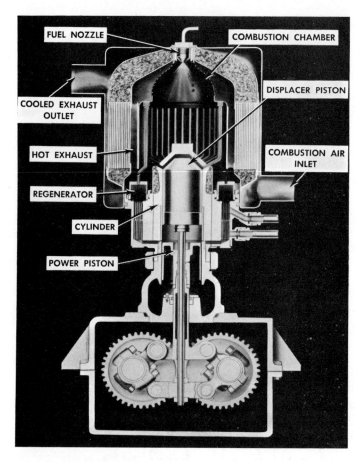

Fig. 12.1 Sterling engine.
The Sterling engine is an interesting engine that can be considered as an alternative for use in the automobile. It is considerably different from the Wankel. The cross-sectional view of a single-cylinder engine shows its complexity. The engine has rather low hydrocarbon and carbon monoxide emissions. It tends to be rather heavy and expensive to produce at this time. (*Courtesy General Motors Corporation*)

since it is only through practice that the designer acquires the art of continually providing new and novel ideas. In developing the design, the engineer or engineering technologist must apply his knowledge of engineering and material sciences while taking into account related human factors, reliability, visual appearance, manufacturing methods, and sale price. It may therefore be said that the ability to design is both an art and a science.

A creative person will almost never follow a set pattern of action in developing an idea. To do so would tend to structure his thinking and might limit the creation of possible solutions. The design process calls for unrestrained creative ingenuity and continual decision making by a free-wheeling mind. However, the total development of an idea, from recognition of a need to the final product, does appear to proceed loosely in stages that are recognized by authors and educators (Fig. 12.2).

Creative thinking usually begins when a design team, headed by a project leader, has been given an assignment to develop something that will satisfy a particular need. The need may have been suggested by a salesman, a housewife, or even an engineer from another company now using a product or a machine produced by the design team's own company. Most often the directive will come down from top management, as was the case with the development of the electric knife some years ago. Although it is always pleasant for an individual to think about the careers of famous inventors of the past and dream of the fame and fortune that might await the development and marketing of one of his ideas, the fact is that almost all new and improved products, from food choppers to aircraft engines, represent a team effort.

12.2
Design Synthesis. The process of combining constituent elements in a new or altered arrangement to achieve a unified entity is known as design synthesis. It is a process that involves reasoning from assumed propositions and known principles to arrive at comparatively new design solutions to recognized problems. The synthesis of systems for simple combinations as well as for complicated assemblies requires creative ability of the very highest order. The synthesis of both parts and systems usually requires successive trials to create new arrangements of old components and new features.

Proof of this point is the Land camera, considered by many people to be an entirely new product, although in reality the camera represents a combination of features and principles common to existing cameras to which Mr. Land added several new ideas of his own, including a new type of film and film pack that for the first time made it possible to develop film and pictures within the camera itself. Land's design activity no doubt started with a recognition of the need and desire that people had to take pictures that could be seen almost immediately. The early automobile is another example that bears out the fact that old established products have features that are used as a starting basis for a new product. For example, at the turn of the century the automobile looked

like a horseless carriage; the horse was taken to the barn and a motor was added in its place.

12.3

Design of Systems and Products. In general most design problems may be classified as being either a systems design or a product design, even though it may be quite difficult in many instances to recognize a problem as belonging entirely to one classification or the other. This is due to the fact that there often will be an overlap of identifying characteristics.

A systems design problem involves the interaction of numerous components that together form an operating unit. A complex system such as the climatic-control system for an automobile (heating and cooling), an automatic movie projector, a stadium, an office building, or even a parking lot represents a composition of several component systems that together form the complete composite system (Fig. 12.3). Some of the component systems of an office building, aside from the structural system itself, are: the electrical system, the plumbing system (including sewers), the heating and cooling systems, the elevator system, and the parking facilities system. All of these component systems, when combined, will meet the needs of a total system; but usually the total system design will involve more than just a technological approach. In the design of composite systems for use by the public, as in the design of a highway, an office building, or a stadium, the designer must adhere to the availability of funds. He must build into his design the safety features that are required by law and, at the same time, give due consideration to human factors, trends, and even present-day social problems. In many cases, a designer will find that there are political and special-interest groups that limit his freedom of decision making. Under such conditions he must be willing to compromise in the best interests of all concerned. This means that a successful designer must have experience in dealing with people. Social studies courses taken along with and in addition to the scientific courses offered in engineering colleges and technical schools will prove to be helpful.

Product design is concerned with the design of some appliances, systems components, and other similar unit-type items for which there appears to be a market. Such a product may be an electric lock, a power car-jack, a lawn sprinkler, a food grinder, a toy, a piece of specialized furniture, an electrical component, a valve, or other item that can be readily marketed as a commercial unit. Such products are designed to perform a specific function and to satisfy a particular need. Total product design includes not only the design of an item, but the testing, manufacture, and distribution of the product as well. Design does not end with the solution of a problem through creative thinking. The design phase for a product to be sold on the open market ends only when the item has received wide acceptance by the public. Of course, if the item being designed is not for the consumer market, the designer will have less reason to be concerned with details of marketing. This would be true in the case of components for systems and for items to be used in products produced in finished form by other companies.

After initially recognizing a need or desire for a new product, the creative individual enters the next step in the design procedure—the research and exploration phase. During this phase, the possible ways and the feasibility of fulfilling the need are investigated (Fig. 12.4). That is, the question is raised at this point as to whether the contemplated product can be marketed at a competitive price or at a price the public will be willing to pay. Except in the case of the public sector, where a system is being designed under the direction of a government unit (Fig. 12.5),

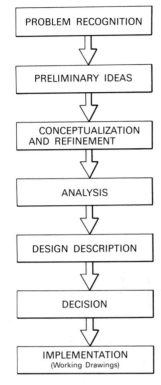

PROBLEM RECOGNITION

PRELIMINARY IDEAS

CONCEPTUALIZATION AND REFINEMENT

ANALYSIS

DESIGN DESCRIPTION

DECISION

IMPLEMENTATION (Working Drawings)

Fig. 12.2 The design process.

Fig. 12.3 Xerox 400 Telecopier transceiver.
This transceiver marks the beginning of a new era in facsimile transmission. It sends and receives, over regular telephones, a letter-size document or drawing in approximately 4 minutes. (*Courtesy Business Products Group—Xerox Corporation*)

Fig. 12.4 Minigap System.
Creative thinking produces new ways of life for man. The Minigap System links cars together into a caravan that is led by a specially built leader vehicle. Computers inside the cars take over control of brakes, accelerator, and steering to free the motorist from the task of driving as long as he remains hooked-up. Caravans can mix in the traffic flow with other vehicles. (*Courtesy Ford Motor Company*)

profit is the name of the game. A search of literature and patent records will often show that there have been previous developments with the possibility of infringements on the design and patents of other companies. The discovery that others already may have legal protection for one or more of the possible solutions will tend to control the paths that innovative thinking may take in seeking an acceptable design solution. Although freewheeling innovative thinking is the basic element of successful design, the experienced engineer never fails to consider known restraints when he is brainstorming possible solutions. In addition to avoiding patent infringements, the designer must be fully aware of possible changes in desires and needs and give

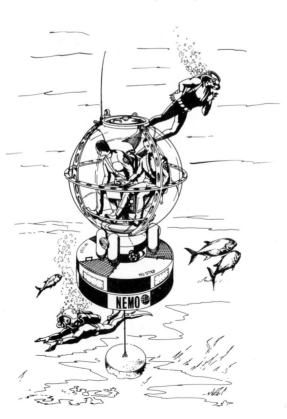

Fig. 12.5 NEMO on the ocean floor.
NEMO is a 168-cm-diameter sphere of acrylic plastic. The sphere is constructed from 12 identical curved pentagons. The capsule, with 63.5-mm walls, is bonded with acrylic adhesive. One of the first uses of NEMO will be as a diver control center at points of Seabee underwater construction sites. In operation, the manned observatory is lowered into the sea from a Navy support vessel with the crewmen flooding the ballast tanks for the descent. NEMO operates independently, controlled by an array of push buttons (41) linked to a solid-state circuitry control system. NEMO has been designed for a normal stay underwater of 8 hours. Under emergency conditions it may remain submerged for as long as 24 hours. (*Official Navy photograph. Courtesy* The Military Engineer Magazine)

full consideration to the production processes. Also not to be overlooked in the decision-making stage are the visual design (styling aspects) and the price range within which the product must be sold.

The ultimate success of a design is judged on the basis of its acceptability in the marketplace and on how well it satisfies the needs of a particular culture (Fig. 12.9). In these latter decades of the twentieth century, we have added to these judgments a requirement that what has been created must not damage the physical environment around us, an environment which has been rapidly deteriorating because of some past technological advances. New requirements are being prepared and approved in rapid order that set standards to reduce if not eliminate pollution. These restrictions were almost unknown to the creative designers of the first half of this century. Today, it has become almost mandatory for a designer selecting a material or an energy source to ask himself if the material can be recycled and whether any residue discharged into the atmosphere is damaging to life.

12.4
Innovative Design—Individuals and Groups.

In the past there have been a number of creative individuals, working almost alone, who have created products that have advanced our culture, products for which people still have a great desire. Examples of such products are the printing press, the steam engine, the gasoline engine, the automobile, the telephone, the phonograph, the motion picture camera and projector, the radio, the television, and many more. These individuals were keen observers of their culture and they possessed inquisitive attitudes that led to experimentation. Some called them dreamers. If so, their eyes were wide open, they worked long hours, and they persisted in making a new try after each failure. Their fame and their fortune, although rightly due them, resulted from the fact that each brought forth a new product or system at a time when it would be readily accepted and when it could be more or less mass-produced. These people possessed a good sense of timing and they were not afraid of failure or ridicule. They belonged in the age in which they lived.

At present, it is the practice of large industrial organizations to use group procedures in order to stimulate the imagina-

tions of the individual members of the group and thereby benefit from their combined thinking about a specific problem (Fig. 12.6). Within the group, the innovative idea of one individual stimulates another individual to present an alternative suggestion. Each idea forms the basis for still other ideas until a great number have been listed that hopefully will lead to a workable solution to the problem at hand. This group attack on a problem produces a long list of ideas that would be difficult for a single individual to assemble in so short a time (Fig. 12.7). A single person attempting to solve the problem alone would find it necessary to conduct extensive searches of current literature and patent records and to hold numerous discussions with knowledgeable experts in seeking the guidance needed for making necessary decisions.

The most widely used of the group-related procedures is known as *brainstorming*. In applying this procedure to seek the solution to a design problem, a group of optimal size meets in a room

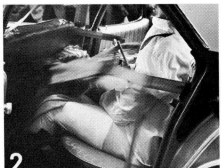

Fig. 12.6 Passive restraint system for rear-seat passengers now being developed by Firestone is designed to provide accident protection in one-thirtieth of a second. It is shown in a stop-action sequence: (1) The rear passenger section with safety blanket stored in a compartment built into the back of the front seat. (2) The blanket about to be deployed. (3) The blanket in place for restraint. (4) This shows how the blanket may be released so that the passengers can leave the car. (*Courtesy The Firestone Rubber Company*)

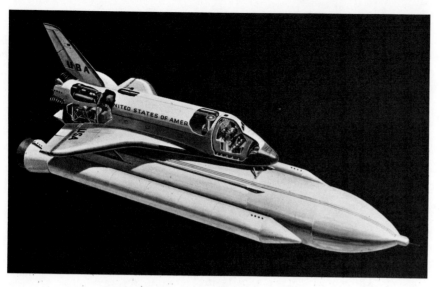

Fig. 12.7 Artist's concept of space shuttle system.
Shown is the complete system, including the airplane-like orbiter, large external propellant tank for the orbiter's three liquid hydrogen/liquid oxygen main engines, flanked by duel solid-rocket boosters.
(*Courtesy National Aeronautics and Space Administration*)

where they will be relatively free of interruptions and distractions. The people selected should be knowledgeable and there should be no one assigned to the group who may take a strong negative attitude toward the design problem to be considered. Usually, group size ranges from six to fifteen persons. If the group consists of less than six people, the back-and-forth interchange of ideas is reduced and the length of the list of ideas, which will be the basis for eventual decision making, is shorter. On the other hand, should the group be larger than fifteen, some individuals who may be capable of making excellent suggestions will have little chance to talk. Also, very large groups tend to be dominated by a few individuals.

During the brainstorming phase of design, no appraisals or judgments should be made nor should criticisms or ridicule of any nature be permitted. The group leader, if he is a capable person, will encourage positive thinking and stimulating comments. He will discourage those who may want to dominate the discussion.

The ideas and suggestions of the members of the group should be listed on a chalk-board and, no matter how long the list, none should be omitted. All ideas should be welcomed and recorded. After the meeting, all the ideas presented may be typed and reproduced for the information of the group and for reference at later discussions.

As the list of suggestions lengthens, several possible solution patterns usually emerge (Fig. 12.8). These in turn lead to still more suggestions for other possible combinations and for improvements to likely solutions to the design problem.

One or two lengthy brainstorming sessions can result in several acceptable design solutions that represent the combined suggestions of the individuals in the group. However, making a group evaluation of the ideas that have evolved from brainstorming is another matter, since no one member feels completely responsible for the results. The major weakness of any group procedure, such as brainstorming, is that individual motivation is dampened to some extent. However, since group procedures are productive and have become widely used in industry, there have been studies made that will hopefully lead to changes that will minimize this recognized weakness and improve group effectiveness. Means must be found to raise the motivation of each participant to the highest possible level.

12.5
Creation of New Products. A product or a system may be said to have been produced either through evolutionary change or by what appears to be pure innovation. The word *appears* has been used appropriately, since few, if any, products are ever entirely new in every respect. Most products that appear to have drawn heavily on innovation usually combine both old and new ideas in a new and more workable arrangement. A product of evolutionary change, however, develops rather slowly, over a long period of time, and with slight improvements being made only now and then. Such a product may be reliable and virtually free of design and production errors, but the small amount of design work involved, done at infrequent intervals, will never really challenge a creative person.

In today's competitive world, when products are produced for world-wide consumption, evolutionary change is hardly sufficient to ensure either the economic well-being or even the survival of those companies that seem to be willing to let well enough alone. Rapid technological changes coupled with new scientific discoveries have increased the emphasis on the importance of new and marketable products that can gain a greater share of the total market than is possible with the product the company may now be pro-

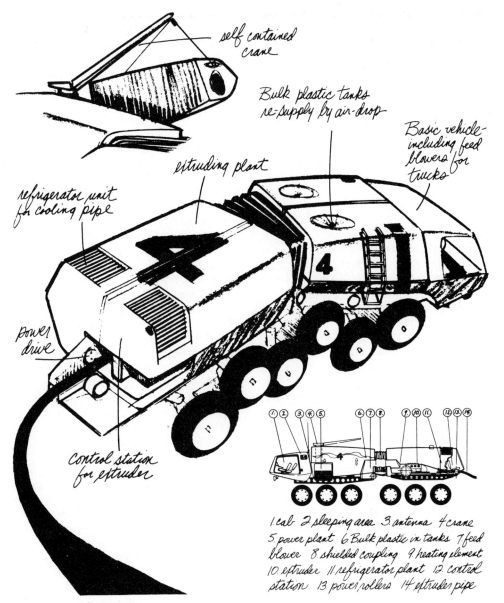

self contained
crane

Bulk plastic tanks
re-supply by air-drop

Basic vehicle
including feed
blowers for
trucks

extruding plant

refrigerator unit
for cooling pipe

power
drive

control station
for extruder

1.cab 2.sleeping area 3.antenna 4.crane
5.power plant 6.Bulk plastic in tanks 7.feed
blower 8.shielded coupling 9.heating element
10.extruder 11.refrigerator plant 12.control
station 13.power rollers 14.extruder pipe

Fig. 12.8 Design of a self-contained pipelayer that could be in use in 1985. It is intended to facilitate irrigation of large tracts of desert land. The equipment as designed is capable of transporting sufficient bulk plastic to lay approximately 2 miles of plastic pipe from each pair of storage tanks. Tanks are to be discarded when empty and replaced by air drop. *(Courtesy Donald Desky Associates, Inc., and Charles Bruning Company)*

moting. In meeting this need for new and marketable ideas, the designer will find that his innovative ability and his experience and knowledge are being taxed to the limit and, since he may in one sense be stepping into the unknown, he will be taking some risks.

The unusual characteristics that seem to be a part of the general makeup of every outstanding designer are: (1) the ability to recognize a problem, (2) the ability to take a questioning approach toward all possible solutions, (3) the possession of an active curiosity about a problem at hand, (4) the innate willingness to take responsibility for what he has done or may do, (5) the ability to make needed decisions and to defend his deci-

sions in writing and orally, and (6) the possession of intellectual integrity.

William Lear, one of the most prominent designers of the last three decades, has spent his entire working life discovering needs and then finding ways to fulfill them. In the case of the development of the eight-track stereo tape, he was working from economic considerations that required more repertoire on tape without adding more tape. This meant either running the tape slower or adding more tracks. The practical answer from Lear's viewpoint was to add more tracks. In addition to the Learoscope, an automatic direction-finder for use on airplanes, Lear is responsible for the development of the car radio, the automatic pilot, the Lear jet plane, and more than 150 other inventions. A few years ago, he began development work on the problem of steam-powered automobile engines. Lear's inventing has been done when he was surrounded by many people with considerable know-how. His work involved the gathering of a maze of information and ideas from which he could pick out the salient facts and discard the unimportant ones, while always keeping the goal in mind and solving the problem at the least possible cost.

12.6
Background for Innovative Designing. Designing should be done by people who have a diversified background and who are not entirely unfamiliar with the problem at hand. As an example, even though the design of a product may be thought of as being in the field of mechanical engineering, a designer with a knowledge of electrical applications and controls will find it a distinct advantage since many of our present-day products use electric current as an energy source. In this case, even though an electrical engineer may be a member of the design group, the mechanical engineer and others should have at least some understanding of his suggestions. This added knowledge will enable them to modify their thinking about a product that is largely a mechanical device. Examples of these products include electric locks, electric food choppers, and electric typewriters.

The background required will vary considerably depending upon the field in which the individual works. For example, a person who may be designing small household appliances would probably never need more than the knowledge acquired from his basic engineering courses, while a designer in the aerospace field would need a background based upon advanced study in chemistry, physics, and mathematics. On the other hand, there are respected and competent designers in industry who have had as little as two years of technical education. With this limited training and several years of on-the-job experience, these men and women have become able to design fairly complex machines.

Because of the increasing complexity of engineering, the rapid development of new materials, and the accumulation of new knowledge at an almost unbelievable rate, it has become absolutely necessary for engineering design to become a team effort in some fields. Under such conditions, the design effort becomes the responsibility of highly qualified specialists. A project requiring designers of varied specialized backgrounds might need, for example, people with experience in mechanical, electrical, and structural design and persons with considerable knowledge of materials and chemical processes (Fig. 12.9). If it is decided that styling is important, then one or more stylists must be added to the team. A complete design group for a major project could include pure scientists, metallurgists, craftsmen, and stylists in addition to the designers.

Finally, graphics must not be overlooked when one considers the background needed to become a successful designer. Anyone who hopes to enter the field of design, other than as a specialist or craftsman, must have a thorough training in this area. He must have a working knowledge of all of the forms of graphical expression that are presented in this text and at the same time be capable of expressing himself well both orally and in written form in his preliminary and final reports. The methods used for the preparation of oral and written reports are discussed in Sec. 12.24.

12.7
History and Background of Human Engineering. Human engineering is a relatively new area in the field of design; it has gained the recognition that it rightly deserves only within the last few decades. In order to simplify matters we might define human engineering as adapting design to the needs of man; that is, we engineer our designs to suit human behavior, human motor activities, and human physical and mental characteristics. The applications of human engineering apply not only to man–machine systems and consumer products but to work methods and to work environments as well. In the early years of its development, human engineering was concerned mainly with the working environment and with the comfort and general welfare of all human beings. As time passed, designers came to realize that more than just safety and comfort should be considered, and that almost all the designs with which they were concerned were in some way related to man's general physical characteristics, his behavior, and his attitudes. At this point, designers began to include these added human factors in their approach to the solution of a design problem so as to

Fig. 12.9 Jet engine.
The ultimate goal of every design must be the production of a product or system useful to man such as the jet engine shown being assembled. (*Courtesy Pratt & Whitney Aircraft Division of United Aircraft Corporation*)

secure the most satisfying and efficient man–machine relationship possible.

It is in the area of human engineering that an engineering designer must adhere to the input of specialists who may be involved in a wide range of disciplines. These disciplines may be industrial engineering, industrial psychology, medicine, physiology, climatology, and statistics. Stimulated to a great extent by the space program, scientists from all of these disciplines have, together and separately, become deeply involved in basic research and laboratory experimentation, which has led to a continuous input of new man–machine information into design. Although in the past human engineering has been associated mainly with industrial engineering, industrial psychology, and industrial design, designers in all fields of engineering must now not only be knowledgeable about the principles of human engineering but they must be capable of utilizing these principles and related information whenever they are developing a product that involves human relationship.

Typical body dimensions, representing an average-size person, are used when a product is being designed for general use.

The measurements of typical adult males and females, as determined from studies made by Henry Dryfuss, may be used for most designs requiring close adaptation to human physical characteristics. Since many designs involve both foot and hand movements these average body dimensions, as tabulated, include arms, hands, legs, and feet along with other parts of the human body. Data relating to body proportions and dimensions are known as anthropometric data. Information relating to body proportions may be obtained from *The Measure of Man*, Whitney Library of Design.

The first known serious studies of the human body were made by Leonardo da Vinci. To record his studies for his own use and for the use of others, he made some of the finest and most accurate detail sketches and drawings of the human body known to man. These drawings show even the intricate details of muscle formation. His work, done in defiance of the laws of his time, is still used today in several textbooks. Da Vinci's studies of the human body mark the beginning of the science of biomechanics. See Fig. 12.10.

From the time of Leonardo da Vinci

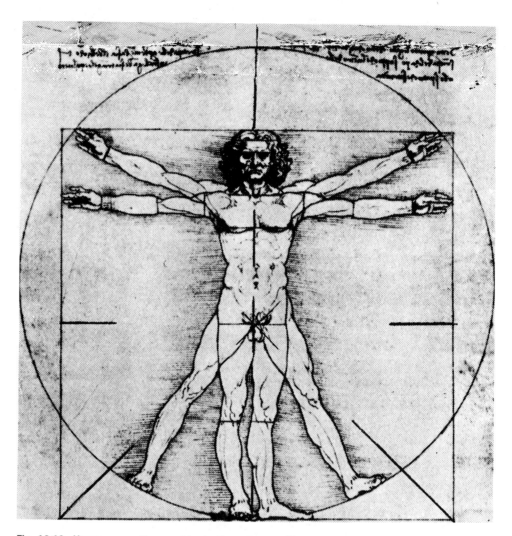

Fig. 12.10 Human proportions and body dimensions as illustrated by Leonardo da Vinci (1452–1519).

until early in the twentieth century, very little work was done toward the development of this science. This lack of interest in people and their relationship with equipment and tasks to be performed was due largely to the fact that from the beginning of the Industrial Revolution (which many people say began with invention of the steam engine patented by James Watt in 1769) until the early 1900s, the interest of designers was centered mainly on the creation of new products and the raising of production efficiency to the high levels needed to compete in world markets. Until about 1911, revolutionary change was the order of the day and there was very little time available for consideration of the human anatomical, physiological, behavior, and attitude factors that have now become the basis of our present man–machine–task systems. Even though our computer-programmed numerically controlled machine tools permit us to do almost any task with only minimal human intervention, man is still needed in most of our man–machine–task systems and he must be taken largely as he has been created. In our designs we must not overlook even the possibility of boredom, for should man become sufficiently bored he might just "pull the plug" and everything would stop.

12.8

Human Engineering in Design. There are a number of factors in human engineering, other than human anthropometric measurements, that must be dealt with in design. These factors include motor activities (Fig. 12.11) and body orientation; the five human senses (sight, hearing, touch, smell, and sometimes even taste); atmospheric environment, temperature, humidity, and light; and, finally, accelerative forces if they are exceptional and are likely to cause undue physical discomfort.

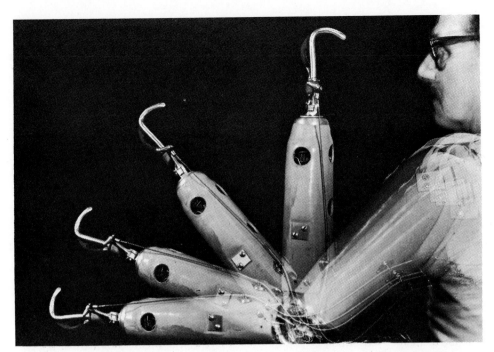

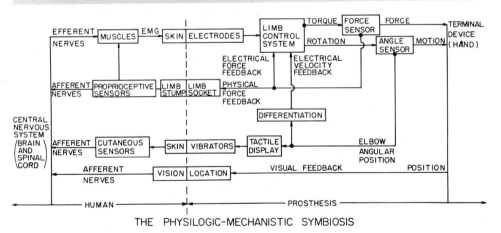

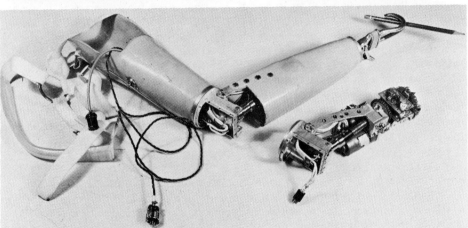

THE PHYSILOGIC-MECHANISTIC SYMBIOSIS

Fig. 12.11 Boston Arm.
The range of movement of the Boston Arm is demonstrated in the upper photograph. The arm, developed as a joint project of Liberty Mutual Insurance Companies, Harvard Medical School, Massachusetts General Hospital, and the Massachusetts Institute of Technology, acts as does a normal arm through thought-impulses transmitted from the brain to existing arm muscles. The design of a product for the handicapped can provide much satisfaction to the designer. (*Courtesy Liberty Mutual*)

During the first half of this century most of the research done in human engineering, aside from the anthropometric studies already mentioned, was directed toward: work areas and the position of controls; physical effort and fatigue; and the speed and accuracy to be expected in the performance of particular tasks. Finally, when it became evident that this was not enough, industrial engineers and industrial psychologists turned their attention to the more complex activities of the average human being. These new studies dealt primarily with receiving information through sight and sound, the making of decisions in response to stimuli, and, finally, the performance in direct response to these decisions.

The study of body motion deals primarily with the effective range of operation of parts of the body, usually the arms and legs, and the amount of body force a human being may reasonably be expected to exert in the performance of an assigned task. For example, many time and motion studies have been made of the range of operation of persons performing given tasks while seated at assigned work areas. At the same time, in many of these studies attention has been given to the location and the amount of force required to operate levers and controls in relation to the size and strength of the operator.

Vision is an important factor in all designs where visual gauges or colored lights on control panels are a part of a control system that involves manual operation or, in the case of numerically controlled machines, monitoring for manual intervention at specified times. Control panels for such equipment must be designed to be within the visual range of the average person, and distinctive colors must be selected and used for the colored lights. The colors selected must be easily recognizable and must be capable of quickly attracting the operator's attention. Not to be overlooked is the fact that vision studies have produced much new information for use in highway design. From these same studies have come new ideas for our highway warning and information signs along with suggestions for their placement.

When a designer must consider the working environment as a part of the total design, he should realize that environment includes (1) temperature, (2) humidity, (3) lighting, (4) color schemes, and (5) sound. These are a few of the factors that also deserve full consideration in the design of a large industrial plant, a particular work area in a plant, the cockpit of an airplane, or the cabin of a space vehicle.

The overall environment and the design of the working areas and the living quarters of the undersea laboratory shown in Fig. 12.12 were based on human requirements. The design of any undersea craft or laboratory involves problems that are similar in many ways to those encountered in the design of space vehicles, in that an artificial living environment must be created and maintained for extended periods of time. This requires a self-contained atmosphere. The members of the crew also must have ample space in which to work and live under climatic conditions that duplicate those on land. Crew members must be able to perform their tasks under normal lighting conditions and they must be able to see to the outside. Aside from the design of features and components that are related directly to the performance of the research assignments, the overall development of the undersea craft can be said to be based on the physical needs and the psychological attitudes and reactions of human beings.

Human engineering is applied to a wide range of consumer items. Automobiles (Fig. 12.13), refrigerators, furniture, office equipment, lawn mowers, hand tools, and like items have long been designed with human factors in mind.

The automotive industry, which has been designing and redesigning cars for more than 75 years with human factors in mind, is now producing cars in accordance with stricter government safety regulations after being accused by a vocal few of having produced American cars for comfort, power, style, and speed with less than full attention being given to the safety of the occupants. Even though the charge in this case is probably unfounded designers must realize that this situation can happen whenever engineers ignore, even to limited extent, some of their professional responsibilities and give in to the whims of stylists and to the suggestions of administrators and sales managers before having made a thorough study of each and every proposal that has been presented.

Under pressure from consumer groups and organizations interested in safety and in protecting our natural environment, there has been a rising tide of governmental laws and regulations. Many of these new laws and directives in a sense provide controls in the field of human

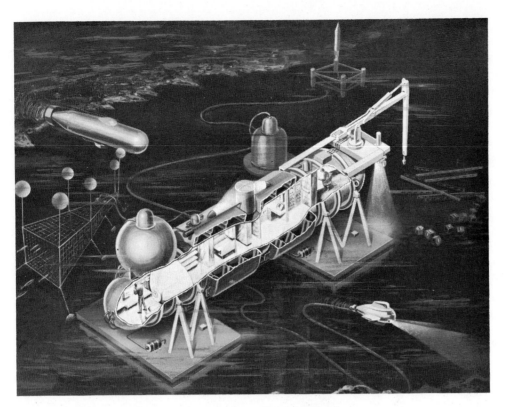

Fig. 12.12 Artist's conception of the Atlantis undersea habitat.
The project, which was to be designed and developed as a joint project of the
University of Miami and the Space Division of Chrysler Corporation, has been
abandoned at this time. The project goal was to explore the continental shelf
along our coasts. (*Courtesy Space Division Chrysler Corporation*)

engineering. In the case of environmental
pollution, automobile companies, acting
to meet regulations set forth by the Environmental Protection Agency, have developed several new devices to reduce
undesirable pollutants at a scheduled
percentage rate to produce an almost pollution-free car by 1980. In the interest of
safety, some type of passive protection
system, such as the air-bag device shown
in Fig. 12.13, must be added by a set date.

These government laws and regulations
have come into existence because of a
growing interest on the part of the general
public in human engineering. Designers in
the years ahead must be fully cognizant of
all such regulations and must be willing to
abide by them or seek to have them
changed should they appear to be unrea-
sonable or impractical.

12.9

Visual Design. Visual design includes the
use of line, form, proportion, texture, and
color to produce the eye-pleasing appearance needed to bring about the acceptance of a consumer item. Without this acceptance there would be no profit, and

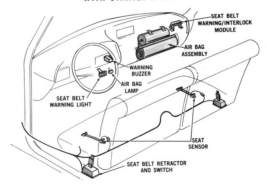

**Fig. 12.13 Major elements of an air-bag/seat-
belt restraint system now under development
by the Ford Motor Company.**
The air-bag assembly and the seat-belt
starter-interlock components are identified. It
should be noted that there are separate sig-
nal lights for the seat-belt and air-bag systems. (*Courtesy Ford Motor Company*)

even though the item might otherwise have been carefully engineered, it would soon disappear from the marketplace. The sketches shown in Fig. 12.14, for the dash panel of an automobile, tastefully combine these visual elements into an attractive design. An illusion of depth has been obtained by means of pencil shading.

Since artistic styling is now recognized as being one of the most important factors in sales, many engineers have come to accept the role of stylists in the development of a consumer product, particularly when they are employed by a company that is small and cannot afford to employ one or more trained stylists. Large companies, such as the Ford Motor Company (Fig. 12.15), General Motors, and the Boeing Company (Fig. 12.16), have styling divisions. Medium-size companies often turn to nationally known organizations to get needed help; the styling is then done

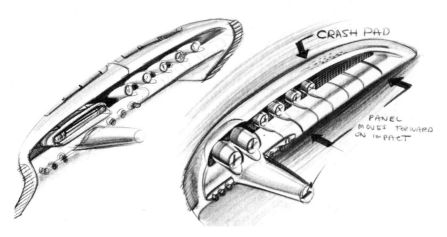

Fig. 12.14 Design sketches for a dashboard panel. (*Courtesy General Motors Corporation*)

Fig. 12.15 Stylist at work. (*Courtesy Ford Motor Company*)

under contract agreement. Many books have been written about aesthetics that have proved to be helpful to engineers (see the Bibliography). At present, design engineers, who have been trained largely to solve technical problems, are reading more about art in design and they are considering the eye-appeal and the overall appearance of products as part of their engineering interests.

12.10
Constructive Criticism. There are people who seem to find it easier to criticize than to mix praise with alternative suggestions. In any group meeting a critic should show respect for good ideas and be able to offer constructive suggestions. If there is to be feedback, which will hopefully lead to the introduction of more ideas, the discussion must be free of any harsh criticism. Harsh criticism may cause a sensitive person to assume a defensive position or to withdraw almost entirely from participation in group action. It is the responsibility of the project leader to prevent this from happening.

12.11
Recognition of a Need. A design project usually begins with the recognition of a need and with the willingness of a company to enter the market with a new product. At other times an idea may be initiated and developed by an individual who either seeks economic benefits for himself or who seeks to solve some social or environmental problem. In either case, the identification of the need in itself represents a high order of creative thinking and the search for a solution to the need requires considerable self-confidence and inner courage. As can be easily observed from reviewing the achievements of our distinguished inventors of the past, those who are closely attuned to life around them become aware of needs or less-than-ideal situations that are worthy of their attention.

It is important that a proposed design activity have clear and definite objectives that will justify the money and effort to be expended in product design and development. The statement defining the objectives should identify the need and state the function the product is to perform in satisfying this need. The identification of the need may be based on the designer's personal observations, suggestions from salesmen in the field, opinion surveys, or on new scientific concepts. The identification of the design problems involved in creating the needed product comes later in the design process.

12.12

Formal Proposal. The statement covering the recognition of need can be used as a basis for a formal proposal that may be either a few short paragraphs or several typewritten pages. A complete proposal may include supporting data in addition to the description of the plan of action that is to be taken to solve the problem as identified (Fig. 12.17). The report should have the same general form as other technical reports and might include a listing of requirements and possible limitations as then recognized. In preparing the report one should keep in mind that the proposal, when approved, gives the broad general parameters of an agreement under which the project will be developed to its conclusion.

12.13

Phases of the Design Process. Many people in the past have prepared outlines of the steps that can be followed in the process of design (Fig. 12.18). They have prepared these outlines in order to give some semblance of order to the total design process from the point of recognition of the need to the point of marketing the product. One must recognize, however, that there are actually many combinations of steps in the overall procedure, with no single listing either the best or the one and only combination. The design procedure required in many cases can be very complex and successful designers have found different ways to achieve their goals. However, the phases of design, as recognized by this author, have been listed here in sequential order to provide some degree of direction to the student who is making his first attempt to design a product under a contrived classroom situation. More experienced persons may find it desirable to alter this outline to make it more suitable to their own method of designing.

The basic phases in the design process can be thought of as being (1) identification of need, (2) task definition (goal), (3) task specifications, (4) ideation, (5) conceptualization, (6) analysis, (7) experimental testing (Fig. 12.26), (8) design (solution) description, and (9) design for production. Phase 9 (implementation) is not usually a primary concern of a designer. Also, the designer may or may not

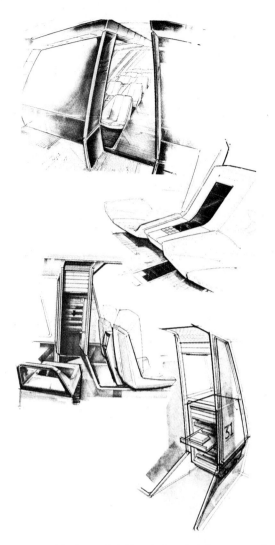

Fig. 12.16 Several styling design sketches prepared for the U.S. supersonic transport (SST).
Visual appeal is important to the success of the total design project. (*Courtesy The Boeing Company*)

give some thought to manufacture, distribution, and consumption of the product. Consideration of these factors may be thought of as being the tenth, eleventh, and twelfth stages of total design.

12.14

Task Definition—Definition of Goals. Briefly stated, task definition is the expression of a commitment to produce either a product or a system that will satisfy the need as identified. By means of broad statements, the product and the goals of the project are identified. The statements, as written, must be clear and concise to avoid at least some of the difficulties (often encountered in design) that can be traced directly back to poorly defined goals.

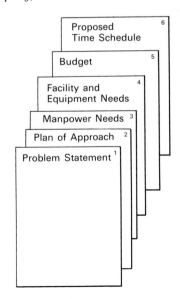

Fig. 12.17 Elements of a proposal.

Even though it is probable that the person who has initiated the project has already gathered some pertinent information and has some preconceived ideas, it is desirable that this material not be included (Fig. 12.19). It is better to present the goals in terms of objectives and then allow the designers to pursue the project in their own way, as free of restrictions as possible. The task definition should be included in the proposal.

12.15

Task Specifications. This is a listing of parameters and data that will serve to control the design. This stage will ordinarily be preceded by some preliminary research to collect information related to the goal as defined. In preparing the task specifications, the designer or design group lists all the pertinent data that can

be gathered from research reports, trade journals, patent records, catalogues, and other sources that possess information relating to the project at hand. Included in this listing should be the parameters that will tend to control the design. Other factors that deserve consideration, such as materials to be used, maintenance, and cost may also be noted.

12.16

Ideation. The ideation phase of design has been discussed for group procedure situations in Sec. 12.4. It is recommended that the reader review this discussion to refresh his memory regarding brainstorming procedures. It is well to remember that often a lasting solution to a problem has resulted from a creative idea selected from a number of alternative ones (Fig. 12.20). The mathematical likelihood of finding an optimum solution is greatly enhanced as the list of possible alternatives grows longer. Truly great creations are possible when one, acting either alone or in a group, lets his imagination soar with little restraint. If an engineer on a design assignment can set aside his engineering know-how and blind himself, at least temporarily, to traditional approaches he is in the right mood to accept almost any challenge. This is the mood and the mental approach to great discoveries. Some call it ideation; others have called it *imagineering*. If we can learn to open up our engineering minds to new approaches to our technical problems as well as to our existing problems of air pollution, industrial waste, transportation, and even unemployment, there is no limit to what we can accomplish.

This open-minded *imagineering* approach to problems must be tempered with a sense of professional responsibility. It is no longer acceptable to solve an immediate problem using a solution that in years to come can endanger the environment. Imagineering is engineering for the total well-being of mankind and all other living things.

12.17

Conceptualization. Conceptualization follows the preliminary idea (ideation) stage when all the rough sketches (Figs. 12.21–12.23) and notes have been assembled and reviewed to determine the one or more apparent solutions that seem to be worthy of further consideration. In evaluating alternative solutions, consideration must be given to any restrictions that

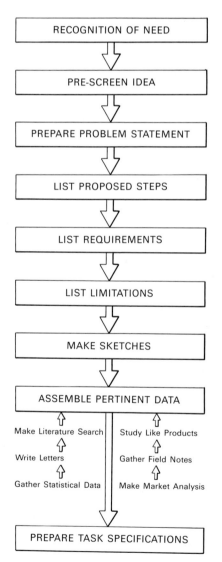

Fig. 12.18 Basic phases of total design.

Fig. 12.19 Initial steps of the design process.

have been placed on the final design. It is at this stage that the preliminary sketches should be restudied to see that all worthwhile ideas are being included and that none have been inadvertently overlooked. At no time during this phase should the designer become so set in his thinking that he does not feel free to develop still another and almost entirely new and different concept, if necessary. He should realize that it is more sensible to alter or even abandon a concept at this stage than later, when considerable money and time will have been invested in the project. The conceptualization stage of design is that stage where alternative solutions are developed and evaluated in the form of concepts. Considerable research may be necessary and task specifications must be continually reviewed. As activity progresses, many idea sketches are made as alternative approaches are worked out; these approaches are evaluated for the best possible chance of product success. It is not necessary at this stage of the design procedure for any of the alternative solutions to be worked out in any great detail.

12.18

Selection of Optimum Concept. As the design of a product or system progresses, a point is reached in the procedure when it becomes necessary to select the best design concept to be presented to the administrators in the form of a proposal. In making this final selection, a more-or-less complete design evaluation is made for each of the alternative concepts under

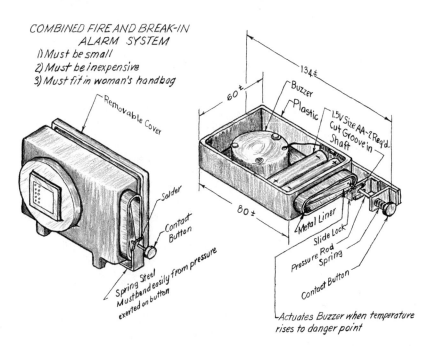

Fig. 12.20 Idea sketches for a small portable safety alarm that will give warning for both fire and an attempted break-in. (Dimension values are in millimeters)

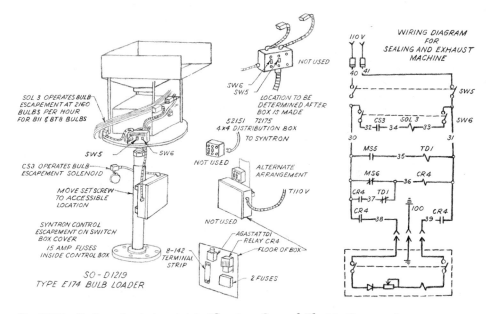

Fig. 12.21 Portion of a design sketch. (*Courtesy General Electric Company*)

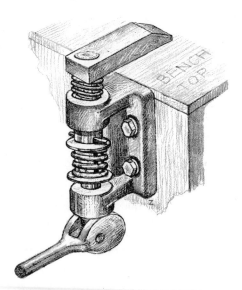

Fig. 12.22 Idea sketch for a quick-acting machine clamp.

Fig. 12.23 Idea sketch showing remote control system for a motor boat.

consideration. These evaluations may reveal ways that costs can be reduced and value improved; means of simplifying the design to reduce costs may also become apparent.

12.19

Design Analysis. After a design concept has been chosen as the best possible solution to the problem at hand, it must be subjected to a design analysis; that is, it must be tested against physical laws and evaluated in terms of certain design factors that are almost certain to be present (Fig. 12.24).

Total analysis of a proposed design will include a review of the engineering principles involved and a study of the materials to be used. In addition, there should also be an evaluation of such design considerations as (1) the environmental conditions under which the device will operate, (2) human factors, (3) possible production methods and production problems, (4) assembly methods, (5) maintenance requirements, (6) cost, and (7) styling and market appeal. If the design is based on newly discovered scientific principles, some research may be in order before a final decision can be reached.

It is at this stage in the design procedure that physics, chemistry, and the engineering sciences are utilized most fully. In making the usual design analysis, the engineer or engineering technologist must depend on the formal training that he has received in school, and although considerable mathematics may be needed in making most of the necessary calculations, he

will find it convenient at times to resort to graphical methods. Over the years, graphical methods have proved to be most helpful in evaluating and developing a design. For example, descriptive geometry methods can be employed for making spatial analyses and critical information can be obtained by scaling accurate drawings.

If the design analysis should prove that the design as proposed is inadequate and does not meet requirements, the designer may then either make certain modifications or incorporate into his designs some new concept that might well be a modification of an earlier idea that was abandoned along the way.

12.20

Experimental Testing. The experimental testing phase of the design process ranges from the testing of a single piece of software or hardware to verify its workability, durability, and operational characteristics through the construction and testing of a full-size prototype of the complete physical system.

A component of a product should be tested in such a way that the designer can predict its durability and performance under the conditions that will be encountered in its actual use. Needed tests may be performed using standard test apparatus or with special devices that have been produced for a particular test.

There are three types of models that may be constructed for the purpose of testing and evaluating a product. These are: (1) the mockup, (2) the scale model, and (3) the prototype.

12.21

Mockups. A mockup is a full-size *dummy* constructed primarily to show the size, shape, component relationships, and styling of the finished design. At this point the designer can see his conception begin to take shape for the first time. Automobile manufacturers customarily produce mockups to evaluate proposed changes in the styling of automobile bodies for new models (Fig. 12.25). Needed modifications in size and body configuration can be determined by studying the mockup and analyzing its overall appearance. Since interior styling is important also, interiors are modeled in clay to reveal the aesthetic appearance the stylist had in mind. In the automobile industry, mockups are made to secure early approval of management for model change. A mockup is more meaningful than a sketch to those whose

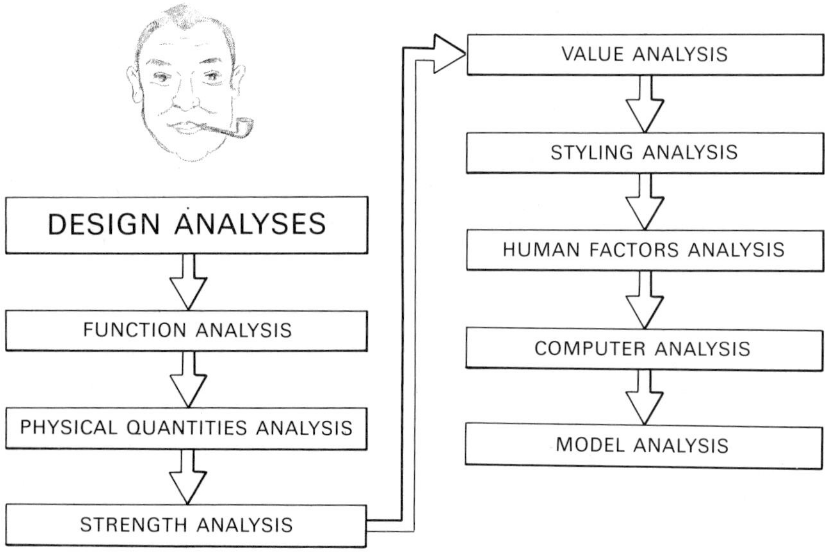

Fig. 12.24 Analysis phase of design.

support and approval is needed. One must realize, however, that numerous sketches and artistic renderings (Fig. 12.16) are made before any work is started on a mockup. These sketches are used as guides. Mockups may be made of clay, wood, plastic, and so forth.

12.22
Preliminary and Scale Models. Models may be made at almost any stage in the design process to assist the designer in evaluating and analyzing his design. Models are made to strengthen three-dimensional visualization, to check the motion and clearance of parts, and to make necessary tests to clear up questions that have arisen in the designer's own mind or in the mind of a colleague.

The designer may prepare a preliminary model to understand more fully what the shape of a component should be, how well it may be expected to operate, and how it might be fabricated most economically. In some cases, the model might be so simple that it could be made of paper, wood, or clay.

Scale and test models may be constructed either for analysis and evaluation or for the purpose of presenting for approval the design as developed in a more or less refined stage. Scale models may be made of balsa wood, plastic, aluminum, wire, steel, or any other material that can be used to a good advantage. The designer should select a scale that will make the model large enough to permit the movement of parts should a demonstration of movement be desirable.

12.23
Prototypes. A prototype is the most expensive form of model that can be constructed for experimental purposes. Yet, since it will yield valuable information that is difficult to obtain in any other way, its cost is usually justified. Since a prototype is a full-size working model of a physical system that has been built in accordance with final specifications, it represents the final step of the experimental stage (Fig. 12.26). In it the designers and stylists see their ideas come to life. From a prototype the designers can gain information needed for mass-production procedures that are to come later. Much can be learned at this point about workability, durability, production techniques, assembly procedures, and, most important of all, performance under actual operating conditions. Since prototype testing offers

Fig. 12.25 Full-size mockup in clay is prepared to evaluate a proposed body configuration for a forthcoming model. (*Courtesy Ford Motor Company*)

the last chance for modification of the design, possible changes to improve the design should not be overlooked, nor should a designer ever be reluctant to make a desirable change.

Since a prototype is a one-of-a-kind working model, it is made by hand using

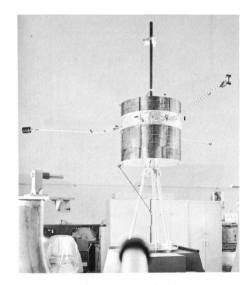

Fig. 12.26 Full-size prototype of the Pioneer Satellite A-B-C-D series.
A prototype provides the best means for an evaluation of a complex design. (*Courtesy TRW Systems Group*)

general purpose machine tools. Although it might be best to use the same materials that will be used for the mass-produced product, this is not always done. Materials that are easier to fabricate by hand are often substituted for hard-to-work materials.

In the development of a design, designers deal first with the mockup; next they work out specific problems relating to single features with preliminary models; then they evaluate the design with scale models; and, finally, they may, if desirable, test the whole concept using a prototype. This order in the use of models maintains a desirable relationship between concept and analysis and represents a logical procedure for the total design process.

12.24

Design (Solution) Description—Final Report. In the solution description the designer describes his design on paper, to communicate his thinking to others. Although the purpose of the final report may be to sell the idea to upper management, it may also be used to instruct the production division on how to construct the product. It usually will contain specific information relating to the product or system. In some cases a process will be described in considerable detail.

A complete design description, prepared as the main part of a formal report, should include: (1) a detailed description of the device or system, (2) a statement of how the device or system satisfies the need, (3) an explanation of how the device operates, (4) a full set of layout drawings, sketches, and graphs, (5) pictorial renderings, if needed, (6) a list of parts, (7) a breakdown of costs, and (8) special instructions to ensure that the intent of the designer is followed in the production stage (Fig. 12.27). After the design description is accepted and approved, there remain only the commercial stages of the total design process, namely, implementation, manufacture, distribution (sales), and consumption.

12.25

Implementation—Design for Production. Implementation is that phase of the total design process when working drawings are prepared for the men in the shops who must fabricate the nonstandard parts and assemble the product. A complete set of working drawings, both detail and assembly, are needed to perimit the manufacture of a product.

Even though in theory the production design phase directly follows the preliminary design phase and the acceptance of the design recommended in the final report, in actual practice there is often no clear dividing line between these two phases. This is because detailed drawings of some components may have been started and to some extent completed well back in the preliminary design stage, along with one or more design layout assemblies (see Fig. 12.28). Furthermore, it is not unusual for detail design drawings to be made for two or more likely solutions in the design-for-production stage and the final selection of the one best solution delayed until the information derived from these detail design drawings and related assemblies can be used in making the final decision. Such a delay in decision making tends to cause the preliminary design and the production design phases to appear to melt into one another. However, there is still a division line even though the designers themselves may not recognize this fact.

12.26

Manufacture (Fig. 12.29). From the time the task specifications are written and through all the stages up to the manufacture, the designer works closely with a

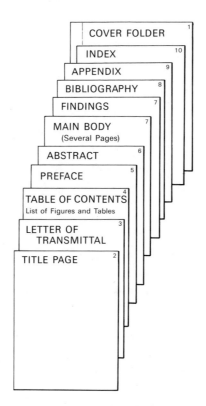

Fig. 12.27 Contents of a design (final) report.

COVER FOLDER 1
INDEX 10
APPENDIX 9
BIBLIOGRAPHY 8
FINDINGS 7
MAIN BODY 7
(Several Pages)
ABSTRACT 6
PREFACE 5
TABLE OF CONTENTS 4
List of Figures and Tables
LETTER OF 3
TRANSMITTAL
TITLE PAGE 2

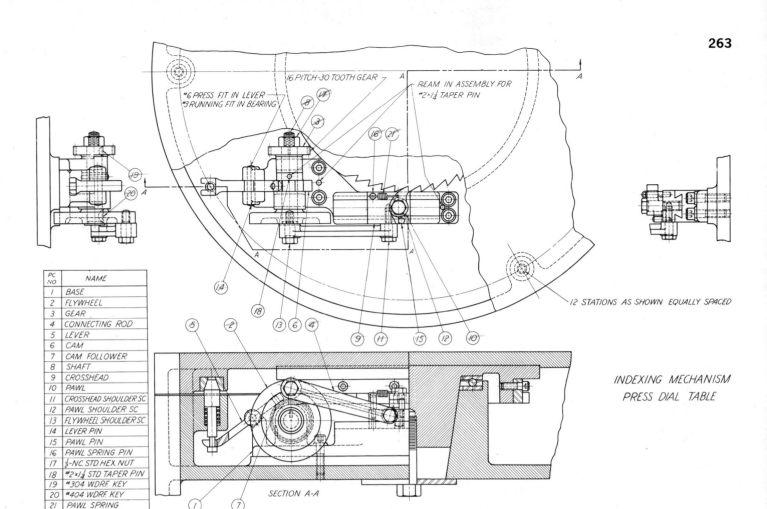

PC. NO	NAME
1	BASE
2	FLYWHEEL
3	GEAR
4	CONNECTING ROD
5	LEVER
6	CAM
7	CAM FOLLOWER
8	SHAFT
9	CROSSHEAD
10	PAWL
11	CROSSHEAD SHOULDER SC
12	PAWL SHOULDER SC
13	FLYWHEEL SHOULDER SC
14	LEVER PIN
15	PAWL PIN
16	PAWL SPRING PIN
17	$\frac{1}{2}$-NC. STD. HEX. NUT
18	#2×1$\frac{1}{4}$ STD TAPER PIN
19	#304 WDRF KEY
20	#404 WDRF KEY
21	PAWL SPRING

SECTION A-A

INDEXING MECHANISM
PRESS DIAL TABLE

Fig. 12.28 Design layout drawing.

production engineer who is familiar with the available shop facilities, production methods, inspection procedures, quality control, and the assembly line. If this is done, the problems encountered in the manufacturing stage will be few in number.

12.27
Distribution. Since a designer usually has little expertise in this area, task problems relating to distribution are passed along to marketing specialists. These specialists have the knowledge and the supporting staff required to decide on the proper release date and to set a competitive price based upon market testing and cost and profit studies. Included among these specialists will be experienced advertising men who prepare the needed advertising and promotional literature. However, these specialists do consult the designer frequently during this stage since he knows more about the product than anyone else. The designer is the one person

who can be depended upon to supply needed technical data and information concerning the capabilities and limitations of the product. Furthermore, the sales promotion people expect ideas from the designer that will lead to a wide and favorable distribution.

12.28
Consumption. There should hopefully be a consumer feedback from this last stage of the design process that will prove useful to the designer when it becomes necessary to alter and improve the product at the time of the next model change. Most of this feedback information will come from the sales force, from distributors, and from service departments. Some of this information will be in the form of complaints made by irate users, but much of it will be received as constructive suggestions. These constructive suggestions take the designer back to the need stage for the next model, and a new round of the design cycle starts all over again.

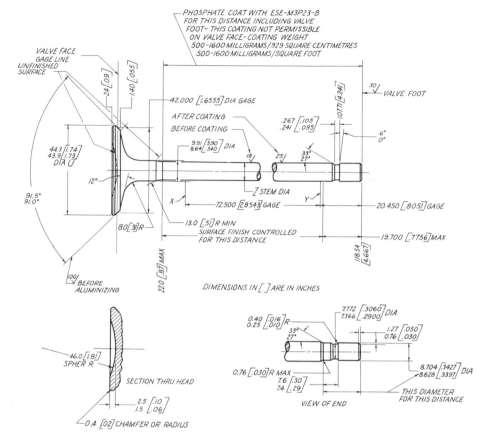

Fig. 12.29 Detail drawing, showing the use of dual-dimensioning.
The reader should note the procedures used when both inch measurements and millimeters are
shown. Many industrial drawings are now being dimensioned in this manner. See Chapter 13.
(*Courtesy Ford Motor Company*)

The consumption stage represents the goal sought by the designer and this is where the design has its ultimate test. At this stage the design will be pronounced a success or a failure by the users and consumers who are and always will be the final judges.

B □ Patents and patent office drawings

12.29
Patent. A patent, when granted to an inventor, excludes others from manufacturing, using, or selling the device or system covered by the patent anywhere in the country for a period of seventeen years. In the patent document, in which the invention is fully described, the rights and privileges of the inventor are set forth and defined. Upon the issuance of the patent, the inventor has the right to either manufacture and sell the product himself with a protected market or he can assign his rights to others and charge them for the manufacture and sale or use of his invention. After the seventeen-year period has expired, the inventor no longer has any protection and the invention becomes public property and may be produced, sold, and used by anyone for the good of all.

To ensure full protection of the patent laws, patented products must be marked *Patent* with the patent number following. Even though an invention is not legally protected until a patent actually has been issued, a product often bears a statement that reads: *Patent Pending.*

In accordance with patent law "any person who has invented any new and useful process, machine, manufacture or composition of matter, or any new and useful improvement thereof, may obtain a patent," subject, however, to restrictions, conditions, and requirements imposed by patent law. To be patentable an invention must be new and original and be uniquely different; it must perform a useful function; and finally the invention must not have been previously described in any publication anywhere nor have been sold or in general use in the United States before the applicant made the invention. It should be noted that an idea in itself is not patentable, since it is a requirement that a specific design and design description of a device accompany an application for a patent.

12.30

Patent Attorney. When an inventor has decided that he has a patentable device, he should engage a patent attorney who will help prepare the necessary application for a patent. The inventor should depend upon this attorney to advise him as to whether or not the product or process that he has conceived may infringe upon the rights of others. Since most patent attorneys are also graduate engineers, the inventor will find that he has allied himself with a person who can guide his application through the searching, investigating, and processing that takes place before a patent is granted. Generally, an application remains pending for as long as four or more years before a patent is finally granted. In this period the pending patent application may well be amended to include engineering changes that have been made in the device or system.

If after some investigation, the attorney–engineer thinks that the device is novel and therefore patentable, an application for a patent should be prepared and filed. A patent application includes a formal portion consisting of the petition, a power of attorney, and an oath or declaration. This is followed by a description of the invention, called the specifications, and a list of claims relating to it. If the device can be illustrated, one or more drawings should be included (Fig. 12.30).

In selecting a patent attorney, the inventor should remember that this man can be of service to him over many years going well beyond the time when the patent is issued. The attorney or his firm may be retained to prepare all agreements covering the sale and leasing of patent rights, to assist in negotiations relating to these rights, and if the need should arise, to handle charges of alleged infringements.

Many of the large corporations have patent divisions that operate as a section of the home office legal staff. An additional office to deal with patent matters may also be maintained in Washington,

D.C. In this case, both offices, staffed with patent attorneys, will have all activities coordinated by a director of patents.

12.31

Role of the Inventor in Obtaining a Patent. Even though an inventor must rely almost entirely upon a patent attorney to take the final steps to secure a patent, the inventor plays a vital role up to the time of filing the application. Very often what he has done or not done determines whether his rights to his invention can be safeguarded by his attorney.

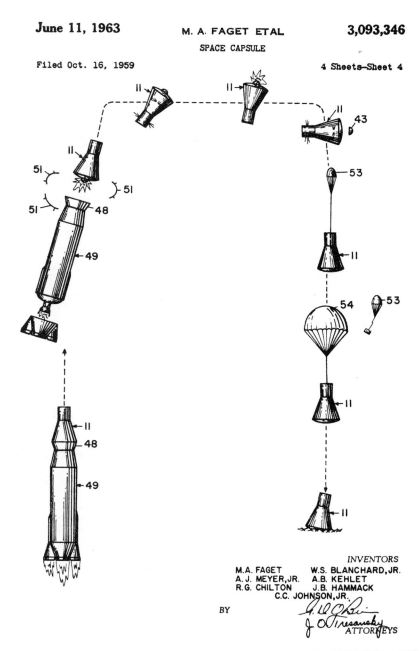

Fig. 12.30 Patent drawing, showing the sequence of events from launch to landing.

Since court decisions in patent suits usually depend upon the inventor's ability to prove that certain design events happened on a specific date, it is always advisable to keep the design records in a hardcover permanently bound notebook so that there will be no question as to whether or not pages have been added (as could be charged if a loose leaf notebook had been used). A well-kept patent notebook is a complete file of information covering a design. It can serve as the basis of project reports; it can spare the designer and his company the unnecessary expense of repeating portions of the experimental work; and, finally, it furnishes indisputable proof of dates of conception and development.

To prepare a legally effective patent notebook the designer should:

1. Use a bound notebook having printed page numbers.

2. Make entries directly using either a black ink pen or an indelible pencil.

3. Keep entries in a chronological order. Do not add retroactive entries.

4. Add references to sources of information.

5. Describe all procedures, equipment, and instruments used for development work.

6. Insert photos of instrument and equipment setups. Add other photos showing models, mockups, and so forth.

7. Sign and date all photos.

8. Have a qualified witness sign and date every completed page. The witness cannot be a co-inventor nor can he have a financial interest in the development of the device. The witness must be a person who is capable of understanding the construction and operation of the device or system. He must also be experienced in reading drawings and understand the specifications.

9. Have the witness write the words "Witnessed and Understood" and then write and date his signature.

10. Have two or more witnesses sign and date every page after reading it.

11. Avoid having blank spaces on any page.

12. Have all new entries witnessed at least once a week.

13. Have the notebook evaluated periodically by a patent attorney. Follow his suggestions.

It takes a considerable time and effort to adhere to these thirteen recommendations for the preparation of an effective patent notebook. However, if an inventor has once become entangled in an infringement lawsuit he realizes that a little extra effort pays off.

12.32
Patent Drawings. A person who has invented a new machine or device, or an improvement for an existing machine, and who applies for a patent is required by law to submit a drawing showing every important feature of his invention. When the invention is an improvement, the drawing must contain one or more views of the invention alone, and a separate view showing it attached to the portion of the machine for which it is intended.

Patent drawings must be carefully prepared in accordance with the strict rules of the U.S. Patent Office. These rules are published in a pamphlet entitled *Rules of Practice in the United States Patent Office,* which may be obtained, without charge, by writing to the Commissioner of Patents, Washington, D.C.

In the case of a machine or mechanical device, the complete application for a patent will consist of a petition, a "specification" (written description), and a drawing. An applicant should employ a patent attorney, preferably one who is connected with or regularly employs competent draftsmen capable of producing well-executed drawings that conform with all of the rules. Ordinary draftsmen lack the skill and experience necessary to produce such drawings. Two sheets of drawings for a patent for a radiation protection system are shown in Figs. 12.31 and 12.32.

Several U.S. Patent Office publications that are available to the general public have been listed in the Bibliography.

12.33
Rules. The following rules (49–55) are quoted verbatim from the pamphlet, *Rules of Practice in the United States Patent Office:*

49. The applicant for a patent is required by law to furnish a drawing of his invention whenever the nature of the case admits of it.

50. The drawing may be signed by the inventor or one of the persons indicated in rule 25, or the name of the applicant

may be signed on the drawing by his attorney in fact. The drawing must show every feature of the invention covered by the claims, and the figures should be consecutively numbered, if possible. When the invention consists of an improvement on an old machine, the drawing must exhibit, in one or more views, the invention itself, disconnected from the old structure, and also in another view, so much only of the old structure as will suffice to show the connection of the invention therewith.

51. Two editions of patent drawings are printed and published—one for office use, certified copies, etc., of the size and character of those attached to patents, the work being about 6 by 9½ inches; and one reduction of a selected portion of each drawing for the official Gazette.

52. This work is done by the photolithographic process, and therefore the character of each original drawing must be brought as nearly as possible to a uniform standard of excellence, suited to the requirements of the process, to give the best results, in the interests of inventors, of the office, and of the public. The following rules will therefore be rigidly enforced, and any departure from them will be certain to cause delay in the examination of an application for letters patent:

(a) Drawings must be made upon pure white paper of a thickness corresponding to two-sheet or three-sheet Bristol board. The surface of the paper must be calendered and smooth. India ink alone must be used, to secure perfectly black and solid lines.

(b) The size of a sheet on which a drawing is made must be exactly 10 by 15 inches. One inch from its edges a single marginal line is to be drawn, leaving the "sight" precisely 8 by 13 inches. Within this margin all work and signatures must be included. One of the shorter sides of the sheet is regarded as its top, and measuring downwardly from the marginal line, a space of not less than 1¼ inches is to be left blank for the heading of title, name, number, and date.

(c) All drawings must be made with the pen only. Every line and letter (signatures included) must be absolutely black. This direction applies to all lines, however fine, to shading, and to lines representing cut surfaces in sectional views. All lines must be clean, sharp, and solid, and they must not be too fine or crowded. Surface shading, when used, should be open. Sectional shading should be made by oblique parallel lines, which may be about one-twentieth of an

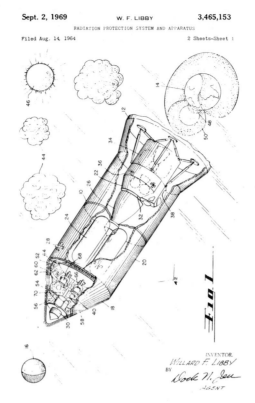

Fig. 12.31 **Patent drawing showing a radiation protection system.** See also Fig. 12.33.

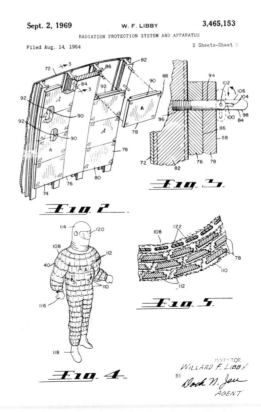

Fig. 12.32 **Drawings showing the details that are related to the claims.**

inch apart. Solid black should not be used for sectional or surface shading. Free-hand work should be avoided wherever it is possible to do so.

(*d*) Drawings should be made with the fewest lines possible consistent with clearness. By the observance of this rule the effectiveness of the work after reduction will be much increased. Shading (except on sectional views) should be used only on convex and concave surfaces, where it should be used sparingly, and may even there be dispensed with if the drawing be otherwise well executed. The plane upon which a sectional view is taken should be indicated on the general view by a broken or dotted line, which should be designated by numerals corresponding to the number of the sectional view. Heavy lines on the shade sides of objects should be used, except where they tend to thicken the work and obscure letters of reference. The light is always supposed to come from the upper left-hand corner at an angle of 45°.

(*e*) The scale to which a drawing is made ought to be large enough to show the mechanism without crowding, and two or more sheets should be used if one does not give sufficient room to accomplish this end; but the number of sheets must never be more than is absolutely necessary.

(*f*) The different views should be consecutively numbered. Letters and figures of reference must be carefully formed. They should, if possible, measure at least one-eighth of an inch in height, so that they may bear reduction to one twenty-fourth of an inch; and they may be much larger when there is sufficient room. They must be so placed in the close and complex parts of drawings as not to interfere with a thorough comprehension of the same, and therefore should rarely cross or mingle with the lines. When necessarily grouped around a certain part they should be placed at a little distance, where there is available space, and connected by lines with the parts to which they refer. They should not be placed upon shaded surfaces, but when it is difficult to avoid this, a blank space must be left in the shading where the letter occurs, so that it shall appear perfectly distinct and separate from the work. If the same part of an invention appears in more than one view of the drawing, it must always be represented by the same character, and the same character must never be used to designate different parts.

(*g*) The signature of the applicant should be placed at the lower right-hand corner of each sheet, and the signatures of the witnesses, if any, at the lower left-hand corner, all within the marginal line, but in no instance should they trespass upon the drawings. The title should be written with pencil on the back of the sheet. The permanent names and title constituting the heading will be applied subsequently by the office in uniform style.

(*h*) All views on the same sheet must stand in the same direction and must, if possible, stand so that they can be read with the sheet held in an upright position. If views longer than the width of the sheet are necessary for the proper illustration of the invention, the sheet may be turned on its side. The space for heading must then be reserved at the right and the signatures placed at the left, occupying the same space and position as in the upright views and being horizontal when the sheet is held in an upright position. One figure must not be placed upon another or within the outline of another.

(*i*) As a rule, one view only of each invention can be shown in the Gazette illustrations. The selection of that portion of a drawing best calculated to explain the nature of the specific improvement would be facilitated and the final result improved by the judicious execution of a figure with express reference to the Gazette, but which must at the same time serve as one of the figures referred to in the specification. For this purpose the figure may be a plan, elevation, section, or perspective view, according to the judgment of the draftsman. All its parts should be especially open and distinct, with very little or no shading, and it must illustrate the invention claimed only, to the exclusion of all other details. (See specimen drawing.) When well executed it will be used without curtailment or change, but any excessive fineness or crowding or unnecessary elaborateness of detail will necessitate its exclusion from the Gazette.

(*j*) Drawings transmitted to the office should be sent flat, protected by a sheet of heavy binder's board; or should be rolled for transmission in a suitable mailing tube, but should never be folded.

(*k*) An agent's or attorney's stamp, or advertisement, or written address will not be permitted upon the face of a drawing, within or without the marginal line.

53. In reissue applications the drawings upon which the original patent was issued may be used upon the filing of suitable permanent photographic copies thereof, if no changes are to be made in the drawings.

54. The foregoing rules relating to drawings will be rigidly enforced. A drawing not executed in conformity thereto may

United States Patent Office

3,465,153
Patented Sept. 2, 1969

269

*The design process
and graphics*

1

**3,465,153
RADIATION PROTECTION SYSTEM AND
APPARATUS**
Willard F. Libby, Los Angeles, Calif., assignor to
McDonnell Douglas Corporation, Santa Monica,
Calif., a corporation of Maryland
Filed Aug. 14, 1964, Ser. No. 389,734
Int. Cl. G21f 1/12, 3/02, 7/00
U.S. Cl. 250—108   16 Claims

My present invention relates generally to astronautics, the science of space flight, and more particularly to a system and apparatus and method for the protection of astronauts from the hazards of suddenly encountered radiation fields of extreme intensity in space.

Manned space flights have now been successfully achieved by both the United States and the Soviet Union. Such flights will be followed by manned space probes including lunar and interplanetary missions for the manned exploration of the Moon, Mars and Venus. The manned space program of the United States is directed towards manned exploration first of the moon and then initially only of the two planets Mars and Venus of the solar system since all of its other planets appear to be barren and lifeless. These and other probes will, of course, eventually lead to interstellar journeys over vast distances to other stellar systems for the purpose of conducting explorations aimed at discovering new worlds which are susceptible to colonization by the human race. The propulsion systems require

2

by using a greater number of stages are offset by the additional complexity involved, and it is very difficult to increase the propellant-weight ratio much beyond a certain value in the present concepts of vehicle systems.

The only feasible alternative remaining is to reduce the inert weight which is not useful for propulsion in the vehicle system so that a greater payload weight can be obtained without the need to increase initial launch weight of the vehicle system. This is an important consideration since any unnecessary inert weight in the various stages of the vehicle system imposes a heavy, additional demand on required engine thrust which is functionally related to launch weight. In a large, three stage booster system to be used on a lunar flight, for example, any change in weight of the final stage will be reflected in a similar change in the total launch weight multiplied, however, by a growth factor which may easily number in the hundreds.

In undertaking manned, space exploration missions, the astronauts may be exposed to radiation fields of high intensity in space. Biological damage is done by the ionization produced by radiation and high energy charged particles which pass through the tissues of the astronauts. A lethal action arises when the radiation dosage is excessive such that changes in living cells result in their death when they attempt division. Of course, extremely high radiation dose rates which may be lethal to an astronaut after a relatively short exposure period are encountered except rarely in ordinary

---

Fig. 12.33 Portion of the first of six sheets of specifications (petition) for the radiation protection system illustrated in Figs. 12.31 and 12.32. Copies of patents may be purchased from the Patent Office (50 cents each).

be admitted for purposes of examination if it sufficiently illustrates the invention, but in such case the drawing must be corrected or a new one furnished before the application will be allowed. The necessary corrections will be made by the office, upon applicant's request and at his expense.

55. Applicants are advised to employ competent draftsmen to make their drawings.

The office will furnish the drawings at cost, as promptly as its draftsmen can make them, for applicants who cannot otherwise conveniently procure them.

Design projects

Note to Instructor: Design projects may be included within traditional courses covering engineering drawing. Either two or three weeks of a semester may be set aside for design projects, or the design tasks may be spread out over a much longer time to parallel the work on the regularly assigned classroom drawing problems. In developing the design of a simple product, a student working in a group can acquire much of the basic knowledge deemed essential.

The following problems are offered as suggestions to stimulate creativity and give some additional experience in both pictorial and multiview sketching. Students who have an inclination to design useful mechanisms should be encouraged to select a problem for themselves, for the creative mind works best when directed to a task in which it already has some interest. However, the young beginner should confine his activities to ideas for simple mechanisms that do not require extensive training in machine design and the engineering sciences.

It is suggested that the group leader for the design project prepare a design event and activities schedule using a form that is similar to the one shown in Fig. 12.34. A carefully prepared record should be kept of the progress of the design work.

1. Prepare design sketches (both pictorial and multiview) for an open-end wrench to fit the head of a bolt (regular series) having a body diameter of 1 in. Give dimensions on the multiview sketch and specify the material.

2. Prepare design sketches (both pictorial and multiview) for a wrench having a head with four or more fixed openings to fit the heads of bolts having nominal diameters of $\frac{5}{8}$, $\frac{3}{4}$, $\frac{7}{8}$, and 1 in. Dimension the multiview sketch and specify the material.

DESIGN SCHEDULE & PROGRESS REPORT

Project:_____   **Work Periods:**_____

Design Team:_____   ***Due Dates***

Leader:_____   **Investigation Report:**_____

Member:_____   **Proposal:**_____

_____   **Preliminary Report:**_____

_____   **Final Report:**_____

_____

ASSIGNMENT		EST. HRS	START DATE	PERCENT COMPLETE				ACTUAL HOURS
	Members			25%	50%	75%	100%	
1								
2								
3								
4								
5								

Fig. 12.34 Design schedule and progress record.

3. Prepare a series of design sketches for a bumper hitch (curved bumper) for attaching a light two-wheel trailer to a passenger automobile. Weight of trailer is 135 kg.

4. Prepare sketches for a hanger bracket to support a $\frac{1}{2}$-in. control rod. The bracket must be attached to a vertical surface to which the control rod is parallel. The distance between the vertical surface and the center line of the control rod is 4 in.

5. Prepare sketches for a mechanism to be attached to a two-wheel hand truck to make it easy to move the truck up and down stairs with a heavy load.

6. Prepare sketches for a quick-acting clamp that can be used to hold steel plates in position for making a lap weld.

7. Prepare a pictorial sketch of a bracket that will support an instrument panel at an angle of 45° with a vertical bulkhead to which the bracket will be attached. The bracket should be designed to permit the panel to be raised or lowered a height distance of 100 mm as desired.

8. Prepare sketches for an adjustable pipe support for a $1\frac{1}{2}$-in. pipe that is to carry a chemical mixture in a factory manufacturing paint. The pipe is overhead and is to be supported at 10-ft intervals where the adjustable supports can be attached to the lower chords of the roof trusses. The lower chord of a roof truss is formed by two angles $2\frac{1}{2} \times 2\frac{1}{2} \times \frac{5}{16}$ that are separated by $\frac{3}{8}$-in.-thick washers.

9. Prepare sketches and working drawings for an easy-to-operate fast-release glider hitch. It is suggested that the student talk with several members of a local glider club to determine the requirements for the hitch and what improvements can be made for the hitch that is being used. Follow the stages of design listed in Sec. 12.13.

10. Prepare design sketches for a camera mounting that may be quickly attached and removed from any selected surface on an automobile, boat, or other type of moving vehicle. The device is to sell for not more than $9.95. Standard parts are to be used if possible. Follow through the several stages of design listed in

Sec. 12.13 as required by the instructor. Make either a complete set of working drawings or drawings of selected parts.

11. Design a tire pump that will be more efficient and easier to operate than the ones now on the market.

12. Design an electric door lock that can be opened by pushing buttons in a special pattern.

13. Design a thermostatically controlled, electrically heated sidewalk upon which snow and ice will not accumulate.

14. Design an automatic automobile theft alarm.

15. Design a weather-controlled window-closing system.

16. Design a child's toy.

17. Design an automatic pet food dispenser.

18. *Class Projects.* Each student in the class is to prepare a description (on a single sheet of paper) of an innovational idea of a needed design that he deems to be suitable for a group project. From among the ideas collected several of the most worthy will be assigned to the class for development, one idea to each group. Each group of students is to be considered as a project group and shall be headed by one student member who may be thought of as being the project engineer. The instructor will assume the role of coach for all of the groups. About midway in the total time period assigned for the development of the design, each group is to submit a written preliminary design report that shall be accompanied by sketches. A final design report (see Sec. 12.24) will be due when the project has been completed. Each report will be judged on the following: (1) evidence of good group organization, (2) quality of technical work, (3) function and appearance, (4) economic analysis, (5) manufacturing methods and requirements, and (6) over-all effectiveness of communication (written and/or oral). (*Note:* The instructor will decide whether or not finished shop drawings are to be made for all of the parts. Information needed for the preparation of finished drawings may be found in Chapters 13–16.)

Part IV

Graphics for design and communication

The Mole was used for rock tunneling in constructing the 98-mile Metro System in our nation's capital. This massive machine has a rotating cutter head 19 feet in diameter with 45 disc cutters. The cutting head, which may turn from 2 to 4 revolutions per minute, is driven by six 150-horsepower electric motors. Hydraulic rams force the cutting head against the rock face. The Mole weighs 285 tons and is 300 feet long. Many well-dimensioned drawings were needed for the construction of this giant machine. *(Courtesy The Robbins Co. and Washington Metropolitan Transit Authority)*

Part IV

Graphics for design and communication

The Mole was used for rock tunneling in constructing the 98-mile Metro System in our nation's capital. This massive machine has a rotating cutter head 19 feet in diameter with 45 disc cutters. The cutting head, which may turn from 2 to 4 revolutions per minute, is driven by six 150-horsepower electric motors. Hydraulic rams force the cutting head against the rock face. The Mole weighs 285 tons and is 300 feet long. Many well-dimensioned drawings were needed for the construction of this giant machine. *(Courtesy The Robbins Co. and Washington Metropolitan Transit Authority)*

13

Dimensions, notes, limits, and geometric tolerances

A □ Fundamentals and techniques

13.1

Introduction. A detail drawing, in addition to giving the shape of a part, must furnish information such as the distances between surfaces, locations of holes, kind of finish, type of material, number required, and so forth. The expression of this information on a drawing by the use of lines, symbols, figures, and notes is known as *dimensioning*.

Intelligent dimensioning requires engineering judgment and a thorough knowledge of the practices of patternmaking, forging, and machining.

13.2

Theory of Dimensioning. Any part may be dimensioned easily and systematiclally by dividing it into simple geometric solids. Even complicated parts, when analyzed, are usually found to be composed principally of cylinders and prisms and, frequently, frustums of pyramids and cones. The dimensioning of an object may be accomplished by dimensioning each elemental form to indicate its size and relative location from a center line, base line, or finished surface. A machine drawing requires two types of dimensions: *size dimensions* and *location dimensions*.

13.3

Size Dimensions (Fig. 13.2). Size dimensions give the size of a piece, component part, hole, or slot.

Figure 13.1 should be carefully analyzed, as the placement of dimensions

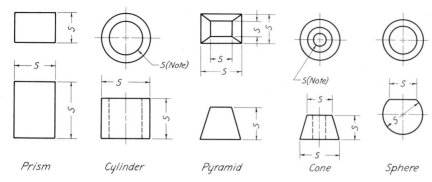

Fig. 13.1 Dimensioning geometric shapes.

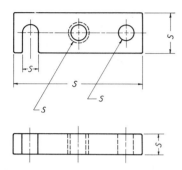

Fig. 13.2 Size dimensions.

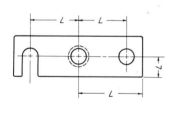

Fig. 13.3 Location dimensions.

shown is applicable to the elemental parts of almost every piece.

The rule for placing the three principal dimensions (width, height, and depth) on the drawing of a prism or modification of a prism is as follows: *Give two dimensions on the principal view and one dimension on one of the other views.*

The circular cylinder, which appears as a boss or shaft, requires only *the diameter and length, both of which are shown preferably on the rectangular view.* It is better practice to dimension a hole (negative cylinder) by giving the diameter and operation as a note on the contour view with a leader to the circle (Figs. 13.1 and 13.54).

Cones are dimensioned by giving *the diameter of the base and the altitude on the same view.* A taper is one example of a conical shape found on machine parts (Fig. 13.49).

Pyramids, which frequently form a part of a structure, are dimensioned by giving *two dimensions on the view showing the shape of the base.*

A sphere requires only the diameter.

13.4
Location Dimensions. Location dimensions fix the relationship of the component parts (projections, holes, slots, and

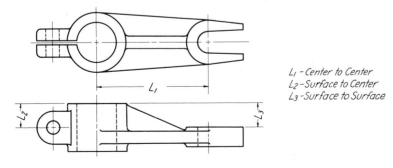

Fig. 13.4 Types of location dimensions.

L_1 – Center to Center
L_2 – Surface to Center
L_3 – Surface to Surface

other significant forms) of a piece or structure (Fig. 13.3). Particular care must be exercised in the selection and placing of location dimensions because on them depends the accuracy of the operations in making a piece and the proper mating of the piece with other parts. To select location dimensions intelligently, one must first determine the contact surfaces, finished surfaces, and center lines of the elementary geometric forms and, with the accuracy demanded and the method of production in mind, decide from what other surface or center line each should be located. Mating location dimensions must be given from the same center line or finished surface on both pieces.

Location dimensions may be from center to center, surface to center, or surface to surface (Fig. 13.4).

13.5
Procedure in Dimensioning. The theory of dimensioning may be applied in six steps, as follows:

1. Mentally divide the object into its component geometrical shapes.

2. Place the size dimensions on each form.

3. Select the locating center lines and surfaces after giving careful consideration to mating parts and to the processes of manufacture.

4. Place the location dimensions so that each geometrical form is located from a center line or finished surface.

5. Add the overall dimensions.

6. Complete the dimensioning by adding the necessary notes.

13.6
Placing Dimensions. Dimensions must be placed where they will be most easily understood—in the locations where the reader will expect to find them. They generally are attached to the view that shows the contour of the features to which they apply, and a majority of them usually will appear on the principal view (Fig. 13.14). Except in cases where special convenience and ease in reading are desired, or when a dimension would be so far from the form to which it referred that it might be misinterpreted, dimensions should be placed outside a view. They should appear directly on a view only when clarity demands.

All extension and dimension lines should be drawn before the arrowheads have been filled in or the dimensions, notes, and titles have been lettered. Placing dimension lines not less than .50 in. (13 mm) from the view and at least .38 in. (10 mm) from each other will provide spacing ample to satisfy the one rule to which there is no exception: *Never crowd dimensions*. If the location of a dimension forces a poor location on other dimensions, its shifting may allow all to be placed more advantageously without sacrificing clarity. Important location dimensions should be given where they will be conspicuous, even if a size dimension must be moved.

The person dimensioning a drawing must make certain that every feature has been completely dimensioned and that no dimension has been repeated in a second view.

13.7

Dimensioning Practices. A generally recognized system of lines, symbols, figures, and notes is used to indicate size and location. Figure 13.5 illustrates dimensioning terms and notation.

A *dimension line* is a lightweight line that is terminated at each end by an arrowhead. A numerical value, given along the dimension line, specifies the number of units for the measurement that is indicated (Fig. 13.6). When the numerals are in a single line, the dimension line is broken near the center, as shown in (*a*) and (*b*). Under no circumstances should the line pass through the numerals. When the numerals are in two lines the dimension line may be drawn without a break and one line of numerals may be placed above the dimension line and the other below, as in (*c*).

Extension lines are light, continuous lines extending from a view to indicate the extent of a measurement given by a dimension line that is located outside a view. They start .06 in. (1.5 mm) from the view and extend .12 in. (3 mm) beyond the dimension line (Fig. 13.5).

Arrowheads are drawn for each dimension line, before the figures are lettered. They are made with the same pen or pencil used for the lettering. The size of an arrowhead, although it may vary with the size of a drawing, should be uniform on any one drawing. To have the proper proportions, the length of an arrowhead must be approximately three times its spread. This length for average work is usually .12 in. (3 mm). Figure 13.7 shows enlarged drawings of arrowheads of correct

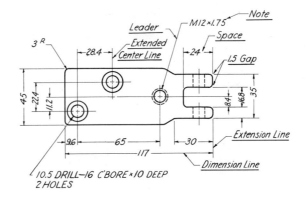

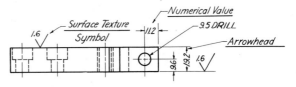

Fig. 13.5 Terms and dimensioning notation. (Dimension values are in millimeters.)

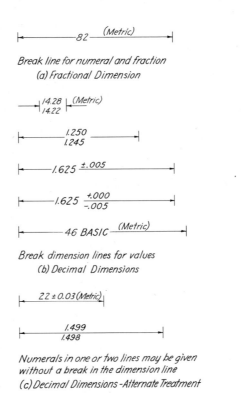

Fig. 13.6 Dimension line.

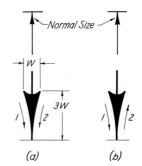

Fig. 13.7 Formation of arrowheads.

proportions. Although many draftsmen draw an arrowhead with one stroke, the beginner will get better results by using two slightly concave strokes drawn toward the point (*a*) or, as shown in (*b*), one stroke drawn to the point and one away from it.

A *leader* or *pointer* is a light, continuous line (terminated by an arrowhead) that extends from a note to the feature of a piece to which the note applies (Fig. 13.5). It should be made with a straightedge and should not be curved or made freehand.

A leader pointing to a curve should be radial, and the first .12 in. (3 mm) of it should be in line with the note (Fig. 13.5).

Finish marks indicate the particular surfaces of a rough casting or forging that are to be machined or "finished." They are placed in all views, across the visible or invisible lines that are the edge views of surfaces to be machined (Fig. 13.8).

The modified italic *f*, shown in Fig. 13.8(*b*), is the form of finish mark that is still used to some extent even though it has not been accepted by ANSI and does not appear in ANSI Y14.5 or any related standard. At this time, an effort is being made to come to an agreement on a new and different symbol that will be accepted worldwide. In the meantime, a number of U.S. companies are using a symbol that is a modification of the surface-texture symbol. As can be noted in (*a*), the check-mark (✓) portion of the texture symbol is used with a horizontal bar added at the top of the short leg of the check-mark.

It is not necessary to show finish marks on holes. They are also omitted, and a title note, "finish all over," is substituted, if the piece is to be completely machined. Finish marks are not required when limit dimensions are used.

Dimension figures should be lettered either horizontally or vertically with the whole numbers equal in height to the cap-ital letters in the notes and guidelines and slope lines must be used. The numerals must be legible; otherwise, they might be misinterpreted in the shop and cause errors, which would be embarrassing to the draftsman.

13.8
Fractional Dimensioning. For ordinary work, where accuracy is relatively unimportant, shopmen work to nominal dimensions given as common fractions of an inch, such as $\frac{1}{2}$, $\frac{1}{4}$, $\frac{1}{8}$, $\frac{1}{16}$, $\frac{1}{32}$, and $\frac{1}{64}$. When dimensions are given in this way, many large corporations specify the required accuracy through a note on the drawing that reads as follows: *Permissible variations on common fraction dimensions to machined surfaces to be $\pm.010$ unless otherwise specified.* It should be understood that the allowable variations will differ among manufacturing concerns because of the varying degree of accuracy required for different types of work.

13.9
Decimal System. At present, the fractional system, which dominated the whole of American industry until the Ford Motor Company adopted the decimal-inch some 40 years ago, is still used in those fields that are not under the direct influence of the automotive and aircraft companies. How long the inch system of fractions and decimals will last cannot be foretold at this time. In the long run this matter may be of little importance, since it may be only a few years before the millimeters of the metric system completely replace the old English inch system. In the United States more and more multinational companies are making metric drawings (Figs. 5.1 and 16.3).

However, this need not be of great concern to the student, for he should be primarily interested in the selection and placement of dimensions. At a later time he will find it easy to use either the fractional or decimal inch or metric systems as required in his field of employment. To assist one to use any system, a standard conversion table has been provided in the Appendix (Table 1).

In Fig. 13.9 a drawing is shown that illustrates decimal dimensioning.

The following practices are recommended for decimal-inch dimensioning.

(*a*) Two-place decimals should be used for dimensions where tolerance limits of $\pm.01$ or more can be allowed (Fig. 13.9).

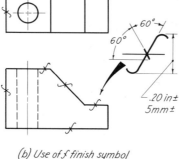

(a) Use of ✓ finish symbol   (b) Use of *f* finish symbol

Fig. 13.8 Finish marks.

13

Dimensions, notes, limits, and geometric tolerances

A □ Fundamentals and techniques

13.1

Introduction. A detail drawing, in addition to giving the shape of a part, must furnish information such as the distances between surfaces, locations of holes, kind of finish, type of material, number required, and so forth. The expression of this information on a drawing by the use of lines, symbols, figures, and notes is known as *dimensioning*.

Intelligent dimensioning requires engineering judgment and a thorough knowledge of the practices of patternmaking, forging, and machining.

13.2

Theory of Dimensioning. Any part may be dimensioned easily and systematiclally by dividing it into simple geometric solids. Even complicated parts, when analyzed, are usually found to be composed principally of cylinders and prisms and, frequently, frustums of pyramids and cones. The dimensioning of an object may be accomplished by dimensioning each elemental form to indicate its size and relative location from a center line, base line, or finished surface. A machine drawing requires two types of dimensions: *size dimensions* and *location dimensions*.

13.3

Size Dimensions (Fig. 13.2). Size dimensions give the size of a piece, component part, hole, or slot.

Figure 13.1 should be carefully analyzed, as the placement of dimensions

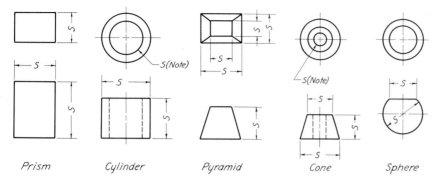

Fig. 13.1 Dimensioning geometric shapes.

Prism   Cylinder   Pyramid   Cone   Sphere

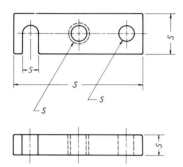

Fig. 13.2 Size dimensions.

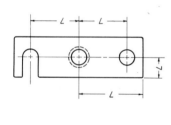

Fig. 13.3 Location dimensions.

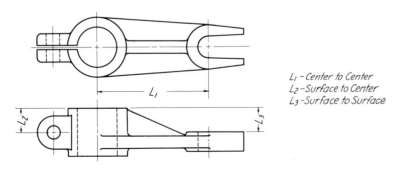

L_1 – Center to Center
L_2 – Surface to Center
L_3 – Surface to Surface

Fig. 13.4 Types of location dimensions.

shown is applicable to the elemental parts of almost every piece.

The rule for placing the three principal dimensions (width, height, and depth) on the drawing of a prism or modification of a prism is as follows: *Give two dimensions on the principal view and one dimension on one of the other views.*

The circular cylinder, which appears as a boss or shaft, requires only *the diameter and length, both of which are shown preferably on the rectangular view.* It is better practice to dimension a hole (negative cylinder) by giving the diameter and operation as a note on the contour view with a leader to the circle (Figs. 13.1 and 13.54).

Cones are dimensioned by giving *the diameter of the base and the altitude on the same view.* A taper is one example of a conical shape found on machine parts (Fig. 13.49).

Pyramids, which frequently form a part of a structure, are dimensioned by giving *two dimensions on the view showing the shape of the base.*

A sphere requires only the diameter.

13.4

Location Dimensions. Location dimensions fix the relationship of the component parts (projections, holes, slots, and other significant forms) of a piece or structure (Fig. 13.3). Particular care must be exercised in the selection and placing of location dimensions because on them depends the accuracy of the operations in making a piece and the proper mating of the piece with other parts. To select location dimensions intelligently, one must first determine the contact surfaces, finished surfaces, and center lines of the elementary geometric forms and, with the accuracy demanded and the method of production in mind, decide from what other surface or center line each should be located. Mating location dimensions must be given from the same center line or finished surface on both pieces.

Location dimensions may be from center to center, surface to center, or surface to surface (Fig. 13.4).

13.5

Procedure in Dimensioning. The theory of dimensioning may be applied in six steps, as follows:

1. Mentally divide the object into its component geometrical shapes.

2. Place the size dimensions on each form.

3. Select the locating center lines and surfaces after giving careful consideration to mating parts and to the processes of manufacture.

4. Place the location dimensions so that each geometrical form is located from a center line or finished surface.

5. Add the overall dimensions.

6. Complete the dimensioning by adding the necessary notes.

13.6

Placing Dimensions. Dimensions must be placed where they will be most easily understood—in the locations where the reader will expect to find them. They generally are attached to the view that shows the contour of the features to which they apply, and a majority of them usually will appear on the principal view (Fig. 13.14). Except in cases where special convenience and ease in reading are desired, or when a dimension would be so far from the form to which it referred that it might be misinterpreted, dimensions should be placed outside a view. They should appear directly on a view only when clarity demands.

All extension and dimension lines should be drawn before the arrowheads

(*b*) Decimals to three or more places should be used for tolerance limits less than ±.010 [see Fig. 13.6(*b*) and (*c*)].

(*c*) In the case of a two-place decimal, the second decimal place should preferably be an even digit such as .02, .04, and .06 so that when it is divided by 2, the result will remain a two-place decimal. Odd two-place decimals may be used where necessary for design reasons.

(*d*) Common fractions may be used to indicate standard nominal sizes for materials and for features produced by standard tools as in the case of drilled holes, threads, keyways, and so forth.

(*e*) When desired, decimal equivalents of nominal commercial sizes may be used for materials and for features such as drilled holes, threads, and so forth that are produced by standard tools.

13.10

Use of the Metric System (Fig. 13.10). Since the metric system has worldwide acceptance and has been legalized for use in this country (approved by act of Congress in 1866), engineers and draftsmen in the United States will find it advantageous to be able to interpret and quite possibly prepare detail drawings using metric dimensioning. At present, there are persons who argue strongly that American industry must change over to this system if the United States is to retain leadership in world trade. In any case, metric dimensioning is appearing more frequently than in the past on drawings made in the United States. The unit of measurement is the millimeter (mm) and, since this fact is generally known by all persons using the system, no indicating marks are needed.

Under the metric system, drawings are prepared to scales based on divisions of 10, such as 1 to 2, 1 to 5, 1 to 10, and so forth. A millimeter is one-thousandth part of a meter, which our government has established as being 39.37 in. in length.

Many of the illustrations in this chapter have been dimensioned using millimeters.

B □ General Dimensioning Practices

13.11

Selection and Placement of Dimensions and Notes. The reasonable application of the selected dimensioning practices that

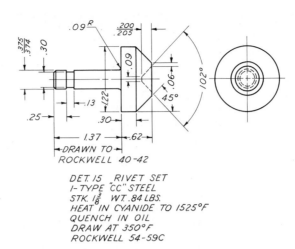

DET. 15   RIVET SET
1- TYPE "CC" STEEL
STK. 1⅜   WT. .84 LBS.
HEAT IN CYANIDE TO 1525°F
QUENCH IN OIL
DRAW AT 350°F
ROCKWELL 54-59C

Fig. 13.9 Decimal dimensioning. (*Courtesy Ford Motor Company*)

follow should enable a student to dimension acceptably. The practices in boldface type should never be violated. In fact, these have been so definitely established by practice that they might be called rules.

When applying the dimensioning practices that follow, the student should realize that each company has its own drafting standards and thus the practices set forth in this chapter may not be exactly the same as the practices followed in all corporations throughout the United States. However, the practices given are in general agreement with recommendations set forth in ANSI standards.

1. Place dimensions using either of two recognized methods—aligned or unidirectional.

a. *Aligned method.* Place the numerals for the dimension values so that they are readable from the bottom and right side of the drawing. An aligned expression is placed along and in the direction of the dimension line (Fig. 13.11). Make the values for oblique dimensions readable from the directions shown in Fig. 13.31.

b. *Unidirectional method.* Place the numerals for the dimension values so that they can be read from the bottom of the drawing (see Fig. 13.12). The fraction bar for a common fraction should be parallel to the bottom of the drawing.

2. Place dimensions outside a view, unless they will be more easily and quickly understood if shown on the view (Figs. 13.12 and 13.13).

3. Place dimensions between views unless the rules, such as the contour rule,

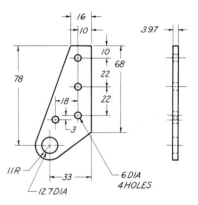

Fig. 13.10 Metric dimensioning. (Dimensions are in millimeters.)

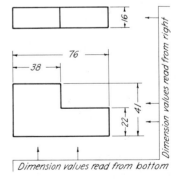

Fig. 13.11 Reading dimensions aligned.

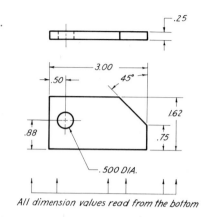

All dimension values read from the bottom

**Fig. 13.12 Reading dimensions—
unidirectional.**

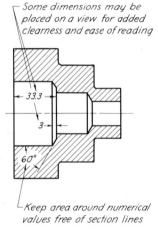

⌐Some dimensions may be
placed on a view for added
clearness and ease of reading

⌐Keep area around numerical
values free of section lines

Fig. 13.13 Dimensions on the view.

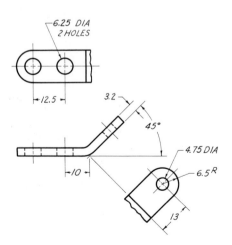

Fig. 13.16 Dimensioning an angle bracket.
(Dimension values are in millimeters.)

the rule against crowding, and so forth,
prevent their being so placed.

**4. Do not use an object line or a center
line as a dimension line.**

5. Locate dimension lines so that they
will not cross extension lines.

6. If possible, avoid crossing two dimension lines.

7. A center line may be extended to
serve as an extension line (Fig. 13.14).

8. Keep parallel dimensions equally
spaced [usually .38 in. (10 mm) apart]
and the figures staggered (Fig. 13.15).

**9. Always give locating dimensions to
the centers of circles that represent
holes, cylindrical projections, or bosses**
(Figs. 13.5 and 13.14).

10. If possible, attach the location dimensions for holes to the view on which
they appear as circles (Fig. 13.16).

11. Group related dimensions on the
view showing the contour of a feature
(Fig. 13.14).

12. Arrange a series of dimensions in a
continuous line (Fig. 13.17).

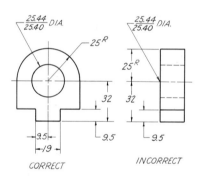

Fig. 13.14 Contour principle of dimensioning.

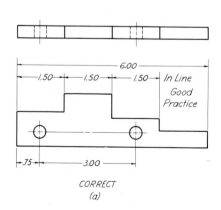

Fig. 13.17 Consecutive dimensions.

13. Dimension from a finished surface,
center line, or base line that can be
readily established (Figs. 13.38 and
13.50).

14. Stagger the figures in a series of
parallel dimension lines to allow sufficient space for the figures and to prevent confusion (Fig. 13.15).

15. Place longer dimensions outside
shorter ones so that extension lines will
not cross dimension lines.

16. Give three overall dimensions located outside any other dimensions
(unless the piece has cylindrical
ends—see 42 and Fig. 13.42).

17. When an overall is given, one intermediate distance should be omitted
unless noted (REF) as being given for
reference (Figs. 13.18 and 13.19).

18. Do not repeat a dimension. One of
the duplicated dimensions may be
missed if a change is made. Give only
those dimensions that are necessary to
produce or inspect the part.

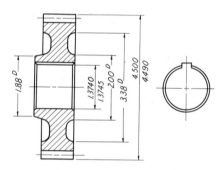

Fig. 13.15 Parallel dimensions.

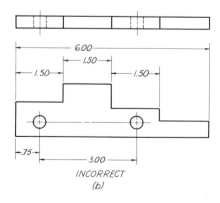

19. Make decimal points of a sufficient size so that dimensions cannot be mis-read.

20. When dimension figures appear on a sectional view, show them in a small uncrosshatched portion so that they may be easily read. This may be accomplished by doing the section lining after the dimensioning has been completed (Fig. 13.20).

21. When an arc is used as a dimension line for an angular measurement, use the vertex of the angle as the center [Fig. 13.21(*a*)]. It is usually undesirable to terminate the dimension line for an angle at lines that represent surfaces. It is better practice to use an extension line [Fig. 13.21(*b*)].

22. Place the figures of angular dimensions so they will read from the bottom of a drawing, except in the case of large angles (Fig. 13.22).

23. Dimension an arc by giving its radius followed by the abbreviation R, and indicate the center with a small cross. [Locate the center by dimensions. (Fig. 13.23).]

24. TRUE R is added after the radius value, where the radius is dimensioned in a view that does not show the true shape of the arc (Fig. 13.24).

25. Show the diameter of a circle, never the radius. If it is not clear that the dimension is a diameter, the figures should be followed by the abbreviation D or DIA (Figs. 13.25 and 13.26). Often this will allow the elimination of one view.

26. When dimensioning a portion of a sphere with a radius the term SPHER R is added (Fig. 13.27).

27. Letter all notes horizontally (Fig. 13.50).

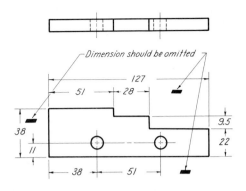

Fig. 13.18 Omit unnecessary dimensions.

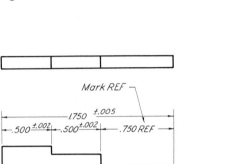

Fig. 13.19 Mark dimensions REF, if given for reference.

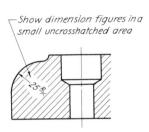

Fig. 13.20 Dimension figures on a section view.

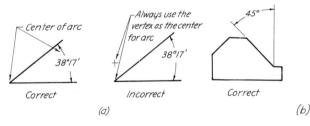

Fig. 13.21 Dimensioning an angle.

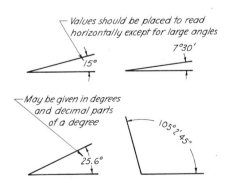

Fig. 13.22 Angular dimensions.

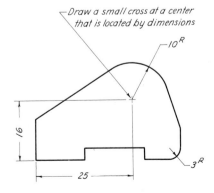

Fig. 13.23 Dimensioning a circular arc. (Dimension values are in millimeters.)

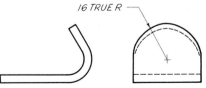

Fig. 13.24 Dimensioning a circular arc— true R.

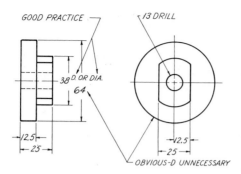

Fig. 13.25 Dimensioning a cylindrical piece.

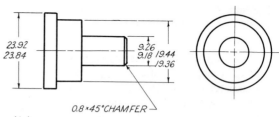

Note:
*Although it is better practice to use a minimum of two views,
a cylindrical part may be completely described in one view (no
end view) by using the abbreviation DIA with the dimension.*

Fig. 13.26 Dimensioning machined cylinders.

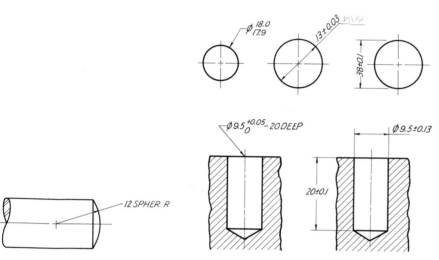

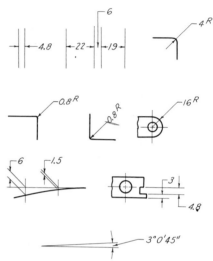

**Fig. 13.27 Dimensioning a piece with a
spherical end.**

Fig. 13.28 Dimensioning holes.

Fig. 13.29 Dimensioning in limited spaces.

**28. Make dimensioning complete, so
that it will not be necessary for a work-
man to add or subtract to obtain a desired dimension or to scale the drawing.**

29. Give the diameter of a circular hole,
never the radius, because all hole-form-
ing tools are specified by diameter. If
the hole does not go through the piece,
the depth may be given as a note (Fig.
13.28).

30. Never crowd dimensions into small

spaces. Use the practical methods suggested in Fig. 13.29.

31. Avoid placing inclined dimensions
in the shaded areas shown in Fig. 13.30.
Place them so that they may be conveniently read from the right side of the
drawing. If this is not desirable, make
the figures read from the left in the
direction of the dimension line [Fig.
13.31 (*b*)]. The unidirectional method is
shown in (*a*).

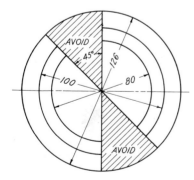

Fig. 13.30 Areas to avoid.

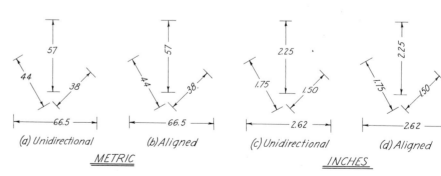

Fig. 13.31 Reading horizontal, vertical, and oblique dimensions.

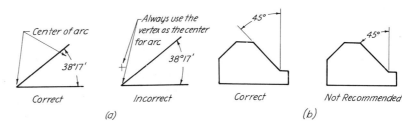

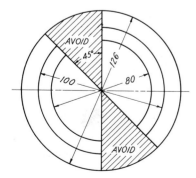

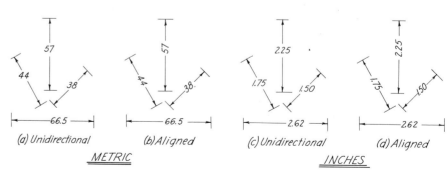

32. Omit superfluous dimensions. Do not supply dimensional information for the same feature in two different ways.

33. Give dimensions up to 72 in. in inches, except on structural and architectural drawings (Fig. 13.32). Omit the inch marks when all dimensions are in inches.

34. Show dimensions in feet and inches as illustrated in Fig. 13.33. Note that the use of the hyphen in (*a*) and (*b*) and the cipher in (*b*) eliminates any chance of uncertainty and misinterpretation.

35. If feasible, design a piece and its elemental parts to such dimensions as .10, .40 and .50 in. or 4, 10, 12 and 20 mm. Except for critical dimensions and hole sizes, dimension values in millimeters should be given in a full number of millimeters, preferably to an even number (12, 16, 20, 40, etc.) that, when divided by 2, will remain a full number. Metric drill sizes are given in Appendix Tables 35 and 36.

36. Dimension a chamfer by giving the angle and length as shown in Fig. 13.34. For a 45° angle only, it is permissible to give the needed information as a note. The word "chamfer" may be omitted in the note form.

37. Equally spaced holes in a circular flange may be dimensioned by giving the dimeter of the bolt circle, across the circular center line, and the size and number of holes, in a note (Fig. 13.35).

38. When holes are unequally spaced on a circular center line, give the angles as illustrated in Fig. 13.36.

39. Holes that must be accurately located should have their location established by the coordinate method. Holes arranged in a circle may be located as shown in Fig. 13.37 rather than through

the use of angular measurements. Figure 13.38 shows the application of the coordinate method to the location of holes arranged in a general rectangular form. The method with all dimensions referred to datum lines is sometimes called *base-line dimensioning*.

40. Dimension a curved line by giving offsets or radii.

a. A noncircular curve may be dimen-

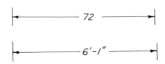

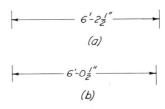

Fig. 13.32　**Dimension values.**

Fig. 13.33　**Feet and inches.**

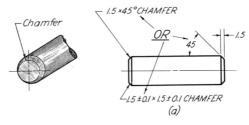

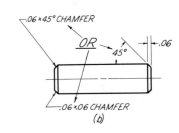

Fig. 13.34　**Dimensioning a chamfer.**

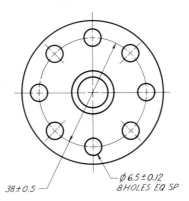

Ø 6.5 ± 0.12
8 HOLES EQ SP

38 ± 0.5

Fig. 13.35　**Equally spaced holes.**

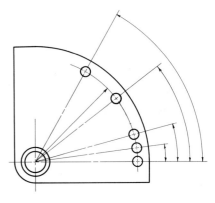

Fig. 13.36　**Locating holes on a circle by polar coordinates.**

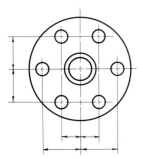

Fig. 13.37　**Accurate location dimensioning of holes.**

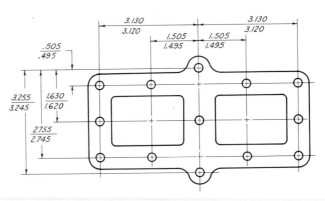

Fig. 13.38　**Location dimensioning of holes.**

sioned by the coordinate method illustrated in Fig. 13.39. Offset measurements are given from datum lines.

b. A curved line, which is composed of circular arcs, should be dimensioned by giving the radii and locations of either the centers or points of tangency (Figs. 13.40 and 13.41).

41. Show an offset dimension line for an arc having an inaccessible center (Fig. 13.41). Locate with true dimensions the point placed in a convenient location that represents the true center.

42. Dimension, as required by the method of production, a piece with cylindrical ends. Give the diameters and center-to-center distance (Fig. 13.42). No overall is required.

43. The method to be used for dimensioning a piece with rounded ends is determined by the degree of accuracy required and the method of production (Figs. 13.43–13.45).

a. It has been customary to give the radii and center-to-center distance for parts and contours that would be laid out and/or machined using centers and radii. A link (Fig. 13.43) or a pad with a slot is dimensioned in this manner to satisfy the requirements of the pattern-maker and machinist. An overall dimension is not needed.

b. Overall dimensions are recommended for parts having rounded ends when considerable accuracy is required. The radius is indicated but is not dimensioned when the ends are fully rounded. In Fig. 13.44, the center-to-center hole distance has been given because the hole location is critical.

c. Slots are dimensioned by giving length and width dimensions. They are located by dimensions given to their longitudinal center line and to either one end or a center line (Fig. 13.45).

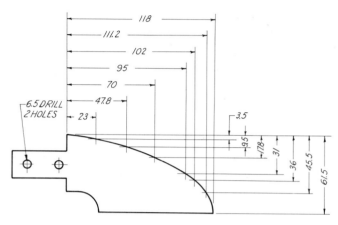

Fig. 13.39 Dimensioning curves by offsets.

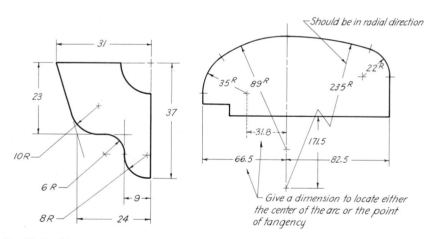

Fig. 13.40 Dimensioning curves consisting of circular arcs.

Fig. 13.41 Dimensioning curves by radii.

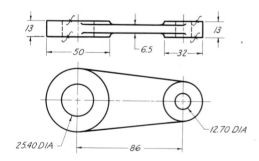

Fig. 13.42 Dimensioning a part with cylindrical ends—link.

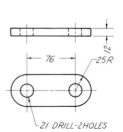

Fig. 13.43 Link with rounded ends.

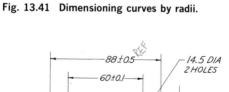

Fig. 13.44 Dimensioning a part with rounded ends.

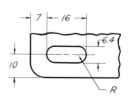

Fig. 13.45 Dimensioning a slot.

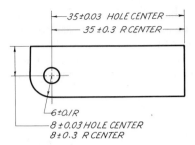

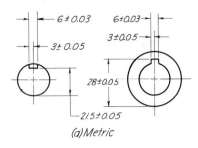

(a) Metric

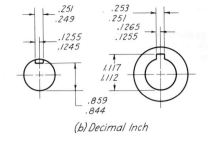

(b) Decimal Inch

Fig. 13.46 Dimensioning a piece with a hole and rounded end.

Fig. 13.47 Dimensioning keyways and keyslots.

d. When the location of a hole is more critical than the location of a radius from the same center, the radius and the hole should be dimensioned separately (Fig. 13.46).

44. A keyway on a shaft or hub should be dimensioned as shown in Fig. 13.47. Woodruff keyslots are dimensioned as shown in (b).

45. When knurls are to provide a rough surface for better grip, it is necessary to specify the pitch and kind of knurl, as shown in Fig. 13.48(a) and (b). When specifying knurling for a press fit, it is best practice to give the diameter before knurling with a tolerance and include the minimum diameter after knurling in the note that gives the pitch and type of knurl, as shown in (c).

46. Dimension special tapers as illustrated in Fig. 13.49. Standard tapers require one diameter, the length, and a note specifying the taper by number. The usual practice is to give the diameter at the large end.

Special tapers may be dimensioned in different ways, as illustrated in Fig. 13.49. As can be observed, the following dimensions may be used in different combinations to specify the size and form of tapered surfaces of conical tapers.

1. The diameter at an end of the tapered feature.

2. The length of the tapered portion of the piece.

3. The diameter at a selected cross-sectional plane. The position of this plane, which may or may not be within the length of the tapered portion of the piece, is shown with a basic dimension.

4. A dimension locating a cross-sectional plane at which a basic diameter is specified.

5. The rate of taper (note).

6. The included angle.

The dimensions and notes given in (b) and (e) are considered adequate for tapers that engage one another permanently or intermittently.

Flat tapers may be dimensioned as shown in (c) or (f).

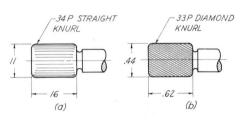

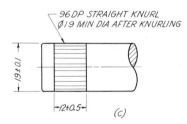

(a)   (b)   (c)

Fig. 13.48 Dimensioning knurls.

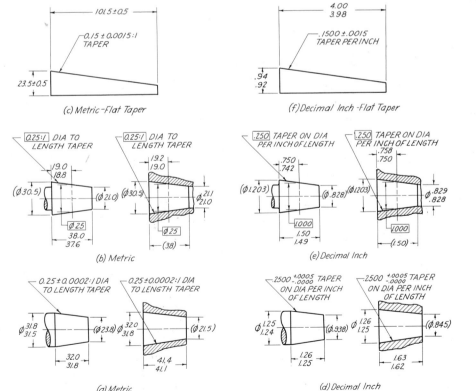

Fig. 13.49 Dimensioning tapers.

47. A half section may be dimensioned through the use of hidden lines on the external portion of the view (Fig. 13.50).

48. The fact that a dimension is out of scale may be indicated by a straight line placed underneath the dimension value (Fig. 13.51).

49. In sheet-metal work, mold lines are used in dimensioning instead of the centers of the arcs (see Fig. 13.52). A mold line (construction line) is the line at the intersection of the plane surfaces adjoining a bend.

13.12

Dimensions from Datum. When it is necessary to locate the holes and surfaces of a part with a considerable degree of accuracy, it is the usual practice to specify their position by dimensions given from a datum (Fig. 13.53) in order to avoid cumulative tolerances. By this method the different features of a part are located with respect to carefully selected datums and not with respect to each other (Fig. 13.38). Lines and surfaces that are selected to serve as datums must be easily recognizable and accessible during production. Corresponding datum points, lines, or surfaces must be used as datums on mating parts.

13.13

Notes (Fig. 13.54). The use of properly composed notes often adds clarity to the presentation of dimensional information involving specific operations. Notes also are used to convey supplementary instructions about the kind of material, kind

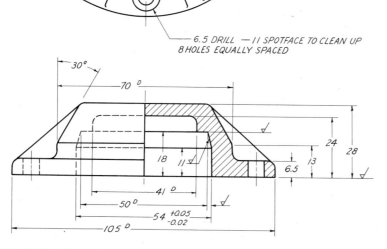

6.5 DRILL — 11 SPOTFACE TO CLEAN UP
8 HOLES EQUALLY SPACED

Fig. 13.50 Dimensioning a half section.

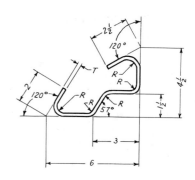

Fig. 13.52 Profile dimensioning.

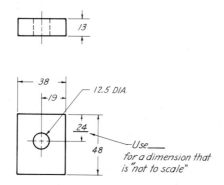

Fig. 13.51 Dimension out of scale.

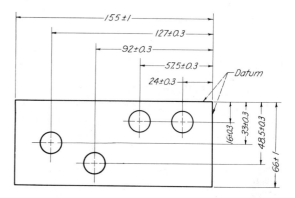

Fig. 13.53 Dimensions from datum lines.

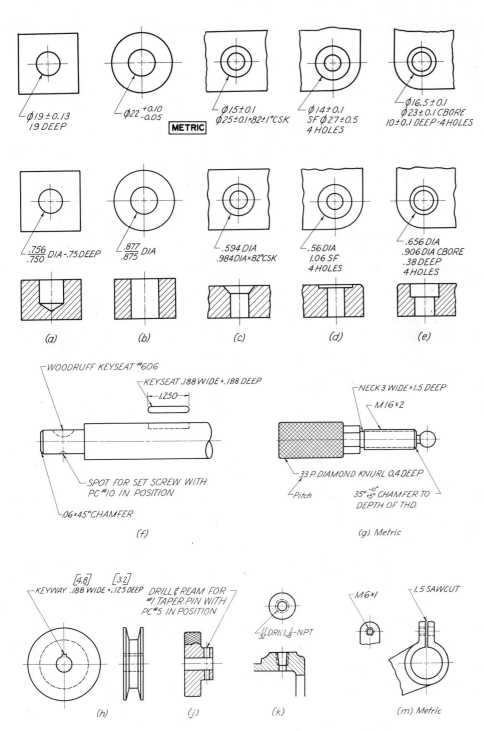

Fig. 13.54 Shop notes.

of fit, degree of finish, and so forth. Brevity in form is desirable for notes of general information or specific instruction. In the case of threaded parts one should use the terminology recommended in Chapter 14.

13.14
Dual Dimensioning. A dual-dimensioning procedure has been adopted recently by several of our large industrial organizations; the procedure calls for showing both U.S. inch units and the SI (Système International) metric units of measurements on the same drawing. See Fig. 13.55.

Dimension values, given in both inch and SI units, must insure the interchangeability of parts. The interchangeability of a part is determined by the number of decimal places retained in rounding off a converted dimension and by the

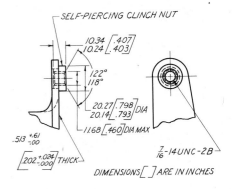

Fig. 13.55 Drawing showing dual-dimensioning. (*Courtesy Ford Motor Company*)

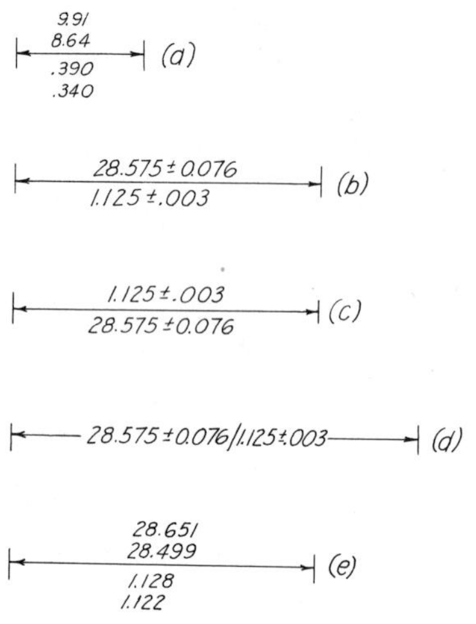

Dimensions, notes, limits, and geometric tolerances

$$\frac{9.91}{8.64}$$
$$\frac{.390}{.340}$$ (a)

$$\frac{28.575 \pm 0.076}{1.125 \pm .003}$$ (b)

$$\frac{1.125 \pm .003}{28.575 \pm 0.076}$$ (c)

28.575±0.076/1.125±.003 (d)

$$\frac{\frac{28.651}{28.499}}{\frac{1.128}{1.122}}$$ (e)

Fig. 13.56 Position method applied to dual-dimensioning.

0.5 [.02] CHAMFER 8.0 [.31] R (a)

$$\frac{9.91}{8.64} \left/ \frac{.390}{.340} \right.$$ (b)

$$\frac{[28.575 \pm 0.076]}{1.125 \pm .003}$$ (c)

$$\frac{\left[\frac{28.651}{28.499}\right] \frac{1.128}{1.122}}{}$$ (d)

[28.575±0.076]/1.125±.003 (e)

$$\frac{\left[\frac{28.651}{28.499}\right]}{\frac{1.128}{1.122}}$$ (f)

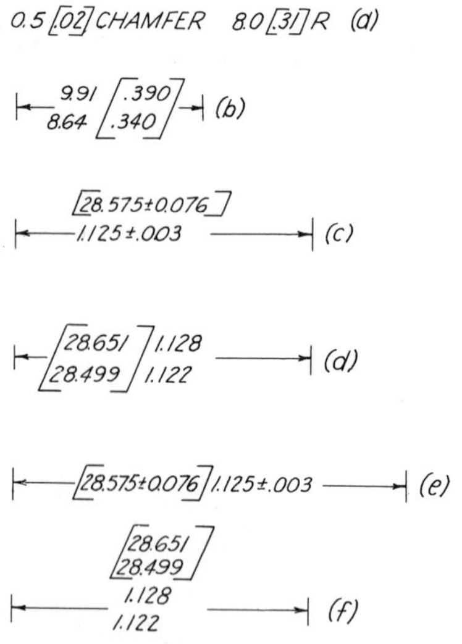

Fig. 13.57 Bracket method applied to dual-dimensioning.

degree that the tolerance limits of the conversion have been permitted to violate the limits of the original dimension. There are several reliable sources for information on conversion principles. These are: (1) Rules for Conversion and Rounding—ASTM E380 Standard Metric Practice Guide and (2) Conversion of Toleranced Linear Dimensions—SAE 1390 Dual Dimensioning Standard.

Each drawing should have noted on it how the dimension values given in inch units and those given in SI units may be identified. For linear dimensions, the SI unit is the millimeter. Two methods have been commonly employed on drawings to distinguish between values given in these different units. These methods are: (1) the position method and (2) the bracket method.

When the position method is used, the millimeter dimension may be placed either above the inch dimension as shown in Fig. 13.56(a) or to the left of the inch dimension with a slash line separating the values (d). Another method of display is similar except that the inch dimension value is placed above or to the left of the metric value.

In using the bracket method, the inch dimension and millimeter dimension values may also be displayed in two ways but only one method should be used on a single drawing. In the first method, the millimeter dimension value is enclosed in square brackets as shown in Fig. 13.57. The bracketed millimeter dimension value may be placed either above or below or to the right or left of the inch value. However, the practice that is adopted must be followed for the entire drawing. The second and alternate method is similar except that the inch values (instead of the milli-

meter dimensions) are enclosed in brackets.

Many of the drawings that are made during this decade will be dual-dimensioned since it is expected that U.S. industry will be changing over to the metric system in order to remain competitive in world trade.

C □ Limit dimensioning and cylindrical fits

13.15

Limit Dimensions. Present-day competitive manufacturing requires quantity production and interchangeability for many closely mating parts. The production of each of these mating parts to an exact decimal dimension, although theoretically possible, is economically unfeasible, since the cost of a part rapidly increases as an absolute correct size is approached. For this reason, the commercial draftsman specifies an allowable error (tolerance) between decimal limits (Fig. 13.58). The determination of these limits depends on the accuracy and clearance required for the moving parts to function satisfactorily in the machine. Although manufacturing experience is often used to determine the proper limits for the parts of a mechanism, it is better and safer practice to adhere to the fits recommended by the American National Standards Institute in ANSI B4.1–1967. This standard applies to fits between plain cylindrical parts. Recommendations are made for preferred sizes, allowances, tolerances, and fits for use where applicable. Up to a diameter of 20 in. the standard is in accordance with

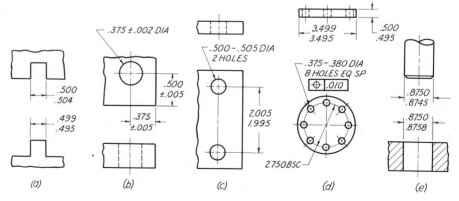

Fig. 13.58 Limit dimensioning for the production of interchangeable parts.

ABC (American–British–Canadian) conference agreements.

There are many factors that a designer must take into consideration when selecting fits for a particular application. These factors might be the bearing load, speed, lubrication, materials, and length of engagement. Frequently temperature and humidity must be taken into account. Considerable practical experience is necessary to make a selection of fits or to make the subsequent adjustments that might be needed to satisfy critical functional requirements. In addition, manufacturing economy must never be overlooked.

Those interested in the selection of fits should consult texts on machine design and technical publications, for coverage of this phase of the dimensioning of cylindrical parts is not within the scope of this book. However, since it is desirable to be able to determine limits of size following the selection of a fit, attention in this section will be directed to the use of Table 34 in the Appendix. Whenever the fit to be used for a particular application has not been specified in the instructions for a problem or has not been given on the drawing, the student should consult his instructor after a tentative choice has been made based on the brief descriptions of fits as given in this section.

To compute limit dimensions it is necessary to understand the following associated terms.

Nominal size. Nominal size is the approximate size used for the purpose of identification. In Fig. 13.59 the nominal size of the hole and shaft is $\frac{1}{2}$, or .50 in.

Basic size. Basic size is the theoretical size from which the limits of size are determined by the application of allowances and tolerances. Values to be used for plus-and-minus tolerancing are given in Tables 34A, B, C, D, and E in the Appendix.

Allowance. An allowance is the intentional difference between the maximum material limits (clearance or maximum interference) of mating parts. For clearance fits the allowance is positive. For interference fits it will be negative. In Fig. 13.59, the allowance is positive and is equal to the difference between the smallest hole and the largest shaft (.5000 − .4988 = .0012).

Tolerance. Tolerance is the permissible variation of a size that has been prescribed for a part. It is the difference between the specified limits. In Fig. 13.59,

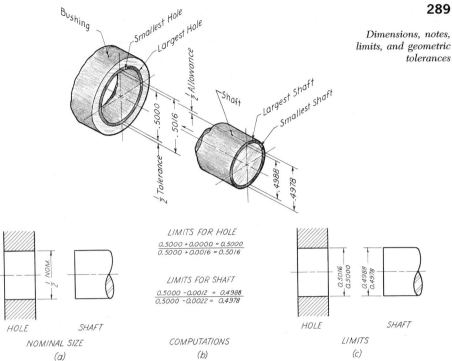

Fig. 13.59 Computation of limits (basic hole system).

the tolerance for the shaft is .0010, the difference between the upper limit (.4988) and the lower limit (.4978).

Limits of size. The limits of size are the extreme maximum and minimum sizes specified by a toleranced dimension. In Fig. 13.59, the limits of size for the hole are .5016 and .5000.

Fit. Fit is the term commonly used to signify the tightness or looseness that may result from the application of a specific combination of allowances and tolerances in the design of mating parts. The fit may be either a clearance fit, an interference fit, or a transition fit.

Clearance fit. A clearance fit results in limits of size that assures clearance between assembled mating parts. In Fig. 13.59, the RC6 fit provides clearance under maximum material conditions. The allowance or minimum clearance is .0012.

Interference fit. An interference fit has limits of size that result in an interference between two mating parts. In the case of a hole and shaft, the shaft will be larger than the hole, to give an actual interference of metal that will result in either a force or a press fit. This has the effect of producing an almost permanent assembly for two assembled parts.

Transition fit. A transition fit has limits of size that can lead to either clearance or

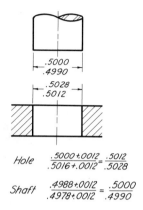

Hole $\dfrac{.5000+.0012}{.5016+.0012} = \dfrac{.5012}{.5028}$

Shaft $\dfrac{.4988+.0012}{.4978+.0012} = \dfrac{.5000}{.4990}$

Fig. 13.60 Computation of limits (basic shaft system).

interference. A shaft may be either larger or smaller than the hole in a mating part.

Basic hole system. For the widely used basic hole system, the design size of the hole (smallest size) is the basic size. The allowance (if any) is applied to the shaft. In Fig. 13.59, the basic hole size is .5000, the smallest diameter. The allowance of .0012 can be subtracted from this diameter to obtain the diameter of the largest shaft (.4988). In design, the basic hole system is used whenever conditions permit because holes are usually formed and checked using standard tools and gauges while cold-rolled shafting can be much more easily machined to any specified size.

Basic shaft system. For the basic shaft system the design size (maximum diameter) of the shaft is the basic size. The allowance, if any, is applied to the hole. In Fig. 13.60, the basic shaft size is .5000. The basic shaft system should be used only when several parts having different fits are to be mounted on standard-size cold-finished shafting of fixed diameter. For each part to be assembled on the shaft the allowance can be applied to the basic shaft size to obtain the smallest diameter of the hole.

Tables 34A, B, C, D, and E in the Appendix cover three general types of fits: running fits, locational fits, and force fits. For educational purposes standard fits may be designated by means of letter symbols, as follows:

RC—Running or Sliding Clearance Fit

LC—Locational Clearance Fit

LT—Transition Clearance or Interference Fit

LN—Locational Interference Fit

FN—Force or Shrink Fit

It should be understood that these letters are not to appear on working drawings. Only the limits for sizes are shown.

When a number is added to these letter symbols a complete fit is represented. For example, FN4 specifies, symbolically, a class 4 force fit for which the limits of size for mating parts may be determined from use of Table 34E. The minimum and maximum limits of clearance or interference for a particular application may be read directly from this table.

Classes of fits as given in these tables are as follows:

Running and Sliding Fits—Classes RC1–RC9

Locational Clearance Fits—Classes LC1–LC11

Locational Transition Fits—Classes LT1–LT6

Locational Interference Fits—Classes LN1–LN3

Force and Shrink Fits—Classes FN1–FN5

Running and sliding fits. Running and sliding fits (RC1-RC9) provide a similar running performance, with suitable lubrication allowance, throughout the range of sizes. The clearances for the RC1 and RC2 classes, which are used mainly for slide fits, increase more slowly with diameter than the other classes, so that accurate location can be maintained even at the expense of free relative motion.

Locational fits. These fits are intended for mating parts requiring some degree of accuracy of location. Locational fits provide for parts requiring rigidity and accurate alignment (interference fits—LN) as well as fits for mating parts where some freedom of location is permissible (locational clearance—LC). Locational fits are divided into three groups: locational clearance fits—LC, locational transition fits—LT, and locational interference fits—LN. Locational transition fits offer a compromise between LC and LN fits and are used where either a small amount of clearance or interference can be permitted.

Force fits. Force fits and shrink fits constitute a special type of interference fit that is normally characterized by maintenance of constant bore pressures throughout the range of sizes. Force fits may be light drive fits—FN1, medium drive fits—FN2, heavy drive fits—FN3, and force fits—FN4 and FN5. Medium drive fits are suitable for ordinary steel parts. FN4 and FN5 force fits are for parts that can be highly stressed. Normally, FN2 fits are used for shrink fits on light sections, while FN3 is applicable for shrink fits in medium sections.

13.16
Computation of Limits of Size for Cylindrical Parts. To obtain the correct fit between two engaging parts, compute limit dimensions that modify the nominal size of both. Numerical values of the modifications necessary to obtain the proper allowance and tolerances for various diameters for all fits mentioned previously are

given in Tables 34A, B, C, D, and E in the Appendix.

The two systems in common use for computing limit dimensions are (1) the basic hole system, and (2) the basic shaft system. The same ANSI tables may be used conveniently for both systems.

13.17

Basic Hole System. Because most limit dimensions are computed on the basic hole system, the illustrated example shown in Fig. 13.59 involves the use of this system. If, as is the usual case, the nominal size is known, all that is necessary to determine the limits is to convert the nominal size to the basic hole size and apply the figures given under "standard limits," adding or subtracting (according to their signs) to or from the basic size to obtain the limits for both the hole and the shaft.

Example

Suppose that a $\frac{1}{2}$-in. shaft is to have a class RC6 fit in a $\frac{1}{2}$-in. hole [Fig. 13.59(a)]. The nominal size of the hole is $\frac{1}{2}$ in. The basic hole size is the exact theoretical size .5000.

From Table 34A it is found that the hole may vary between +.0000 and +.0016, and the shaft between −.0012 and −.0022. As can be readily observed, these values result in a variation (tolerance) of .0016 between the upper and lower limits of the hole, while the variation (tolerance) for the shaft will be .0010. The allowance (minimum clearance) is .0012, as given in the table.

The limits of the hole are

$$\frac{.5000 + .0000}{.5000 + .0016} = \frac{.5000}{.5016}$$

The limits on the shaft are

$$\frac{.5000 - .0012}{.5000 - .0022} = \frac{.4988}{.4978}$$

It is recommended that the maximum limit always be placed directly above the minimum limit where dimensions are associated with dimension lines (ANSI Y14.5–1973). However, when both limits are given in a single line in association with a leader or note, the minimum limit is to be given first [Fig. 13.58(d)].

13.18

Basic Shaft System. When a number of parts requiring different fits but having the same nominal size must be mounted on a shaft, the basic shaft system is used because it is much easier to adjust the

limits for the holes than to machine a shaft of one nominal diameter to a number of different sets of limits required by different fits.

For basic shaft fits the maximum size of the shaft is basic. The limits of clearance or interference are the same as those shown in Tables 34A, B, C, D, and E for the corresponding fits. The symbols for basic shaft fits are identical with those used for the standard fits with a letter S added. For example, LC4S specifies a locational clearance fit, class 4, as determined on a basic shaft basis.

Basic shaft system— clearance fits

To determine the needed limits, increase each of the limits obtained, using the basic hole system, by the value given for the upper shaft limit. For example, if the same supposition (nominal size and fit requirement) is made as for the preceding illustration, the limits shown in Fig. 13.60 can be most easily obtained by adding .0012 to each of the limits shown for the hole and shaft in Fig. 13.59.

Basic shaft system— interference and transition fits

To determine the needed limits, subtract the value shown for the upper shaft limit from the basic hole limits.

Example

The basic shaft limits are to be determined using an FN2 fit and the same nominal diameter as for the two previous illustrations.

$$\text{Hole:} \quad \frac{.5000 - .0016}{.5007 - .0016} = \frac{.4984}{.4991}$$

Shaft:

$$\frac{.5016 - .0016}{.5012 - .0016} = \frac{.5000 \text{ (basic size)}}{.4996}$$

In brief, it can be stated that the limits for hole and shaft, as given in Tables 34A–E, are increased for clearance fits or decreased for transition or interference fits by the value shown for the upper shaft limit, which is the amount required to change the maximum shaft to basic size.

13.19

Tolerances. Necessary tolerances may be expressed by general notes printed on a drawing form or they may be given with definite values for specific dimensions as in the case shown in Fig. 13.65, where the part has been dimensioned using millimeters. When expressed in the form of a printed note, the wording might be as follows: ALLOWABLE VARIATION ON ALL FRACTIONAL DIMENSIONS IS $\pm$.010 UNLESS OTHERWISE SPECIFIED. A general note for tolerance on decimal dimensions might read: ALLOWABLE VARIATION ON DECIMAL DIMENSIONS IS $\pm$.001. This general note would apply to all decimal dimensions where limits were not given.

The general notes on tolerances should be allowed to apply to all dimensions where it is not necessary to use specific tolerances.

13.20

Tolerancing Methods. A tolerance applied directly to a dimension may be expressed by giving limits (maximum and minimum values) or by means of plus-and-minus tolerancing. When limit dimensioning is used, the high limit is given above the low limit, as: $.750 \atop .742$ When given in a single line, the low limit precedes the high limit, with a dash separating the values, as: .742–.750.

When the metric system is used, the same practices are followed. It should be noted that when the high and low limits are placed one above the other, they are not separated by a line.

Examples

$31.8 \atop 31.5$ mm   31.5–31.8 mm

13.21

Plus-and-Minus Tolerancing. When plus-and-minus tolerancing is used, the specified size is given first, followed by a plus-and-minus expression of the tolerance as shown in Fig. 13.61.

Unilateral plus-and-minus tolerancing is illustrated in (*a*) for both the metric and English systems of measurement. It should be noted for metric dimensioning that the nil value is shown by a zero with no plus-or-minus symbol being given. When the tolerance is a plus value, it is placed above the zero. When the tolerance has a minus value, it is placed below the zero.

Bilateral tolerancing is illustrated in (*b*). For equal plus-and-minus values, the mean value is preceded by a $\pm$ symbol. The plus value is shown above when unequal plus-and-minus tolerancing is used. Both tolerance values should be given using the same number of decimal places.

13.22

Cumulative Tolerances. An undesirable condition may result when either the location of a surface or an overall dimension is affected by more than one tolerance dimension. When this condition exists, as illustrated in Fig. 13.62(*a*), the tolerances are said to be cumulative. In (*a*) surface B is located from surface A and surface C is related in turn to surface B. With the tolerances being additive, the tolerance on C is the sum of the separate tolerances ($\pm$.002). In respect to A, the position of C may vary from 1.998 to 2.002. This tolerance, as illustrated by the shaded rectangle, is .004 in. When consecutive dimensioning is used, one dimension should always be omitted to avoid serious inconsistency. The distance omitted should be the one requiring the least accuracy. To avoid the inconsistency of cumulative tolerances it is the preferred practice to locate the surfaces from a datum plane, as shown in (*b*), so that each surface is affected by only one dimension. The use of a datum plane makes it possible to take full advantage of permissible variations in size and still satisfy all requirements for the proper functioning of the part.

13.23

Specification of Angular Tolerances. Angular tolerances may be expressed in de-

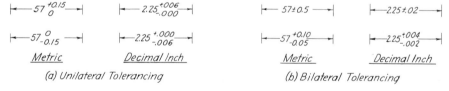

(a) *Unilateral Tolerancing*   (b) *Bilateral Tolerancing*

Fig. 13.61 Plus-and-minus tolerancing.

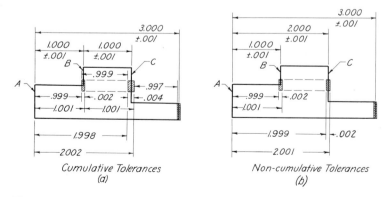

Cumulative Tolerances (a)   *Non-cumulative Tolerances* (b)

Fig. 13.62 Cumulative tolerances.

grees, minutes, or seconds (see Fig. 13.63). If desired, an angle may be given in degrees and decimal parts of a degree with the tolerance in decimal parts of a degree.

Fig. 13.63 Angular tolerances.

D □ Geometric dimensioning and tolerancing

13.24

Geometric Tolerancing. Geometric tolerances specify the maximum variation that can be allowed in form or position from true geometry. Actually, a geometrical tolerance is either the width or diameter of a tolerance zone within which a surface or axis of a hole or cylinder can lie with the resulting part satisfying the necessary standards of accuracy for proper functioning and interchangeability. Whenever tolerances of form are not specified on a drawing for a part, it is understood that the part as produced will be acceptable regardless of form variations. Expressions of tolerances of form control straightness, flatness, parallelism, squareness, concentricity, roundness, angular displacement, and so forth.

13.25

Symbols for Tolerances of Position and Form. The characteristic symbols shown in Fig. 13.64 have been adopted for use in lieu of notes to express positional and

form tolerances. In general, these symbols are the same as those given in Mil. Std. 8C–1962 for use by the armed services. Figure 13.65 shows typical feature control symbols applied to a drawing. After making a careful study of this drawing, the reader is urged to relate and compare the symbolic callouts used with callouts and their illustrated significance as given in Figs. 13.66–13.73 inclusive.

Use of Symbols

Geometric characteristic symbols. These symbols denoting geometric characteristics are shown in Fig. 13.64. Figures 13.65–13.73 show their use.

Datum identifying symbol. The datum identifying symbol consists of a frame (box) containing the datum reference letter. It may be associated with the datum feature in one of the ways shown in Figs. 13.65–13.73.

Basic dimension symbol. In order that a basic dimension may be identified, it is enclosed in a frame as shown in Fig. 13.65.

Supplementary symbols—M, MC, RFS, and diameter. The symbols Ⓜ and Ⓢ are used to designate "maximum material condition" and "regardless of feature size" (Fig. 13.64). In notes, the abbreviations MMC and RFS are used.

CHARACTERISTIC SYMBOLS		
Form Tolerances / Single Features	Straightness	—
	Flatness	▱
	Contour of any surface	⌒
Related Features	Roundness – Circularity	○
	Cylindricity	⌭
	Perpendicularity-Squareness	⊥
	Parallelism	∥
	Angularity	∠
	Runout	⟋
Positional Tolerances	True Position	⊕
	Concentricity	◎
	Symmetry	≡
SUPPLEMENTARY SYMBOLS		
MMC	Maximum Material Conditions	Ⓜ
RFS	Regardless of Feature Size	Ⓢ
	Diameter	⌀

Examples:
─ | 0 0.04   ⌐Symbol   ○ | 0.05
⊥ | 0.15 | A   ⌐Datum Reference   ⌭ | 0.04
∥ | 0.05 | A   ⌐Tolerance   ∠ | 0.04 | A
⊕ | ⌀0.12 Ⓜ | A | D Ⓜ | B   ⌐Diameter Symbol

Fig. 13.64 Geometric characteristic symbols.*

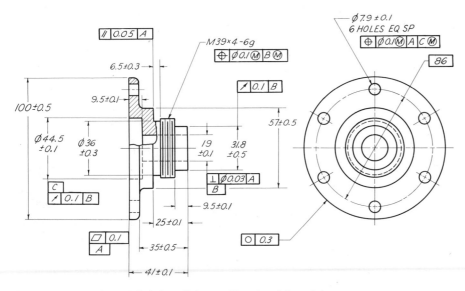

Fig. 13.65 Use of symbols in specifying positional and form tolerances.

*Compiled from ANSI Y14.5–1973.

The symbol φ is used to designate a diameter (Figs. 13.64 and 13.65). On a drawing it is used to replace DIA or D. It always precedes the associated dimension on the drawing and the specified tolerance in a feature control symbol.

Combined symbols. Individual symbols (characteristic and supplementary), datum reference letters, and the needed tolerance may be combined in one frame, as shown in the lower portion of Fig. 13.64, to express a callout for a tolerance symbolically.

In using the universally recognized symbols given in Fig. 13.64, a position or form tolerance is expressed by means of a feature control symbol consisting of a frame that contains the appropriate geometric characteristic symbol followed by the allowable tolerance. It should be noted in Fig. 13.65 that a vertical line separates the symbol from the tolerance. Where applicable, the tolerance should be proceded by the diameter symbol and followed by the symbol Ⓜ or Ⓢ. See Fig. 13.65.

Where a tolerance of position of form is related to a datum, this relationship shall be indicated in the feature control symbol by placing the datum reference letter(s) following the tolerance. As can be noted in the several examples shown in Fig. 13.64, vertical lines separate these entries. Each datum reference letter entered is supplemented by the symbol for MMC or RFS when applicable (Fig. 13.65).

Placement of feature control symbols. A symbol is related to the feature to which it applies by one of the four methods listed.

(*a*) By adding the symbol to a note (Fig. 13.65).

(*b*) By a leader from the feature to the symbol (Fig. 13.65).

(*c*) By attaching the symbol frame to an extension line from the feature (Fig. 13.65).

(*d*) By attaching a side or end of the symbol frame to a dimension line (Fig. 13.65).

The symbols shown in Fig. 13.64 provide the best means for specifying geometric characteristics and tolerances. Notes have been extensively used in the past but notes can be inconsistent and require more space. In addition, these symbols are understood internationally and surmount language barriers. This has been the principal reason for their quick adoption by multinational companies for use on drawings.

13.26

Tolerance Specification for Straightness. Straightness is specified by symbol as shown in Fig. 13.66. The given example illustrates straightness control of individual longitudinal surface elements of a cylindrical part. The symbol meaning is set forth in (*b*). In (*a*) the tolerance zone has been given as 0.05 mm, while in (*c*) it is .002 in. These values are equivalent.

13.27

Tolerance Specification for Flatness. The callout used to control flatness specifies that all points of the actual surface must lie between two parallel planes that are a distance apart equal to the specified tolerance (see Fig. 13.67). The symbol in (*a*) is

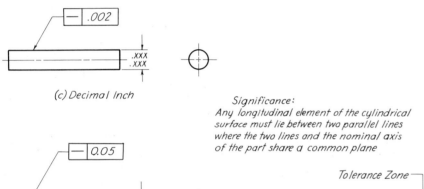

(*c*) Decimal Inch

(*a*) Metric

Significance:
Any longitudinal element of the cylindrical surface must lie between two parallel lines where the two lines and the nominal axis of the part share a common plane

(*b*)

Fig. 13.66 Specification for straightness.

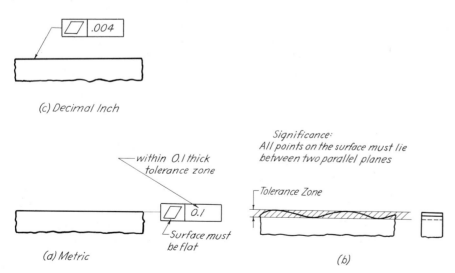

(*c*) Decimal Inch

(*a*) Metric

Significance:
All points on the surface must lie between two parallel planes

(*b*)

Fig. 13.67 Specification for flatness.

interpreted to read: The entire actual surface shall be flat within 0.1 (mm) total tolerance zone. When necessary, the expressions MUST NOT BE CONCAVE or MUST NOT BE CONVEX may be added below the feature control symbol.

13.28
Tolerance Specification for Perpendicularity. Perpendicularity is the condition of a surface, median plane, or axis that is at a 90° angle to a datum plane or axis. In Fig. 13.68, the perpendicularity tolerance specifies a cylindrical tolerance zone perpendicular to a datum plane A within which the axis of the considered feature (hole) must lie. From the feature control symbol we learn that when the feature is at MMC, the maximum perpendicularity tolerance is 0.10 (mm). The ϕ symbol was used to indicate a cylindrical tolerance zone.

The example in Fig. 13.69 illustrates perpendicularity tolerance applied to a surface. The symbol in (*a*) when interpreted specifies that the surface indicated must be perpendicular to datum plane A within an 0.1 (mm)-wide tolerance zone.

13.29
Tolerance Specification for Parallelism. Parallelism is the condition of a surface or axis equidistant at all points from a datum plane or axis. For the example in Fig. 13.70, the datum is considered as a plane established by the high points of surface A. All points of the other surface must lie between two planes that are parallel to the datum. In (*a*) the two planes are 0.05 (mm) apart.

13.30
Tolerance Specification for Concentricity (Fig. 13.71). When the cylindrical or conical features of a part must be basically concentric, it is the practice to specify permissible eccentricity in terms of the maximum permissible deviation from concentricity. Concentricity tolerance, as illustrated in Fig. 13.71(*b*), is the diameter of the cylindrical tolerance zone within which the axis of the feature must lie. The axis of this tolerance zone must coincide with the axis of the datum feature that has been indicated.

13.31
Tolerance Specification for Angularity. The feature control symbol that is commonly used to specify the tolerance for control of angularity is shown in Fig.

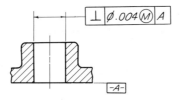

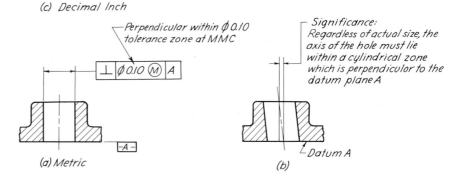

Fig. 13.68 Specification for perpendicularity.

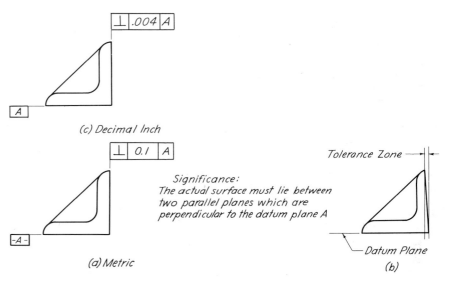

Fig. 13.69 Specification for perpendicularity.

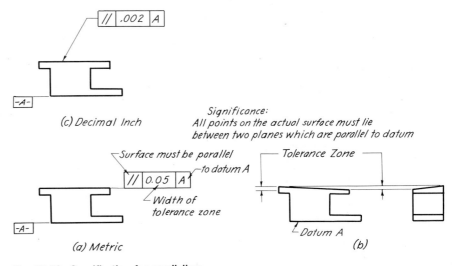

Fig. 13.70 Specification for parallelism.

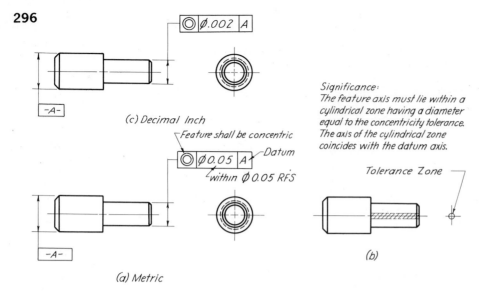

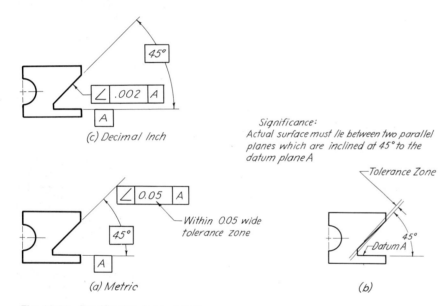

Significance:
The feature axis must lie within a cylindrical zone having a diameter equal to the concentricity tolerance. The axis of the cylindrical zone coincides with the datum axis.

Fig. 13.71 Specification for concentricity.

Significance:
Actual surface must lie between two parallel planes which are inclined at 45° to the datum plane A

Fig. 13.72 Specification for angularity.

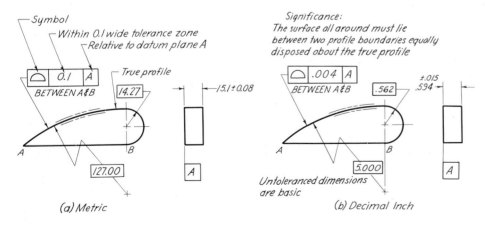

Significance:
The surface all around must lie between two profile boundaries equally disposed about the true profile

Fig. 13.73 Profile (zone) tolerancing.

13.72. Angularity is the condition of a surface or axis that is at some specified angle (other than 90°) from a datum plane or axis. The symbol may be interpreted as: This surface must be at a 45° angle in relation to datum plane *A* within a 0.05 (mm)-wide tolerance zone.

13.32
Profile Tolerancing. Where a uniform amount of variation may be permitted along a profile, a zone tolerance may be specified. The zone is indicated at a conspicuous location by two phantom lines drawn parallel to the profile when the zone is bilateral, that is, symmetrical about the contour line, as shown in Fig. 13.73. Only one phantom line is needed for a unilateral zone that may lie on either side of the true profile. As can be observed from the illustration, the finished surface must lie within the specified tolerance zone. The variation to be permitted and the extent of the tolerance zone must be specified. On the drawing, the applicable feature control symbol should appear with the view where the surface is represented in profile.

13.33
True-position Dimensioning. In the past, it has been the usual practice to locate points by means of rectangular dimensions given with tolerances. A point located in this manner will lie within a square tolerance zone when the positioning dimensions are at right angles to each other, as in Fig. 13.74. Where features are located by radial and angular dimensions with tolerances, wedge-shaped tolerance zones result.

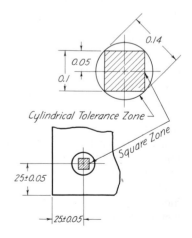

Fig. 13.74 Comparison between coordinate tolerancing and true-position tolerancing.

In making a comparison of coordinate tolerancing and true-position tolerancing of circular features, it can be noted in the case of coordinate tolerancing, as illustrated in Fig. 13.74, that the actual position of the feature can be anywhere within the 0.1 square and that the maximum allowable variation from the desired position occurs along the diagonal of the square. With this allowable variation along the diagonal being 1.4 times the specified tolerance, the diameter of the cylinder for true-position tolerancing of the same feature could be 1.4 times the tolerance that would be used in coordinate tolerancing without any increase in the maximum allowable variation. True-position tolerancing increases the permissible tolerance in all directions, without detrimental effect on the location of the feature. True-position dimensioning takes into full account the relations that must be maintained for the interchangeable assembly of mating parts and it permits the design intent to be expressed more simply and precisely. Furthermore, the true-position approach to dimensioning corresponds to the control furnished by position and receiver gauges with round pins. Such gauges are commonly used for the inspection of patterns of holes in parts that are being mass produced.

Positional tolerancing can be used for specific features of a machine part. When the part contains a number of features arranged in groups, positional tolerances can be used to relate each of the groups to one another as necessary and to tolerance the position of the features within a group independently of the features of the other groups.

The term *true position* denotes the theoretically exact position for a feature. In practice, the basic (exact) location is given with untoleranced dimensions that are excluded from the general tolerances, usually specified near the title block. This is done by enclosing each of the true-position-locating dimensions in a rectangular box to identify them. See Fig. 13.76.

When the alignment between mating parts depends on some functional surface, this surface is selected as a datum for dimensioning, and the datum is identified.

The requirement of true-position dimensioning for a cylindrical feature is illustrated in Fig. 13.75(*a*). It must be understood that the axis of the hole at all points must lie within the specified cylindrical tolerance zone having its center located at true position. This cylindrical tolerance zone also defines the limits within which variations in the squareness of the axis of the hole in relation to the flat surface must be confined.

For noncircular features, such as slots and tabs, the positional tolerance is usually applied only to surfaces related to the center plane of the feature. In applying true-position dimensioning to such features, it will be found that the principal difference is in the geometric form of the tolerance zone within which the center plane of the feature must be contained. See Fig. 13.75(*b*). The center plane of the tolerance zone must be located at true position. It should be noted that this tolerance zone also defines the limits within which variations in the squareness of the center plane of the slot must be contained.

13.34

True-position Dimensioning—Application of the MMC Principle. The least desirable situation exists for the assembly of mating parts when both parts are at their maximum metal condition (designated MMC). The expression *maximum metal condition* by itself, as applied to an internal feature of a finished part, means that the internal feature (hole, slot, etc.) is at its minimum allowable size. In the case of external features (shafts, lugs, tabs, etc.), a maximum material condition exists when these features are at their maximum allowable sizes. MMC occurs for mating parts, say, for a hole and a shaft, when the shaft is at its maximum size and the hole is at its smallest size. Thus, at MMC, there is least clearance between these parts. In general, tolerance of position and the MMC of mating features are considered together in relationship to each other. This leads to the situation where the specified limits of location frequently may be exceeded and acceptable parts produced when the mating features are away from their maximum material limits of size. With this latter condition permissible, it becomes desirable to indicate the fact that specified limits of location need be observed only under MMC.

True-position tolerancing on the MMC basis is both practical and economical for the mass production of interchangeable parts. However, the MMC basis should not be applied where it would be inconsistent with functional requirements.

In those relatively few cases where the more economical MMC basis is not appli-

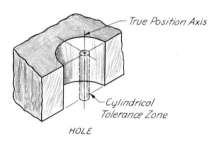

HOLE

(a)

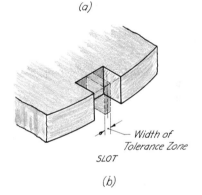

SLOT

(b)

Fig. 13.75 Meaning of true-position dimensioning.

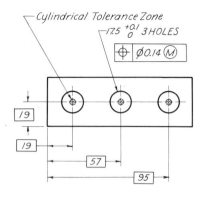

Fig. 13.76 Position tolerancing and use of feature control symbol.

cable and the positional tolerance must be stated without reference to MMC, the more restrictive "regardless of feature size" (RFS) basis is specified. This is accomplished by adding the abbreviation RFS (or S, its symbolic equivalent), to the true-position callout.

Additional information concerning the meaning of MMC as related to positional tolerances may be found in ANSI Y14.5–1973.

The 0.14 position tolerance, illustrated in Fig. 13.74, would usually be based on the MMC size of the hole. Then as the hole size deviates from the MMC size, the position of the hole is permitted to shift off its true position somewhat beyond the specified tolerance zone to offer a realistic bonus to the extent of that departure. In Fig. 13.76, the position tolerance zone would show considerable enlargement as the hole size departs from MMC (17.5 mm) to its high limit size of 17.6 mm. Thus, position tolerancing provides greater production tolerances while meeting fully all design requirements.

Datum planes (surfaces), as the basis for position relationships, may be either specified on a drawing or implied. They have not been specified for the part in Fig. 13.77. In this case, the dimension (19 ± 0.1 mm) from the left edge and a like dimension from the lower edge establish the relationship of the hole pattern to these edges as implied datum surfaces. Datums must be specified on a drawing when feature interrelationships, involving form and position tolerances, must be directly and accurately controlled. See Fig. 13.78.

A circular pattern of six holes located by position tolerancing with respect to the center hole is shown in Fig. 13.78. In (a) the symbol specifies that six holes are to be located at position within ϕ 0.05 at 9.5 MMC size of the holes with respect to datum A and datum hole B at 24 MMC.

Tabs and slots are dimensioned as shown in Fig. 13.79. For such features the positional tolerance is applied only to surfaces related to the center plane of the feature as shown in Fig. 13.75(b). The feature control symbol reads: these features (3 tabs) must be at position within a 0.5 (total)-wide zone with the features at MMC and with respect to datum A at MMC.

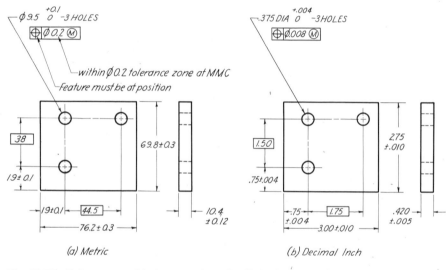

Fig. 13.77 Hole pattern with datum surfaces implied.

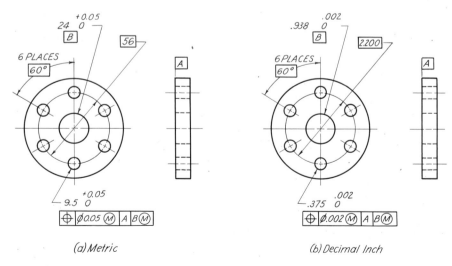

Fig. 13.78 Positional tolerancing with datum references.

E □ Designation of surface texture

13.35

Surface Quality. The improvement in machining methods within recent years coupled with a strong demand for increased life for machined parts has caused engineers to give more attention to the quality of the surface finish. Not only the service life but also the proper functioning of the part as well may depend on obtaining the needed smoothness quality for contact surfaces.

On an engineering drawing a surface may be represented by line if shown in profile or it may appear as a bounded area in a related view. Machined and ground surfaces, however, do not have the perfect smoothness represented on a drawing.

Actually a surface has three dimensions—length, breadth, and curvature (waviness)—as illustrated in Fig. 13.80(*a*). In addition, there will be innumerable peaks and valleys of differing lengths, widths, and heights. An exaggerated profile of surface roughness is shown in (*b*). Combined waviness and roughness is illustrated in (*c*).

The following terms must be understood before the surface symbol shown in Fig. 13.82 can be properly applied:

Surface texture. This term refers to repetitive or random deviations from the nominal surface, which form the pattern of the surface. Included are roughness, waviness, lay, and flaws (Fig. 13.81).

Roughness. Roughness is the relatively finely spaced surface irregularities that are produced by the cutting action of tool edges and abrasive grains on surfaces that are machined.

Roughness height. Roughness height is the average (arithmetical) deviation from the mean line of the profile. It is expressed in microinches or micrometers (Fig. 13.81).

Roughness width. Roughness width is the distance between successive peaks or ridges, which constitute the predominant pattern of roughness. Roughness width is measured in inches or millimeters (Fig. 13.81).

Roughness width cutoff. This term indicates the greatest spacing of repetitive surface irregularities to be included in the measurement of average roughness height. It is measured in inches or millimeters (Fig. 13.81).

Waviness. Waviness is the surface undulations that are of much greater magnitude than the roughness irregularities. Waviness may result from machine or work deflections, vibrations, warping, strains, or similar causes.

Waviness height. Waviness height is the peak-to-valley distance (Fig. 13.81). It is rated in inches or millimeters.

Waviness width. Waviness width (rated in inches or millimeters) is the spacing of successive wave valleys or wave peaks (Fig. 13.81).

Flaws. Flaws are irregularities, such as cracks, checks, blowholes, scratches, and so forth, that occur at one place or at relatively infrequent or widely varying intervals on the surface (Fig. 13.81).

Lay. Lay is the predominant direction of the tool marks of the surface pattern (Fig. 13.84).

Microinch. A *microinch* is one-mil-

lionth (.000001) of an inch (μin.). A micrometer is one-millionth of a meter (μm).

13.36
Designation of Surface Characteristics
(Fig. 13.82). A surface whose finish is to be specified should be marked with the finish mark having the general form of a check mark (**V**) so that the point of the symbol shall be on the line representing the surface, on the extension line, or on a leader pointing to the surface. Good practice dictates that the long leg and the extension shall be to the right as the drawing is read. Figure 13.83 illustrates the specification of roughness, waviness, and lay by listing rating values on the symbol.

Where it is desired to specify only the

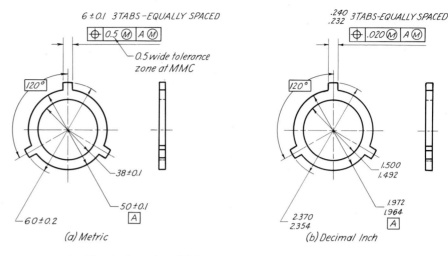

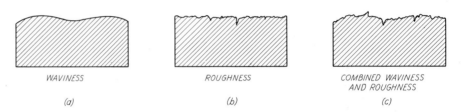

Fig. 13.79 Positional tolerancing of tabs.

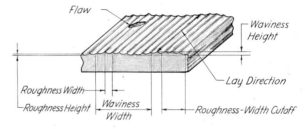

Fig. 13.80 Surface definitions illustrated.

Fig. 13.81 Surface texture definitions.

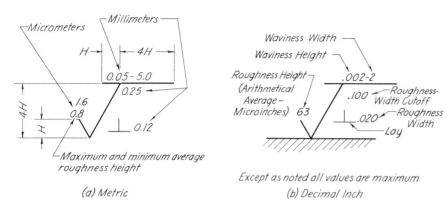

Fig. 13.82 Surface texture symbol.

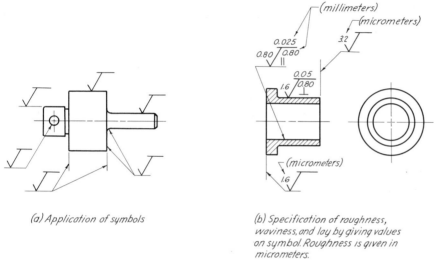

Fig. 13.83 Application of surface texture symbols to a drawing of a machine part.

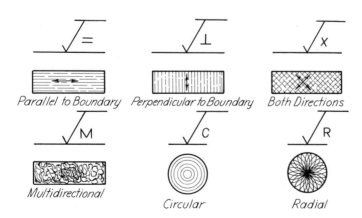

Fig. 13.84 Lay notations.

surface-roughness height, and the width of roughness or direction of tool marks is not important, the simplest form of the symbol may be used. The numerical value is placed in the √, as shown in Fig. 13.82.

Where it is desired to specify waviness height in addition to roughness height, a straight horizontal line must be added to the top of the simple symbol (Fig. 13.82). The numerical value of height waviness would be shown above this line.

If the nature of the preferred lay is to be shown in addition to these two characteristics, it will be indicated by the addition of a combination of lines, as shown in Fig. 13.82. Parallel or perpendicular lines indicate that the dominant lines on the surface are parallel or perpendicular to the boundary line of the surface in contact with the symbol (Fig. 13.84).

Roughness width is placed to the right of the lay symbol, as shown in Fig. 13.82.

The use of only one number to specify the height or width of roughness or waviness will indicate the arithmetical average.

The chart in Fig. 13.85 shows the expected surface roughness in microinches and micrometers for surfaces produced by common production methods.

The surface-quality symbol, which is used only when it is desirable to specify surface smoothness, should not be confused with a finish mark, which indicates the removal of material. A surface-quality symbol might be used for a surface on a die casting, forging, or extruded shape where the surface is to have a natural finish and no material is to be removed.

Surface finish should be specified only by experienced persons because the function of many parts does not depend on the smoothness quality of a surface or surfaces. In addition, surface quality need not be necessarily indicated for many parts that are produced to close dimensional tolerances because a satisfactory surface finish may result from the required machining processes. It should be remembered that the cost of producing a part will generally become progressively greater as the specification of surface finish becomes more exacting.

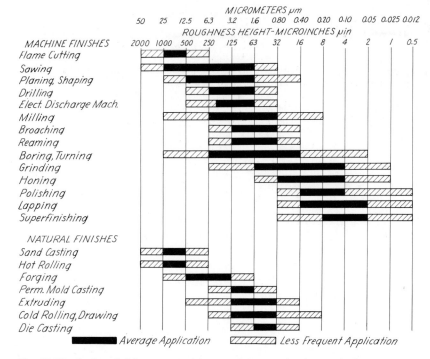

Fig. 13.85 Surface finishes expected from common production methods
(micrometers μm and microinches μin.).

Problems

The following problems offer the student the opportunity to apply the rules of dimensioning given in this chapter. If it is desirable, either millimeters or decimals of an inch may be used in place of fractions. Use Table 1 in the Appendix.

1,2. (Figs. 13.86 and 13.87). Reproduce the given views of an assigned part. Determine the dimensions by transferring them from the drawing to one of the open-divided scales by means of the dividers.

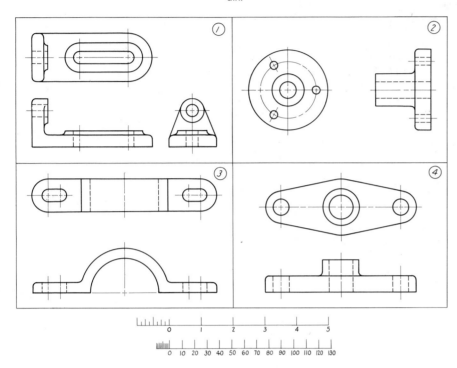

Fig. 13.86 Dimensioning problems.

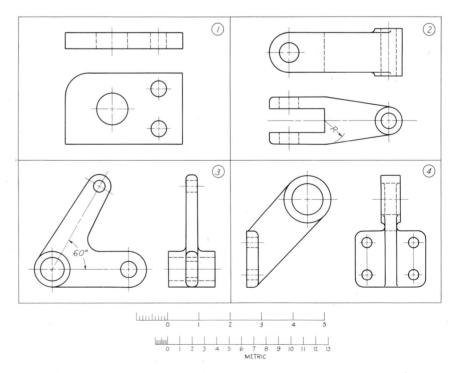

Fig. 13.87 Dimensioning problems.

3–9. (Figs. 13.88–13.94). Make a fully dimensioned multiview sketch or drawing of an assigned part. Draw all necessary views. Give a detail title with suitable notes concerning material, number required, etc. These parts have been selected from different fields of industry—automotive, aeronautical, chemical, electrical, etc.

10. (Fig. 13.95). Make a fully dimensioned multiview sketch or drawing of the rocker arm. Show a detail section taken through the ribs. *Supplementary information:* (1) The distance from the center of the shaft to the center of the hole for the pin in 4.00 in. The distance from the shaft to the threaded hole is 4.50 in. (2) The nominal diameter of the hole for the shaft is

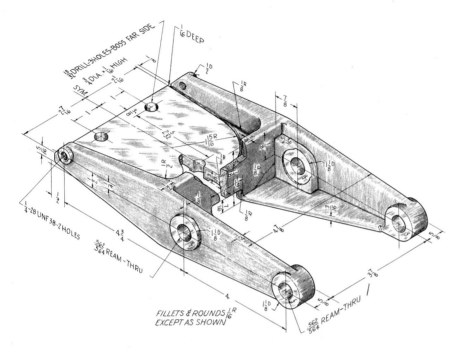

Fig. 13.88 Control pedal—airplane control system.

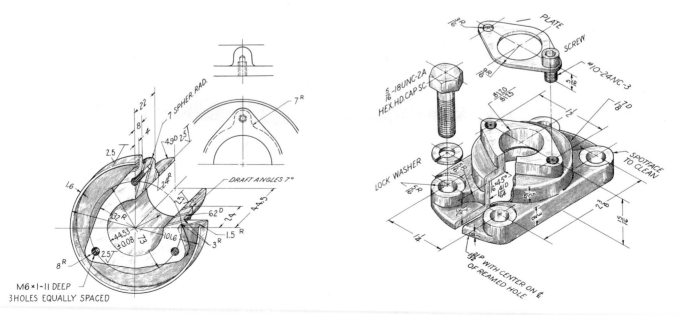

Fig. 13.89 Inlet flange—airplane cooling system.

Fig. 13.90 Cover—mixing machine.

1.875 in. The hole in the rocker arm is to be reamed for a definite fit. Consult your instructor. The diameter of the pin is .969 in. (3) The diameter of the threaded boss is 2.00 in. (4) The diameter of the roller is 2.25 in., and its length is 1.46 in. Total clearance between the roller and finished faces is to be .03 in. (5) The inside faces of the arms are to be milled in toward the hub far enough to accommodate the roller. (6) The rib is .62 in. thick. (7) The lock nut has 1¼–12 UNF thread. (8) Fillets and rounds .12 in. R except where otherwise noted.

11. (Fig. 13.96). Make a fully dimensioned drawing of an assigned part of the shaft support.

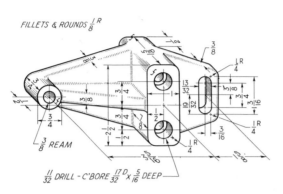

Fig. 13.91 Elevator bracket.

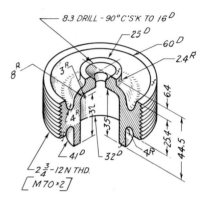

Fig. 13.92 Valve seat.

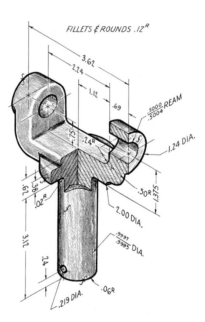

Fig. 13.93 Yoke.

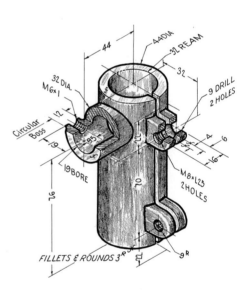

Fig. 13.94 Torch holder—welding.

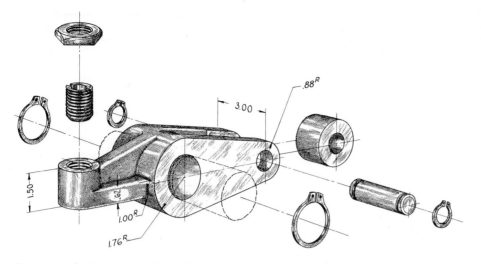

Fig. 13.95 Rocker arm—marine engine.

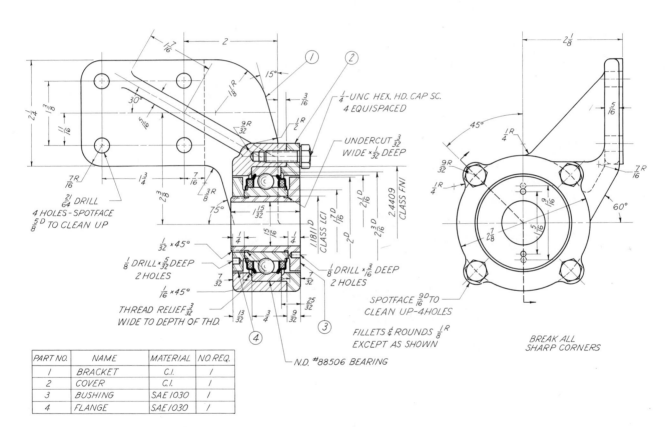

PART NO.	NAME	MATERIAL	NO. REQ.
1	BRACKET	C.I.	1
2	COVER	C.I.	1
3	BUSHING	SAE 1030	1
4	FLANGE	SAE 1030	1

Fig. 13.96 Shaft support.

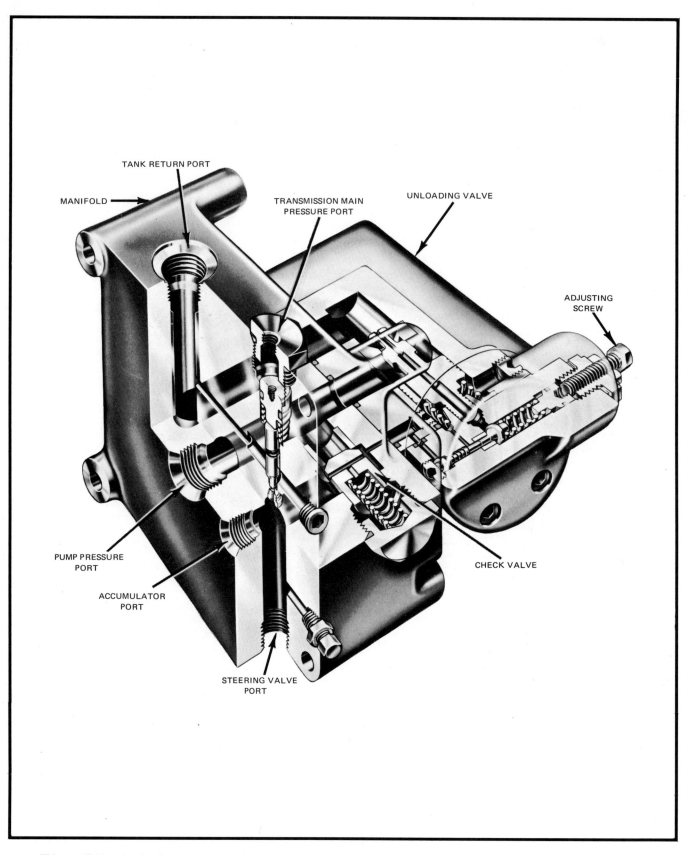

TANK RETURN PORT

MANIFOLD

TRANSMISSION MAIN
PRESSURE PORT

UNLOADING VALVE

ADJUSTING
SCREW

CHECK VALVE

PUMP PRESSURE
PORT

ACCUMULATOR
PORT

STEERING VALVE
PORT

This manifold and unloading control valve is part of the hydraulic steering and braking circuits. It maintains the hydraulic system at a constant pressure. Units of this type require many threaded parts and the use of springs, bearings, and fasteners. (*Courtesy General Motors Corporation*)

14

Threads and standard machine elements

A □ Screw threads

14.1

Introduction. In the commercial field, where the practical application of engineering drawing takes the form of working drawings, knowledge of screw threads and fasteners is important. There is always the necessity for assembling parts either with permanent fastenings such as rivets, or with bolts, screws, and so forth, which may be removed easily.

Engineers, detailers, and draftsmen must be completely familiar with the common types of threads and fastenings, as well as with their use and the correct methods of representation, because of the frequency of their occurrence in structures and machines. Information concerning special types of fasteners may be obtained from manufacturers' catalogues.

Young technologists in training should study Fig. 14.1 to acquaint themselves with the terms commonly associated with screw threads.

14.2

Threads. The principal uses of threads are (1) for fastening, (2) for adjusting, and (3) for transmitting power. To satisfy most of the requirements of the engineering profession, the various forms of threads shown in Fig. 14.2 are used.

The (ANSI) Unified thread form (Figs. 14.2 and 14.3), now recognized as the standard thread form in the United States, Great Britain and Canada, is a modification of the American National (N) thread form in use by American industry since 1935. Since the thread forms are essentially the same, the new *Unified*

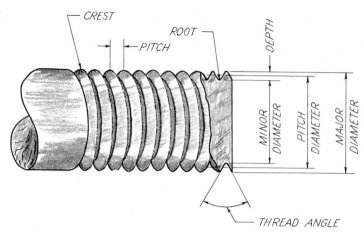

Fig. 14.1 Screw-thread nomenclature.

thread form has found ready acceptance and it can be said to have superseded the old (N) form.

The sharp V is used to some extent where adjustment and holding power are essential.

For the transmission of power and motion, the modified square, Acme, and Brown and Sharpe worm threads have been adopted. The modified square thread, which is now rarely used, transmits power parallel to its axis. A further modification of the square thread is the stronger Acme, which is easier to cut and more readily disengages split nuts (as lead screws on lathes). The Brown and Sharpe worm thread, with similar proportions but with longer teeth, is used for transmitting power to a worm wheel.

The knuckle thread, commonly found on incandescent lamps, plugs, and so on, can be cast or rolled.

The Whitworth and buttress threads are not often encountered by the average engineer. The former, which fulfills the same purpose as the American Standard thread, is used in England but is also frequently found in this country. The buttress or breech-block thread, which is designed to take pressure in one direction, is used for breech mechanisms of large guns and for airplane propeller hubs. The thread form has not been standardized and appears in different modified forms.

The Dardelet thread is self-locking in assembly.

14.3

American–British Unified Thread. The Unified Thread Standard came into existence after the representatives of the United States, Great Britain, and Canada signed a unification agreement on Nov. 18, 1948 in Washington, D.C. This accord, which made possible the interchangeability of threads for these countries, created a new thread form (Fig. 14.3) that is a compromise between our own American National Standard design and the British Whitworth. The external thread of the new form has a rounded root and may have either a flat or rounded crest. The Unified thread is a general-purpose thread for screws, bolts, nuts, and other threaded parts (ANSI B1.1–1960).

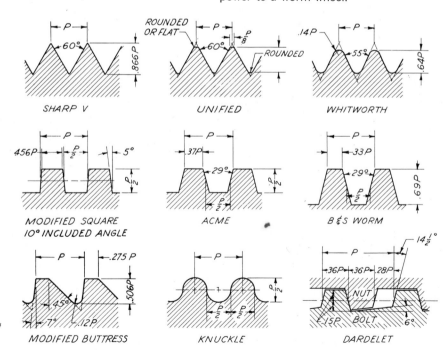

Fig. 14.2 Screw threads.

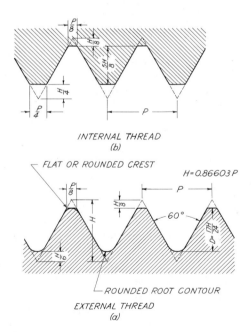

Fig. 14.3 American–British Unified thread.

14.4

Multiple-start Threads. Whenever a quick advance is desired, as on fountain pens, valves, and so on, two or more threads are cut side by side. Two threads form a double (2-START) thread; three, a triple (3-START) thread; and so on. A thread that is not otherwise designated is understood to be a single-start thread. All threads included in ANSI B1.1–1960 are single-start unless specifically identified as being multiple-start.

Fig. 14.4 shows heavy strings wound around a rod for the purpose of demonstrating single-start and double (2-START) threads. The center line of the single string, representing the single thread, assumes the form of a helix. In the case of the double (2-START) thread, it should be noted that there are two strings side by side that are shaded differently for clarity. On a double (2-START) thread, each thread starts diametrically opposite the other one.

In drawing a single or an odd-number multiple-start thread, a crest is always diametrically opposite a root; in a double or other even-number multiple-start thread, a crest is opposite a crest and a root opposite a root.

14.5

Right-hand and Left-hand Threads. A right-hand thread advances into a threaded hole when turned clockwise; a left-hand thread advances when turned counterclockwise. They can be easily distinguished by the thread slant. A right-hand thread on a horizontal shank always slants upward to the left (\) and a left-hand, upward to the right (/). A thread is always considered to be right-hand if it is not otherwise specified. A left-hand thread is always marked LH on a drawing.

14.6

Pitch. The pitch of a thread is the distance from any point on a thread to the corresponding point on the adjacent thread, measured parallel to the axis, as shown in Fig. 14.1.

14.7

Lead. The lead of a screw may be defined as the distance advanced parallel to the axis when the screw is turned one revolution (Fig. 14.5). For a single thread, the lead is equal to the pitch; for a double (2-START) thread, the lead is twice the pitch; for a triple (3-START) thread, the lead is three times the pitch, and so on.

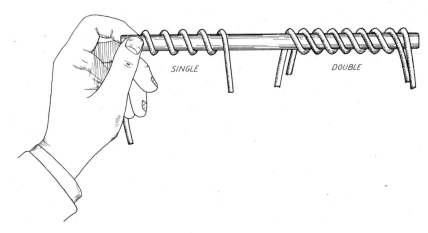

Fig. 14.4 Demonstration of a helix as related to a single-start and double (2-START) screw thread.

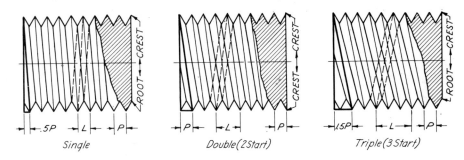

Fig. 14.5 Single-start and multiple-start threads.

14.8

Detailed Screw-thread Representation. The true representation of screw threads by helical curves, requiring unnecessary time and laborious drafting, is rarely used. The detailed representation, closely approximating the actual appearance, is preferred in commercial practice, for it is much easier to represent the helices with slanting lines and the truncated roots and crests with sharp V's (Fig. 14.6). Since

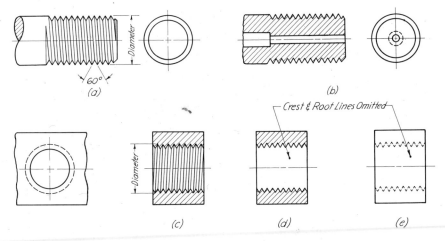

Fig. 14.6 Detailed representation.

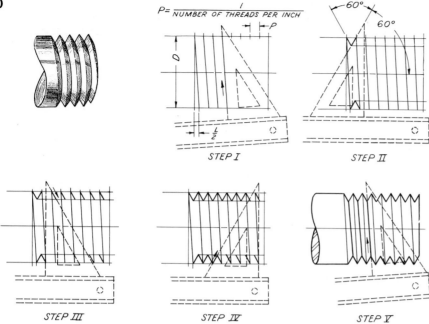

$$P = \frac{1}{\text{NUMBER OF THREADS PER INCH}}$$

STEP I   STEP II

STEP III   STEP IV   STEP V

Fig. 14.7 Detailed representation of Unified and American and sharp V-threads (external).

detailed rendering is also time consuming, its use is justified only in those few cases where appearance and permanency are important factors and when it is necessary to avoid the possibility that confusion might result from the use of one of the symbolic methods. The preparation of a detailed representation is a task that belongs primarily to a draftsman, the engineer being concerned only with specifying that this form be used.

The steps in drawing a Unified thread are shown in Figs. 14.7 and 14.8.

The stages in drawing the detailed representation of modified square and Acme threads are shown in Figs. 14.9, 14.10, and 14.11. All lines of the finished square thread are made the same weight. The root lines of the Acme thread may be made heavier than the other lines.

14.9
American National Standard Conventional Thread Symbols (Fig. 14.12).

To save valuable time and expense in the preparation of drawings, the American National Standards Institute has adopted the "schematic" and "simplified" series of thread symbols to represent threads having a diameter of 1 in. (25 mm) or less.

The root of the thread for the simplified representation of an external thread is

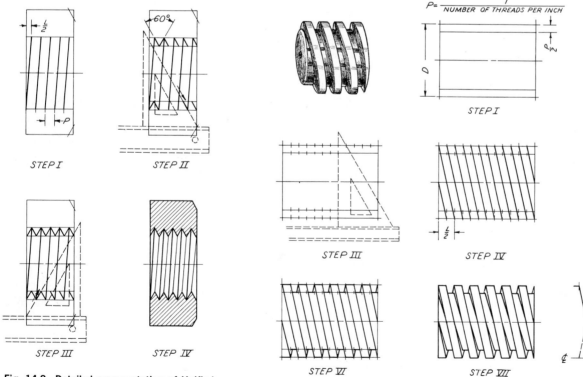

STEP I   STEP II

STEP III   STEP IV

Fig. 14.8 Detailed representation of Unified and American and sharp V-threads (internal).

$$P = \frac{1}{\text{NUMBER OF THREADS PER INCH}}$$

STEP I   STEP II

STEP III   STEP IV   STEP V

STEP VI   STEP VII   ENLARGED VIEW

Fig. 14.9 Detailed representation of square threads (modified).

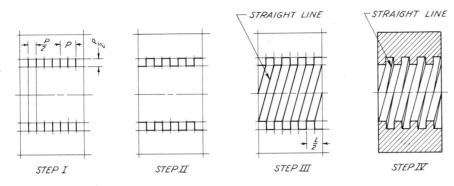

Fig. 14.10 Detailed representation of square threads (internal).

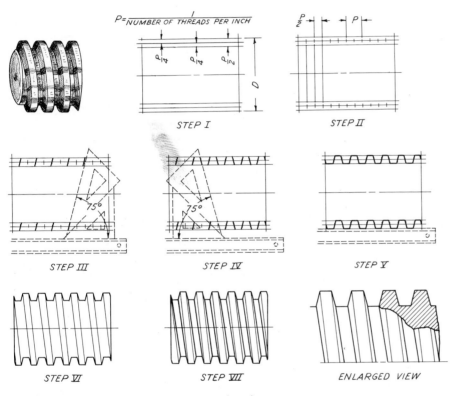

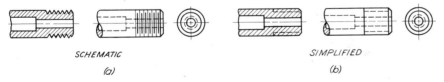

Fig. 14.11 Detailed representation of Acme thread.

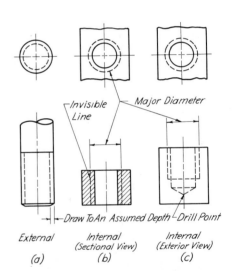

Fig. 14.13 Simplified representation.

Fig. 14.12 External thread representation.

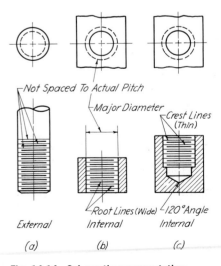

Fig. 14.14 Schematic representation.

shown by invisible lines drawn parallel to the axis [Fig. 14.13(a)].

The schematic representation consists of alternate long and short lines perpendicular to the axis. Although these lines, representing the crests and roots of the thread, are not spaced to actual pitch, their spacing should indicate noticeable differences in the number of threads per inch of different threads on the same working drawing or group of drawings (Fig. 14.14). The root lines are made heavier than the crest lines (Fig. 14.15).

Before a hole can be tapped (threaded), it must be drilled to permit the tap to enter. See Table 10 in the Appendix for tap drill sizes for standard threads. Since the last of the thread cut is not well formed or

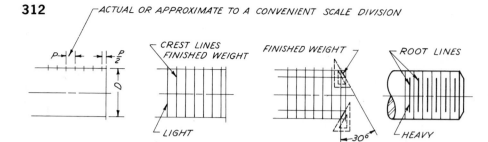

ACTUAL OR APPROXIMATE TO A CONVENIENT SCALE DIVISION

CREST LINES FINISHED WEIGHT — FINISHED WEIGHT — ROOT LINES — LIGHT — HEAVY — 30°

Fig. 14.15 Drawing conventional threads—schematic representation.

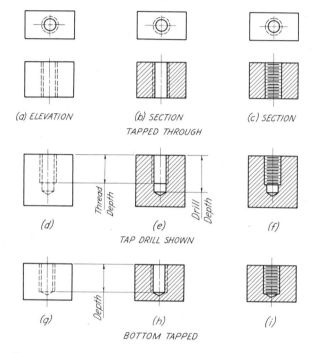

(a) ELEVATION (b) SECTION TAPPED THROUGH (c) SECTION

(d) (e) TAP DRILL SHOWN (f)

Thread Depth Drill Depth

(g) (h) BOTTOM TAPPED (i)

Depth

Fig. 14.16 Representation of internal threads.

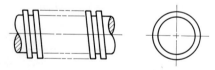

Fig. 14.17 Simplified representation of a square thread.

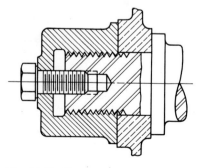

Fig. 14.18 Threads in section.

usable, the hole must be shown drilled and tapped deeper than the screw will enter [Fig. 14.16(d), (e), and (f)]. To show the threaded portion extending to the bottom of the drilled hole indicates the use of a bottoming tap to cut full threads at the bottom. This is an extra and expensive operation not justified except in cases where the depth of the hole and the distance the screw must enter are limited [see Fig. 14.16(g), (h), and (i)].

Figure 14.17 shows a simplified method of representation for square threads.

14.10

Threads in Section. The detailed representation of threads in section, which is used for large diameters only, is shown in Fig. 14.8. Since the far side of an internal thread in section is visible, the crest and root lines incline in the opposite direction

to those of an external thread having the same specifications.

The schematic and simplified representations for threads of small diameter are shown in Figs. 14.12 and 14.16.

A sectioned assembly drawing is shown in Fig. 14.18. When assembled pieces are both sectioned, the detailed representation is used, and the thread form is drawn. Simplified representation could have been used for the thread on the cap screw instead of the schematic representation shown (Fig. 13.96).

14.11

Unified and American Screw-thread Series. The Unified and American screw thread series, as given in ANSI B1.1–1960 consists of six series and a selection of special threads that cover special combinations of diameter and pitch. Each series differs from the other by the number of threads per inch for a specific diameter (see Tables 10 and 11 in the Appendix).

The coarse-thread series (UNC and NC) is designated UNC for sizes above $\frac{1}{4}$ in. in diameter. This series is recommended for general industrial use.

The fine-thread series (UNF and NF), designated UNF for sizes above $\frac{1}{4}$ in., was prepared for use when a fine thread is required and for general use in the automotive and aircraft fields.

The extra-fine-thread series (UNEF and NEF) is used for automotive and aircraft work when a maximum number of threads is required for a given length.

The 8-thread series (8N) is a uniform pitch series for large diameters. It is sometimes used in place of the coarse-thread series for diameters greater than 1 in. This series was originally intended for high-pressure joints.

The 12-thread series (12UN or 12N) is a uniform pitch series intended for use with large diameters requiring threads of medium-fine pitch. This series is used as a continuation of the fine-thread series for diameters greater than $1\frac{1}{2}$ in.

The 16-thread series (16UN or 16N) is a uniform pitch series for large diameters requiring a fine-pitch thread. This series is used as a continuation of the extra-fine-thread series for diameters greater than 2 in.

14.12

Unified and American Screw-thread Classes. Classes of thread are determined by the amounts of tolerance and allowance specified. Under the new unified system, classes 1A, 2A, and 3A apply only to external threads; classes 1B, 2B, and 3B apply to internal threads. Classes 2 and 3 from the former American Standard (ASA) have been retained without change in the Unified and American Thread Standard for use in the United States only, but they are not among the unified classes even though the thread forms are identical. These classes are used with the American thread series (NC, NF, and N series), which covers sizes from size 0 (.060) to 6 in.

Class 1A and class 1B replace class 1 of the old American Standard.

Class 2A and class 2B were adopted as the recognized standards for screws, bolts, and nuts.

Class 3A and class 3B invoke new classes of tolerance. These classes along with class 2A and class 2B should eventually replace class 2 and class 3 now retained from the American (National) Standard. Class 2 and class 3 are defined in the former standard ANSI B1.1–1935 as follows:

Class 2 fit. Represents a high quality of commercial thread product and is recommended for the great bulk of interchangeable screw-thread work.

Class 3 fit. Represents an exceptionally high quality of commercially threaded product and is recommended only in cases where the high cost of precision tools and continual checking is warranted.

14.13

Identification Symbols for Unified Screw Threads. Threads are specified under the unified system by giving the diameter, number of threads per inch, initial letters (UNC, UNF, etc.), and class of thread (1A, 2A, and 3A; or 1B, 2B, and 3B) (see Fig. 14.19).

Unified and American National threads are specified on drawings, in specifications, and in stock lists by thread information given as shown in Fig. 14.20. A multiple-start thread is designated by specifying in sequence the nominal size, pitch, and lead.

14.14

Thread Dimensioning. In general, the thread length dimension shown on a drawing should be the length of the complete (full-form) threads. The incomplete threads should be beyond this dimensioned length.

14.15

Square Threads. Square threads can be completely specified by a note. The nominal diameter is given first, followed by the number of threads per inch and the type of thread (see Fig. 14.20).

14.16

Acme and Stubb Acme Threads. Acme threads have been standardized in a preferred (single) series of diameter–pitch combinations by the American National Standards Institute (B1.5–1952). The standard provides two types of Acme threads, the general-purpose and centralizing. The three classes of general-purpose threads (2G, 3G, and 4G) have clearances on all diameters for free movement. The five classes of centralizing threads (2C, 3C, 4C, 5C, and 6C) have a limited clearance at the major diameters of the external and internal threads to ensure

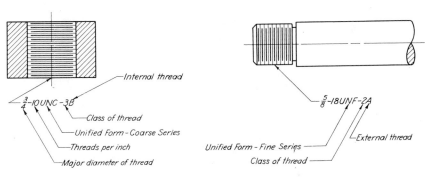

Fig. 14.19 Unified thread identification symbols.

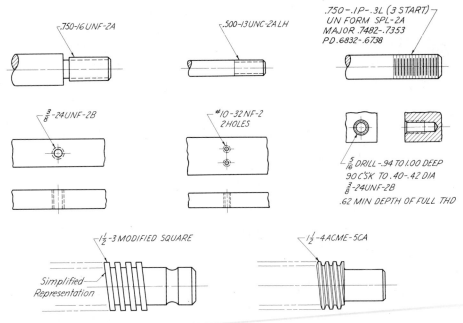

Fig. 14.20 Thread identification symbols.

alignment of mating parts and to prevent wedging on the flanks of the thread.

Acme threads are specified by giving the diameter, number of threads per inch, type of thread, and class (Fig. 14.20). The letter A or B is added to indicate an external or an internal thread.

Examples

$1\frac{3}{8}$–4 ACME–2GA or 1.375–4 ACME–2GA

$1\frac{3}{4}$–4 ACME–2GB–LH

$\frac{3}{4}$–6 ACME–4CB or .750–6 ACME–4CB

$2\frac{3}{4}$–3 ACME–3GA–2-START (2-START indicates a double thread)

14.17

Buttress Threads. The buttress thread is designed for exceptionally high stress along the axis of the thread in one direction only. No pitch-diameter series has been recommended, because of the need for special design of most components. The ANSI B1.9–1953 standard covers three classes of buttress threads. These are 1A, 2A, and 3A for external threads and 1B, 2B, and 3B for internal threads.

Buttress threads are specified by giving, in order, the diameter, threads per inch, type of thread (National Buttress), and class.

Examples

$\frac{5}{8}$–20 BUTTRESS–2A

10–10 BUTTRESS–3B (2-START) (Optionally, these threads may be designated by giving the pitch followed by the letter P instead of the number of threads per inch.)

.625–.05P BUTTRESS–2A

14.18

ISO (International Organization for Standardization) Metric Screw Threads. Some thirty years ago (1946) an ISO Committee was organized to investigate the possibil-

ity for the establishment of a single international system for screw threads. The futility of the idea soon became evident and it was agreed that two systems would be accepted, a new SI Metric Standard System and the Unified Inch System then already developed for use in the United States, Great Britain, and Canada. Fortunately for all concerned, the two thread standards are somewhat alike, with both using essentially the same basic thread profile. See Fig. 14.21. The principal differences between these two recognized standards are in the basic sizes, the magnitude and application of allowances and tolerances, and the method for designating and specifying threads.

When it became evident at the beginning of the 1970s that the United States was about to start a slow conversion to the metric system, the American National Standards Institute realized that the opportunity was at hand to make needed major changes in the many forms of fasteners in use. Performance could be improved and redesign would offer the possibility for more optimum use of materials and for the limitation of products to the fewest number of styles, types, and sizes consistent with present-day requirements of American industry.

In 1971 a special committee was created by ANSI to study and prepare an optimum metric fastener system for use in the United States. This committee was asked to develop a total system of mechanical fasteners that would improve fastener performance through product redesign. All dimensions and properties of the new system were to be given in metric (SI) units. As one of its first tasks the committee made a thorough study of existing screw-thread systems. This project led to the preparation and release of OMFS Recommendations 2, 3, 4, and 6 for threads as an extension of ISO Metric Screw Threads. The ANSI–OMFS committee is continuing with its work so that this interim standard can eventually be replaced by an expanded and complete standard—ANSI–OMFS Metric Screw Threads. Information for the specification of these threads will be found in Sec. 14.21. The diameter–pitch combinations of the ISO Metric Screw Thread Standard Series may be found in Table 8 in the Appendix.

14.19

Designation of ISO Metric Screw Threads (Fig. 14.22). In general, ISO metric

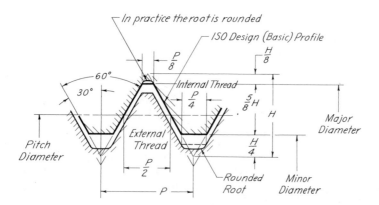

Fig. 14.21 ISO metric internal and external thread design profiles.

threads will be specified in this text using only the basic designation. The basic designation consists of the letter M followed by the nominal size (basic major diameter in millimeters) and then the pitch in millimeters (See Tables 8 and 9 in the Appendix), the nominal size and pitch being separated by the sign ×.

Example

> M8 × 1 (ISO designation)
> M20 × 1.5

For coarse series threads the indication of the pitch may be omitted. However, it is recommended in a number of company engineering standards that this option be disregarded. Therefore, the student should follow the recommendations of his instructor and specify, say, a 10-mm coarse thread as either M10 or M10 × 1.5, whichever is suggested.

The complete designation of an ISO metric thread includes the basic designation previously discussed, followed by an identification for the tolerance class. The tolerance-class designation includes the symbol for the pitch-diameter tolerance followed by the symbol for the crest-diameter tolerance. As can be observed from the examples that follow, each of these symbols consists of a number indicating the tolerance grade followed by a letter indicating the tolerance position. The tolerance-class designation is separated from the basic designation by a dash.

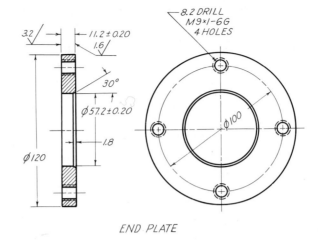

END PLATE

Fig. 14.22 Detail drawing with ISO (metric) thread specification.

For the principal thread elements (pitch diameter and crest diameter) the ISO standard provides a number of tolerance grades that reflect various magnitudes. Basically, three Metric Tolerance Grades are recommended by ISO. These are grades 4, 6, and 8. Grade 6 is commonly used for general-purpose threads with normal lengths of engagement. As might be expected, grade 6 is the closest ISO grade to our Unified 2A and 2B fits.

Under this system, the numbers of the tolerance grades reflect the size of the tolerance. Tolerances below grade 6 are smaller and are specified for fine-quality

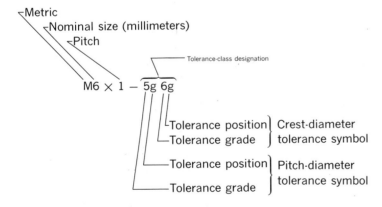

When the pitch-diameter and crest-diameter tolerance symbols are identical, the symbol is only given once. It is not repeated.

Example

> M6 × 1 — 6g
>
> —— Pitch-diameter and crest-diameter
>   tolerance symbols (equal)

requirements or short lengths of engagement. Tolerances above grade 6 are larger and are therefore recommended for coarse quality or long lengths of engagement.

Tolerance positions establish the maximum material limits of the pitch and crest diameters for both internal and external threads. The series of tolerance-position symbols as established by ISO to reflect varying amounts of allowance are as follows:

> For external threads (bolts, etc.):
> Lowercase e—large allowance
> Lowercase g—small allowance
> Lowercase h—no allowance
> For internal threads (nuts, etc.):
> Uppercase G—small allowance
> Uppercase H—no allowance

A desired fit between mating threads may be specified by giving the internal thread tolerance-class designation followed immediately by the external thread tolerance class. The two designations are separated by a slash (/), as shown by the example.

Example

> M6 × 1–6H/6g (or, since this is a coarse thread, as M6–6H/6g)

14.20

ANSI–OMFS Metric Screw Thread Standard Series. For use at this time a series of diameter–pitch combinations for nominal diameters from 1.6 through 36 has been adopted at General Motors for threads considered to be necessary for product design purposes. As ANSI continues its work and new recommendations are forthcoming, this GM interim (one series) standard will be replaced by a complete standard for ANSI–OMFS Metric Screw Threads.

14.21

ANSI–OMFS Metric Threads, Basic Designation. It has been recommended that metric screw threads be designated by giving the nominal diameter and the pitch in millimeters with the two values separated by an uppercase P. Thread class is not specified since there is only one thread class. That is, one class of fit is standard and an allowance has been taken into account.

Example

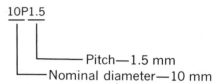

> 10P1.5
> └──── Pitch—1.5 mm
> └──── Nominal diameter—10 mm

When it is desirable to indicate that the basic designation applies to an external thread, an A is added. The letter B added to the designation identifies the thread as an internal thread.

Examples

> 8P1.25A—external thread
> 8P1.25B—internal thread

A detail drawing of a plug having a tapered metric thread is shown in Fig. 14.23. It should be noted that the taper has been specified as 1:16.

B □ Fasteners

14.22

American National Standard Bolts and Nuts (Fig. 14.24). Commerical producers of bolts and fasteners manufacture their products in accordance with the standard specifications given in the American National Standard entitled *Square and Hexagon Bolts and Nuts* (Revised 1965).* See Table 14 in the Appendix.

The ANSI has approved the specifications for three series of bolts and nuts:

1. *Regular series.* The regular series was adopted for general use.

2. *Heavy series.* Heavy bolt heads and nuts are designed to satisfy the special commercial need for greater bearing surface.

3. *Light-series nuts.* Light nuts are used under conditions requiring a substantial savings in weight and material. They are usually supplied with a fine thread.

*ANSI B18.2–1965.

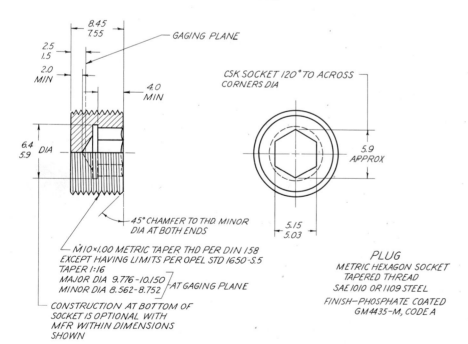

Fig. 14.23 Part with a tapered metric thread. (*Courtesy General Motors Corporation—Standards Section*)

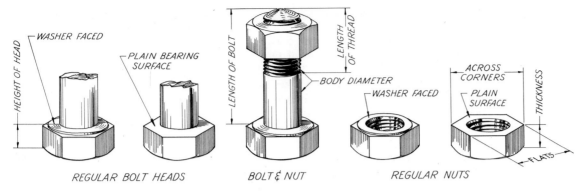

Fig. 14.24 American Standard bolts and nuts.

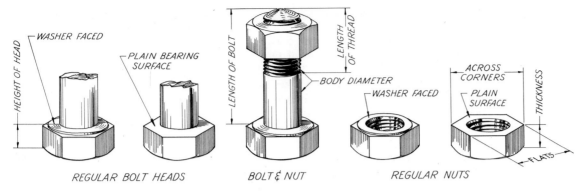

The amount of machining is the basis for further classification of hexagonal bolts and nuts in both the regular and heavy series as unfinished and semifinished.

Square-head bolts and nuts are standardized as unfinished only.

Unfinished heads and nuts are not washer-faced, nor are they machined on any surface.

Semifinished bolt heads and nuts are machined or treated on the bearing surface to provide a washer face for bolt heads and either a washer face or a circular bearing surface for nuts. Nuts, not washer-faced, have the circular bearing surface formed by chamfering the edges.

Bolts and nuts are *always* drawn across corners in all views. This recognized commercial practice, which violates the principles of true projection, prevents confusion of square and hexagonal forms on drawings.

The chamfer angle on the tops of heads and nuts is 30° on hexagons and 25° on squares, but both are drawn at 30° on bolts greater than 1 in. in diameter.

Bolts are specified in parts lists and elsewhere by giving the diameter, number of threads per inch, series, class of thread, length, finish, and type of head.

Example

½–13 UNC–2A × 1¾
SEMIFIN HEX HD BOLT

Frequently it is advantageous and practical to abbreviate the specification thus:

Example

½ × 1¾ UNC SEMIFIN HEX HD BOLT

Although bolt lengths have not been standardized in construction practice, because of the varied requirements in engineering design, length increments for length under the head to the end of a hexagonal bolt can be considered as ⅛ in. for bolts ¼–¾ in. in length, ¼ in. for bolts ¾–3 in. in length, and ½ in. for bolts 3–6 in. in length. Length increments for square-head bolts are ⅛ in. for bolts ¼–¾ in. in length, and ¼ in. for bolts ¾–4¾ in. in length.

The minimum thread length for bolts up to and including 6 in. in length will be twice the diameter plus ¼ in. For lengths over 6 in. the minimum thread length will be twice the diameter of the bolt plus ½ in. (ANSI B18.2–1965).

14.23

To Draw Bolt Heads and Nuts. Using the dimension taken from the tables, draw the lines representing the top and contact surfaces of the head or nut and the diameter of the bolt. Lay out a hexagon about an inscribed chamfer circle having a diameter equal to the distance across the flats (Fig. 14.25) and project the necessary lines to block in the view. Draw in the arcs after finding the centers, as shown in Fig. 14.25.

A square-head bolt or nut may be drawn

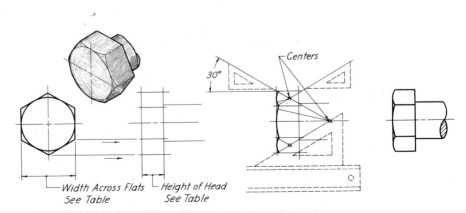

Fig. 14.25 Steps in drawing a hexagonal bolt head.

by following the steps indicated in Fig. 14.26.

The engineer and experienced draftsman wisely resort to some form of template for drawing the views of a bolt head or nut (see Fig. 2.11). To draw the views as shown in Figs. 14.25 and 14.26 consumes valuable time needlessly.

14.24

Studs. Studs, or stud bolts, which are threaded on both ends as shown in Fig. 14.27, are used where bolts would be impractical and for parts that must be removed frequently (cylinder heads, steam chest covers, pumps, and so on). They are first screwed permanently into the tapped holes in one part before the removable member with its corresponding clearance holes is placed in position. Nuts are used on the projecting ends to hold the parts together.

Since studs are not standard they must be produced from specifications given on a detail drawing. In dimensioning a stud, the length of thread must be given for both the stud end and nut end along with an overall dimension. The thread information is given by note.

In a bill of material, studs may be specified as follows:

Example

$\frac{1}{2}$–13 UNC–2A $\times$ 2$\frac{3}{4}$ STUD

It is good practice to abbreviate the specification thus:

Example

$\frac{1}{2}$ $\times$ 2$\frac{3}{4}$ STUD

14.25

Cap Screws (Fig. 14.28). Cap screws are similar to machine screws. They are available in four standard heads, usually in finished form. When parts are assembled, the cap screws pass through clear holes in one member and screw into threaded holes in the other (Fig. 14.29). Hexagonal cap screws have a washer face $\frac{1}{64}$ in. thick with a diameter equal to the distance across flats. All cap screws 1 in. or less in length are threaded very nearly to the head.

Cap screws are specified by giving the diameter, number of threads per inch, series, class of thread, length, and type of head.

Example

$\frac{5}{8}$–11 UNC–2A $\times$ 2 FIL HD CAP SC

It is good practice to abbreviate the specification thus:

Example

$\frac{5}{8}$ $\times$ 2 UNC FIL HD CAP SC

14.26

Machine Screws. Machine screws, which fulfill the same purpose as cap screws, are used chiefly for small work having thin sections (Fig. 14.30). Under the approved American National Standard they range from No. 0 (.060 in. diam.) to $\frac{3}{4}$ in.

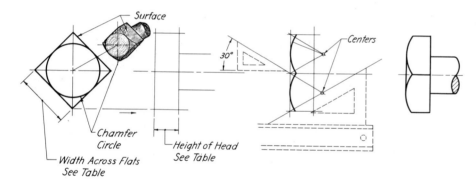

Fig. 14.26 Steps in drawing a square bolt head.

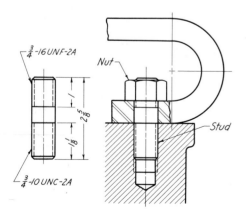

Fig. 14.27 Stud bolt.

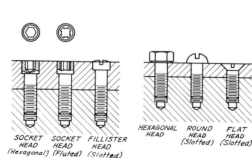

Fig. 14.28 Cap screws.

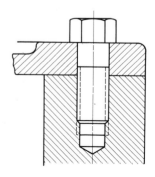

Fig. 14.29 Hexagonal-head cap screw.

(.750 in. diam.) and are available in either the American Standard Coarse or Fine-Threaded Series. The four forms of heads shown in Fig. 14.31 have been standardized.

To specify machine screws, give the diameter, threads per inch, thread series, class of thread, length, and type of head.

Example

No. 12–24 NC–3 × ¾ FIL HD MACH SC

It is good practice to abbreviate by omitting the thread series and class of fit.

Example

No. 12–24 × ¾ FIL HD MACH SC

14.27

Commercial Lengths: Studs, Cap Screws, Machine Screws. Unless a fastening of any of these types carries a constant and appreciable fatigue stress, the usual practice is to have it enter a distance related to its nominal diameter (Fig. 14.32). If the depth of the hole is not limited, it should be drilled to a depth of 1 diameter beyond the end of the fastener to permit tapping to a distance of ½ diameter below the fastener.

The length of the fastening should be determined to the nearest commercial length that will allow it to fulfill minimum conditions. In the case of a stud, care should be taken that the length allows for a full engagement of the nut. Commercial lengths for fasteners increase by the following increments:

Standard length increments

For fastener lengths:
¼–1 in. = ⅛ in.
For fastener lengths:
1–4 in. = ¼ in.

For fastenings and other general-purpose applications, the engagement length should be equal to the nominal diameter (D) of the thread when both components are of steel. For steel external threads in cast iron, brass, or bronze, the engagement length should be $1\frac{1}{2}D$. When assembled into aluminum, zinc, or plastic, the engagement should be $2D$.

14.28

Set Screws. Set screws are used principally to prevent rotary motion between two parts, such as that which tends to occur in the case of a rotating member mounted on a shaft. A set screw is screwed through one part until the point presses firmly against the other part (Fig. 14.33).

The several forms of safety heads shown in Fig. 14.34 are available in combination with any of the points. Headless set screws comply with safety codes and should be used on all revolving parts. The many serious injuries that have been caused by the projecting heads of square-head set screws have led to legislation prohibiting their use in some states [Fig. 14.33(c)].

Set screws are specified by giving the diameter, number of threads per inch, series, class of thread, length, type of head, and type of point.

Example

¼–20 UNC–2A × ½
SLOTTED CONE PT SET SC

The preferred abbreviated form gives the diameter, number of threads per inch, length, type of head, and type of point.

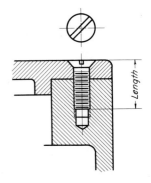

Fig. 14.30 Use of a machine screw.

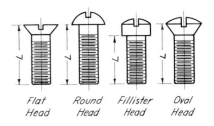

Flat Head   Round Head   Fillister Head   Oval Head

Fig. 14.31 Types of machine screws.

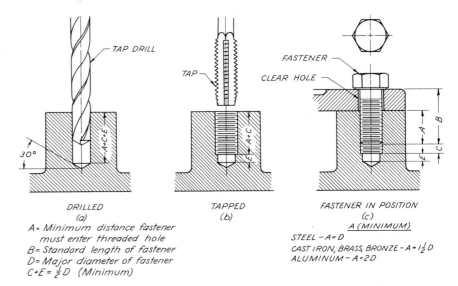

DRILLED (a)
TAP DRILL
30°

TAPPED (b)
TAP

FASTENER IN POSITION (c)
FASTENER
CLEAR HOLE

A = Minimum distance fastener must enter threaded hole
B = Standard length of fastener
D = Major diameter of fastener
C + E = ½ D (Minimum)

A (MINIMUM)
STEEL – A = D
CAST IRON, BRASS, BRONZE – A = 1½ D
ALUMINUM – A = 2D

Fig. 14.32 Threaded hole and fastener.

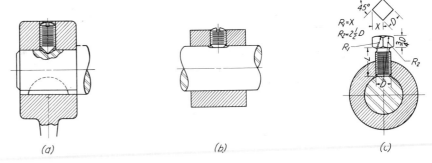

(a)   (b)   (c)

R₁ = X
R₂ = 2½ D
45°

Fig. 14.33 Use of set screws.

320

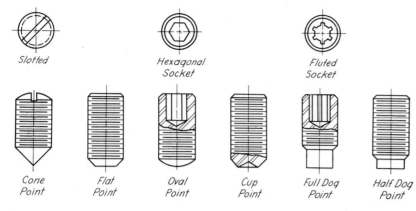

Slotted   Hexagonal Socket   Fluted Socket

Cone Point   Flat Point   Oval Point   Cup Point   Full Dog Point   Half Dog Point

Fig. 14.34   Set screws.

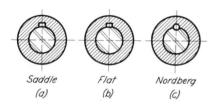

Saddle (a)   Flat (b)   Nordberg (c)

Fig. 14.35   Light-duty keys.

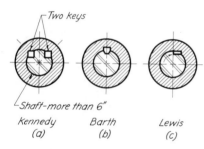

Two keys

Shaft-more than 6"

Kennedy (a)   Barth (b)   Lewis (c)

Fig. 14.36   Heavy-duty keys.

Example

¼–20 × ½ HEX SOCKET CONE PT SET SC

14.29
Keys.   Keys are used in the assembling of machine parts to secure them against relative motion, generally rotary, as is the case between shafts, cranks, wheels, and so on. When the relative forces are not great, a round key, saddle key, or flat key is used (Fig. 14.35). For heavier duty, rectangular keys are more suitable (Fig. 14.36).

The square key (Fig. 14.37) and the Pratt and Whitney key (Fig. 14.38) are the two keys most frequently used in machine design. A plain milling cutter is used to cut the keyway for the square key, and an end mill is used for the Pratt and Whitney keyway. Both keys fit tightly in the shaft and in the part mounted on it.

The gib-head key (Fig. 14.39) is designed so that the head remains far enough from the hub to allow a drift pin to be driven to remove the key. The hub side of the key is tapered ⅛ in. per ft to ensure a fit tight enough to prevent both axial

and rotary motion. For this type of key, the keyway must be cut to one end of the shaft.

14.30
Woodruff Keys.   A Woodruff key is a flat segmental disc with either a flat or a round bottom (Fig. 14.40). It is always specified by a number, the last two digits of which indicate the nominal diameter in eighths of an inch, while the digits preceding the last two give the nominal width in thirty-seconds of an inch.

A practical rule for selecting a Woodruff key for a given shaft is as follows: Choose a standard key that has a width approximately equal to one-fourth of the diameter of the shaft and a radius nearly equal (plus or minus) to the radius of the shaft. Table 26 in the Appendix gives the dimensions for American Standard Woodruff keys.

When Woodruff keys are drawn, it should be remembered that the center of the arc is placed above the top of the key at a distance shown in column E in the table.

14.31
Taper Pins.   A taper pin is commonly used for fastening collars and pulleys to shafts, as illustrated in Fig. 14.41. The hole for the pin is drilled and reamed with the parts assembled. When a taper pin is to be used, the drawing callout should read as follows:

#3 (.213) DRILL AND REAM FOR #4 TAPER PIN WITH PC #6 IN POSITION

Drill sizes and exact dimensions for taper pins are given in Table 28 in the Appendix.

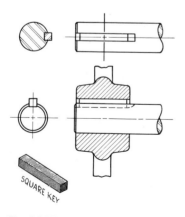

SQUARE KEY

Fig. 14.37   Square key.

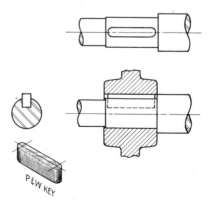

P&W KEY

Fig. 14.38   Pratt and Whitney key.

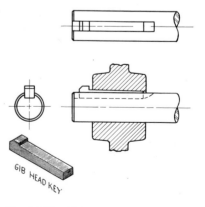

GIB HEAD KEY

Fig. 14.39   Gib-head key.

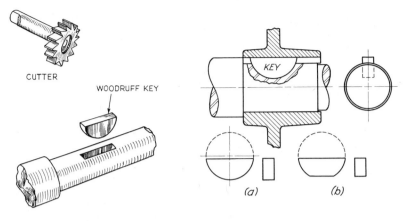

CUTTER

WOODRUFF KEY

(a) *(b)*

Fig. 14.40 Woodruff key.

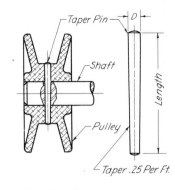

Taper Pin

Shaft

Pulley

Length

Taper .25 Per Ft.

Fig. 14.41 Use of a taper pin.

14.32

Locking Devices. A few of the many types of locking devices that prevent nuts from becoming loose under vibration are shown in Figs. 14.42 and 14.43.

Figure 14.42 shows six forms of patented spring washers. The ones shown in (*D*), (*E*), and (*F*) have internal and external teeth.

In common use is the castellated nut with a spring cotter pin that passes through the shaft and the slots in the top [Fig. 14.43(*a*)]. This type is used extensively in automotive and aeronautical work.

Figure 14.43(*b*) shows a regular nut that is prevented from loosening by an American National Standard jam nut.

In Fig. 14.43(*c*) the use of two jam nuts is illustrated.

A regular nut with a spring-lock washer is shown in Fig. 14.43(*d*). The reaction provided by the lock washer tends to prevent the nut from turning.

A regular nut with a spring cotter pin through the shaft, to prevent the nut from backing off, is shown in Fig. 14.43(*e*).

Special devices for locking nuts are illustrated in Fig. 14.43(*f*) and (*g*). A set screw may be held in position with a jam nut, as in (*h*).

14.33

Areo Thread. The Areo-thread screw-thread system allows the use of high-strength cap screws and studs in light soft metals, such as aluminum and magnesium, through the use of a phosphor bronze or stainless steel coilspring lining in the tapped hole, as shown in Fig. 14.44. This coil (screw bushing) is formed to fit a modified American Standard thread. Special tools are needed for inserting the coil in the tapped hole.

Fig. 14.42 Special lock washers.

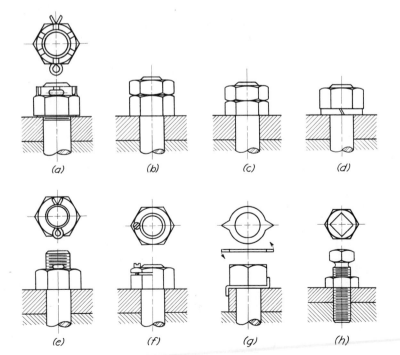

(a) *(b)* *(c)* *(d)*

(e) *(f)* *(g)* *(h)*

Fig. 14.43 Locking schemes.

Fig. 14.44   Aero thread.

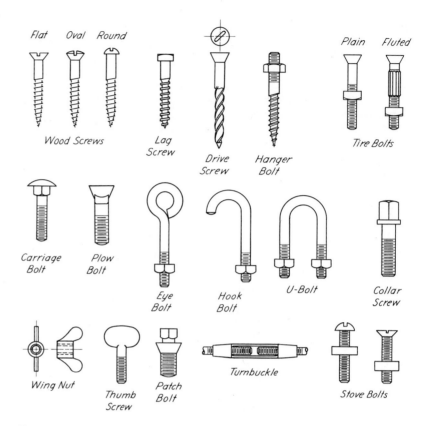

Flat   Oval   Round

Plain   Fluted

Wood Screws

Lag
Screw

Drive
Screw

Hanger
Bolt

Tire Bolts

Carriage
Bolt

Plow
Bolt

Eye
Bolt

Hook
Bolt

U-Bolt

Collar
Screw

Wing Nut

Thumb
Screw

Patch
Bolt

Turnbuckle

Stove Bolts

Fig. 14.45   Miscellaneous bolts, screws, and nuts.

14.34

Miscellaneous Bolts, Screws, and Nuts.
Other types of bolts and screws that have been adopted for commercial use are illustrated in Fig. 14.45.

Wood screws have threads proportioned for the holding strength of wood. They are available with different forms of heads (flat, round, and oval).

Some of the fastenings shown in Fig. 14.45 have been standardized by the American National Standards Institute.

14.35

Phillips Head.   The Phillips head, shown in Fig. 14.46 for a wood screw, is one of various types of recessed heads. Although special drivers are usually employed for installation, an ordinary screwdriver can be used. Machine screws, cap screws, and many special types of fasteners are available with Phillips heads.

14.36

Rivets.   Rivets are permanent fasteners used chiefly for connecting members in such structures as buildings and bridges and for assembling steel sheets and plates for tanks, boilers, and ships. They are cylindrical rods of wrought iron or soft steel, with one head formed when manufactured. A head is formed on the other end after the rivet has been put in place through the drilled or punched holes of the mating parts. A hole for a rivet is generally drilled, punched, or punched and reamed $\frac{1}{16}$ in. larger than the diameter of the shank of the rivet [Fig. 14.47(a)]. Figure 14.47(b) illustrates a rivet in position.

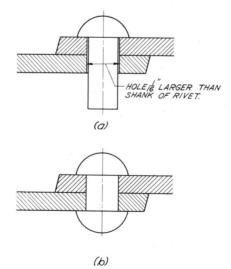

HOLE $\frac{1}{16}$" LARGER THAN
SHANK OF RIVET.

(a)

(b)

Fig. 14.46   Phillips head screw.

Fig. 14.47   Riveting procedure.

Small rivets, less than $\frac{1}{2}$ in. in diameter, may be driven cold, but the larger sizes are driven hot. For specialized types of engineering work, rivets are manufactured of chrome–iron, aluminum, brass, copper, and so on. Standard dimensions for small rivets are given in Table 23 in the Appendix.

The type of rivets and their treatment are indicated on drawings by the American National Standard conventional symbols shown in Fig. 25.13.

14.37

Riveted Joints. Joints on boilers, tanks, and so on, are classified as either lap joints or butt joints (Fig. 14.48). Lap joints are generally used for seams around the circumference. Butt joints are used for longitudinal seams, except on small tanks where the pressure is to be less than 100 lb per sq in.

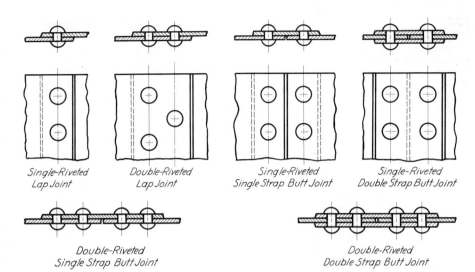

Single-Riveted Lap Joint

Double-Riveted Lap Joint

Single-Riveted Single Strap Butt Joint

Single-Riveted Double Strap Butt Joint

Double-Riveted Single Strap Butt Joint

Double-Riveted Double Strap Butt Joint

Fig. 14.48 Forms of riveted joints.

C □ Springs

14.38

Springs. In production work, a spring is largely a matter of mathematical calculation rather than drawing, and it is usually purchased from a spring manufacturer, with the understanding that it will fulfill specified conditions. For experimental work, and when only one is needed, it may be formed by winding oil-tempered spring wire or music wire around a cylindrical bar. As it is wound, the wire follows the helical path of the screw thread. For this reason the steps in the layout of the representation for a spring are similar to the screw thread. Pitch distances are marked

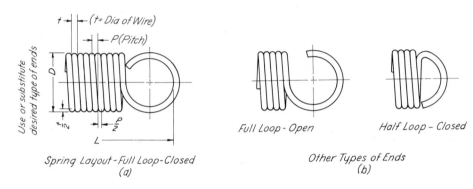

Spring Layout - Full Loop-Closed
(a)

Full Loop - Open

Half Loop - Closed

Other Types of Ends
(b)

Fig. 14.49 Tension springs.

off, and the coils are given a slope of one-half of the pitch. Figure 14.49(a) shows a partial layout of a tension spring. Other types of ends are shown in (b). A compression spring layout, with various types of ends, is illustrated in Fig. 14.50. Single-line symbols for the representation of springs are shown in Fig. 14.51.

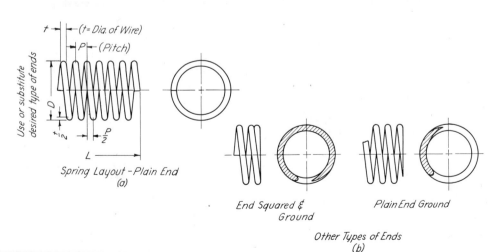

Spring Layout – Plain End
(a)

End Squared & Ground

Plain End Ground

Other Types of Ends
(b)

Compression Spring
(a)

Torsion Spring
(b)

Ends shaped as required

Tension Spring
(c)

Fig. 14.51 Single-line representation of springs.*

Fig. 14.50 Compression springs.

*ANSI Z14.1–1946.

When making a detail working drawing of a spring, it should be shown to its free length. On either an assembly or detail drawing, a fairly accurate representation, neatly drawn, will satisfy all requirements.

It is common practice in industry to rely on a printed spring drawing, accompanied by a filled-in printed form, to convey the necessary information for the production of a needed spring by a reliable manufacturer. The best procedure is to give the spring characteristics along with a list of necessary dimensions and then depend on an experienced spring designer, at the plant where the spring will be produced, to finalize the design.

The method of representing and dimensioning a compression spring is shown in Fig. 14.52. The spring is represented by a rectangle with diagonals (printed or drawn). Pertinent information is added as dimensions and notes. Either an ID (inside diameter) or an OD (outside diameter) dimension is given, depending on how the spring is to be used.

A method of representing and dimensioning an extension spring is shown in Fig. 14.53. Although drawings of this type may be printed forms, it is often necessary to prepare a drawing showing the ends and a few coils, since extension springs may have any one of a wide variety of types of ends. Needed information is presented in tabular form, as shown, with or without a complete spring design. When no printed form is available, a drawing similar to the one shown in Fig. 14.53 must be prepared.

A torsion spring offers resistance to a torque load. The extended ends form the torque arms, which are rotated about the axis of the spring. One method of representing and dimensioning torsion springs is shown in Fig. 14.54. A printed form may be used when there is sufficient uniformity in product requirements to warrant the preparation of a printed form or several printed forms. When a printed form is not available, a drawing similar to the one shown must be prepared.

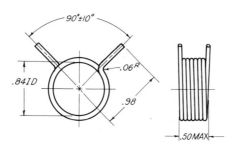

MATERIAL: .105 HARD DRAWN SPRING STEEL WIRE
12 COILS   10 ACTIVE
CLOSED ENDS GROUND
FINISH: PLAIN

Fig. 14.52 Compression spring drawing.

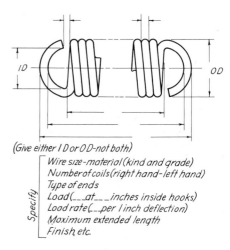

(Give either I D or O D–not both)

Specify {
Wire size–material (kind and grade)
Number of coils (right hand–left hand)
Type of ends
Load (__at___inches inside hooks)
Load rate (__per I inch deflection)
Maximum extended length
Finish, etc.

Fig. 14.53 Extension spring drawing.

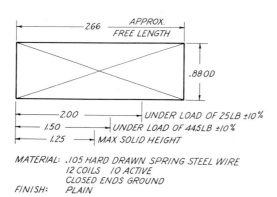

MATERIAL: .059 MUSIC WIRE
6¾ COILS–RIGHT HAND-NO INITIAL TENSION
TORQUE: 2.5 INCH LB AT 155° DEFLECTION SPRING MUST
DEFLECT 180° WITHOUT PERMANENT SET AND
MUST OPERATE FREELY ON .750 DIA SHAFT
FINISH: CADMIUM PLATE

Fig. 14.54 Torsion spring representation and dimensioning.

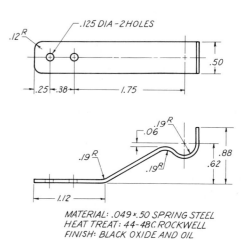

MATERIAL: .049 x .50 SPRING STEEL
HEAT TREAT: 44-48 C. ROCKWELL
FINISH: BLACK OXIDE AND OIL

Fig. 14.55 Flat spring drawing.

The term *flat spring* includes all springs made of a strip material. One method of representing and dimensioning flat springs is shown in Fig. 14.55.

D □ Bearings

14.39

Bearings. A draftsman ordinarily is never called on to make a detail drawing of a ball or roller bearing, because bearings of these two types are precision-made units that are purchased from reliable manufacturers. All draftsmen working on machine drawings, however, should be familiar with the various types commonly used and should be able to represent them correctly on an assembly drawing. An engineer will find it necessary to determine shaft-mounting fits and housing-mounting fits from a manufacturer's handbook, in order to place the correct limits on shafts and housings. Figure 14.56 shows two types of ball bearings. A roller bearing is shown in Fig. 14.57.

Ball bearings may be designed for loads either perpendicular or parallel to the shaft. In the former, they are known as radial bearings and in the latter, as thrust bearings. Other types, designated by various names, are made to take both radial and thrust loads, either light or heavy. In most designs, bearings are forced to take both radial and thrust loads. Ball bearings are designated by a letter and code number, the last number of which represents the bearing bore. They may be extra light, light, medium, or heavy and still have the same bore number. That is, bearings of different capacities and different outer diameters can fit shafts having the same nominal size. Figure 14.58 shows typical mountings in a single mechanism.

Roller bearings are designed for both radial and thrust loads. The bearing consists of tapered rollers that roll between an inner and an outer race. The rollers are enclosed in a retainer (cage) that keeps them properly spaced.

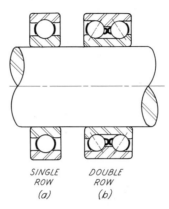

SINGLE ROW (a)   DOUBLE ROW (b)

Fig. 14.56 Ball bearings.

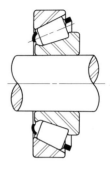

Fig. 14.57 Roller bearing.

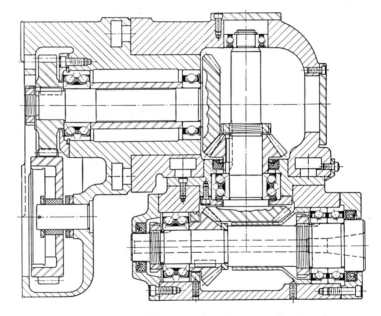

Fig. 14.58 Ball bearings. (*Courtesy New Departure Bearing Co.*)

Problems

Excellent practice in drawing (or sketching) the representations of threads, threaded fasteners, keys, and rivets is provided by the problems of this chapter.

Note: There are no ANSI standards for fasteners in metric sizes at this time. However, metric threads are given in Tables 8 and 9 (see Sec. 14.18).

Use Table 1 in the Appendix if it should be desirable to convert inches and fractions of an inch to millimeters. A metric scale has been given for Prob. 1.

1. Draw or sketch the three layouts shown in Fig. 14.59 to full size, using the given scale to determine the measurements. On layout ① complete the drawing to show a suitable fastener on center line AA. On layout ② show a $\frac{1}{2}$-in. hexagonal-head cap screw on center line BB. On layout ③ show a $\frac{3}{8}$-in. button-head rivet on center lines CC and a No. 608 Woodruff key on center line (CL) DD. Use the schematic symbol for the representation of threads.

2. (Fig. 14.60). Reproduce the views of the assembly of the alignment bearing. On CLs A show $\frac{1}{4}$-in. button-head rivets (four required). On CL B show a $\frac{5}{16} \times \frac{1}{2}$ American Standard square-head set screw. Do not dimension the views.

3. (Fig. 14.61). Reproduce the views of the assembly of the impeller drive. On CLs A show $\frac{1}{4}$-20 UNC $\times \frac{1}{2}$ round-head machine screws and regular lock washers. On CL B show a No. 406 Woodruff key. On CL C show a standard No. 2 $\times$ 1$\frac{1}{2}$ taper pin.

4. (Fig. 14.62). Reproduce the views of the assembly of the bearing head. On CLs A show $\frac{1}{2}$-UNC studs with regular lock washers and regular semifinished hexagonal nuts (four required). On CLs B show $\frac{3}{8}$-UNC $\times$ 1$\frac{1}{4}$ hexagonal-head cap screws (two required). On CL C drill through and tap $\frac{1}{8}$-in. pipe thread.

5. (Fig. 14.63). Reproduce the views of the assembly of the air cylinder. On CL AA at the left end of the shaft show a 1-UNF semifinished hexagonal nut. At the right end show a hole tapped $\frac{3}{4}$-UNF $\times$ 1$\frac{1}{2}$ deep. Between the piston and the (right) end plate draw a spring 3 in. OD, five full coils, $\frac{1}{4}$-in. wire. On CLs B show $\frac{3}{8}$-UNC $\times$ 1$\frac{1}{4}$ hexagonal-head cap screws. On CLs C draw $\frac{1}{4}$-UNC $\times \frac{3}{4}$ flathead cap screws with heads to the left. On CL D show a $\frac{1}{4}$-in. standard pipe thread. On CLs E show $\frac{1}{2}$-UNC $\times$ 1$\frac{3}{4}$ semifinished hexagonal-head bolts. Use semifinished hexagonal nuts. Show visible fasteners on the end view.

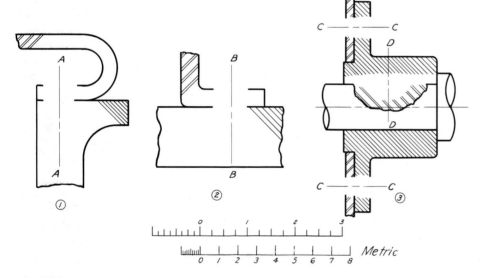

Fig. 14.59

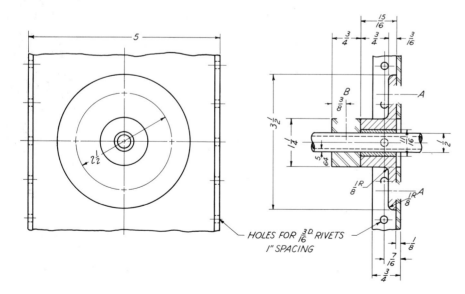

Fig. 14.60 Alignment bearing.

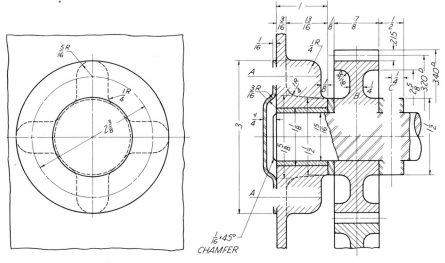

Fig. 14.61   Impeller drive.

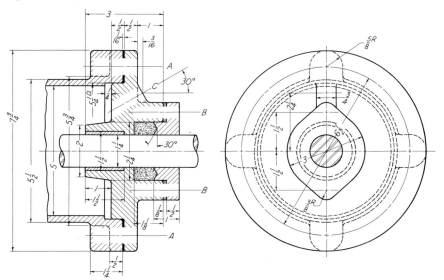

Fig. 14.62   Bearing head.

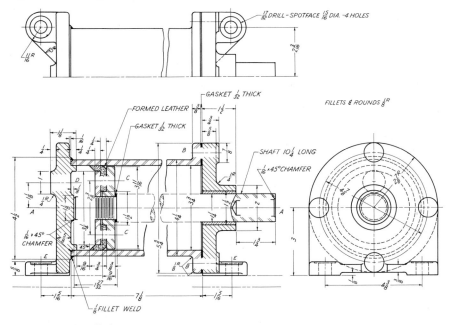

Fig. 14.63   Air cylinder.

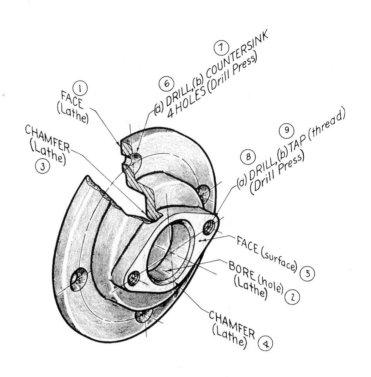

① FACE (Lathe)

③ CHAMFER (Lathe)

⑥ ⑦ (a) DRILL,(b) COUNTERSINK 4 HOLES (Drill Press)

⑧ ⑨ (a) DRILL,(b) TAP (thread) (Drill Press)

⑤ FACE (surface)

② BORE (hole) (Lathe)

④ CHAMFER (Lathe)

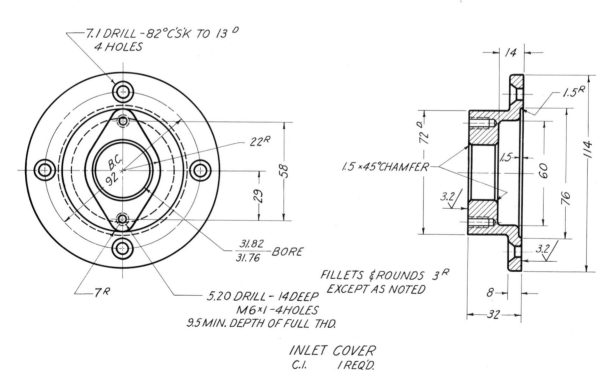

7.1 DRILL -82°C'SK TO 13 D
4 HOLES

22ᴿ

58

29

B.C. 92

7ᴿ

31.82 / 31.76 BORE

5.20 DRILL - 14 DEEP
M6×1 - 4 HOLES
9.5 MIN. DEPTH OF FULL THD.

14

1.5ᴿ

72 D

1.5 × 45° CHAMFER

1.5

60

114

76

3.2

3.2

8

32

FILLETS & ROUNDS 3ᴿ
EXCEPT AS NOTED

INLET COVER
C.I.   1 REQ'D.

Detail drawing of a machined casting.

15

Shop processes, shop terms, and tool drawings

15.1

Shop Processes. An engineering draftsman must be thoroughly familiar with the fundamental shop processes before he is qualified to prepare drawings that will fulfill the requirements of the production shops. In preparing working drawings, he must consider each and every individual process involved in the production of a piece and then specify the processes in terms that the shopman will understand (see the facing page). All too frequently, drawings that specify impractical methods and impossible operations are sent to the shops. Most of these impractical specifications are the result of a lack of knowledge, on the draftsman's part, of what can or cannot be done by skilled craftsmen using modern machines and tools.

Although an accurate knowledge of shop processes can be acquired only through actual experience in the various shops, it is possible for an apprentice draftsman to obtain a working knowledge of the fundamental operations through study and observation. This chapter presents and explains the principal operations in the pattern shop, foundry, forge shop, and machine shop.

15.2

Die Casting. Die casting is an inexpensive method for producing certain types of machine parts, particularly those needing no great strength, in mass production. The castings are made by forcing molten metal or molten alloy into a cavity between metal dies in a die-casting machine. Parts thus produced usually require little or no finishing.

15.3

Sand Casting. Castings are formed by pouring molten metal into a mold or cavity. In sand molding, the molten metal assumes the shape of the cavity that has been formed in a sand mold by ramming a prepared moist sand around a pattern and then removing the pattern. Although a casting shrinks somewhat in cooling, the metal hardens in the exact shape of the pattern used (Fig. 15.1).

A sand mold consists of at least two sections. The upper section, called the *cope,* and the lower section, called the *drag,* together form a box-shaped structure called a *flask.*

When large holes [.75 in. (19 mm) and over] or interior passageways and openings are needed in a casting, dry sand cores are placed in the cavity. Cores exclude the metal from the space they occupy and thus form desired openings. Large holes are cored to avoid an unnecessary boring operation. A dry sand core is formed by ramming a mixture of sand and a binding material into a core box that has been made in the pattern shop. To make a finished core rigid, the coremaker places it in a core oven, where it is baked until it is hard.

The molder when making a mold inserts in the sand a sprue stick that he removes after the cope has been rammed. This resulting hole, known as the *sprue,* con-

ducts the molten metal to the *gate,* which is a passageway cut to the cavity. The adjacent hole, called the *riser,* provides an outlet for excess metal.

15.4

Pattern Shop. The pattern shop prepares patterns of all pieces that the foundry is to cast. Although special pattern drawings are frequently submitted, the patternmaker ordinarily uses a drawing of the finished piece that the draftsman has prepared for both the pattern shop and the machine shop (see the chapter-facing page). The finish marks on such a drawing are just as important to the patternmaker as to the machinist, for he must allow, on each surface to be finished, extra metal, the amount of which depends on the method of machining and the size of the casting. In general, this amount varies from .06 in. (1.5 mm), on very small castings, to as much as .75 in. (19 mm), on large castings.

It is not necessary for the draftsman to specify on his drawing the amount to be allowed for shrinkage, for the patternmaker has available a "shrink rule," which is sufficiently oversize to take care of the shrinkage.

A pattern usually is first constructed of light, strong wood, such as white pine or mahogany, which, if only a few castings are required, may be used in making sand molds. In quantity production, however, where a pattern must be used repeatedly, the wooden one will not hold up, so a metal pattern (aluminum, brass, and so on) is made from it and is used in its place.

Every pattern must be constructed in such a way that it can be withdrawn from each section of the sand mold. If the pattern consists of two halves (split), the plane of separation should be so located that it will coincide with the plane of separation of the cope and the drag (Fig. 15.1). Each portion of the pattern must be slightly tapered, so that it can be withdrawn without leaving a damaged cavity. The line of intersection, where the dividing plane cuts the pattern, is called the *parting line.* Although this line is rarely shown on a drawing, the draftsman should make certain that the design will allow the patternmaker to establish it. Ordinarily, it is not necessary to specify the slight taper, known as *draft,* on each side of the parting line, for the patternmaker assumes such responsibility when constructing the pattern.

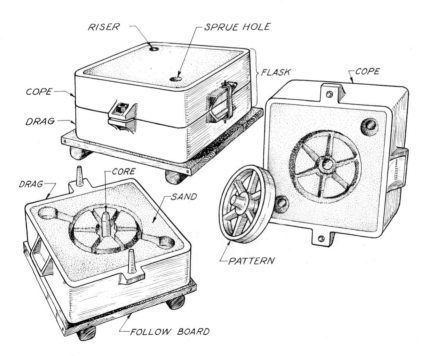

Fig. 15.1 Sand mold.

A "filled-in" interior angle on a casting is called a *fillet*, to distinguish it from a rounded exterior angle, which is known as a *round* (Fig. 5.46). Sharp interior angles are avoided for two reasons: They are difficult to cast; and they are likely to be potential points of failure because the crystals of the cooling metal arrange themselves at a sharp corner in a weak pattern. Fillets are formed by nailing quarter rounds of wood or strips of leather into the sharp angles or by filling the angles with wax.

15.5

Foundry. Although a draftsman is not directly concerned with the foundry, since the patternmaker takes his drawing and prepares the pattern and core boxes for the molder, it is most important that he be familiar with the operations in making a sand mold and a casting. Otherwise, he will find it difficult to prepare an economical design, the cost of which depends on how simple it is to mold and cast.

15.6

Forge Shop. Many machine parts, especially those that must have strength and yet be light, are forged into shape, the heated metal being forced into dies with a drop hammer. Drop forging, since heated metal is made to conform to the shape of a cavity, might be considered a form of casting. However, because dies are difficult to make and are expensive, this method of production is used principally to make parts having an irregular shape that would be costly to machine and could not be made from casting material. Forgings are made of a high-grade steel. Dies are made by expert craftsmen who are known simply as diemakers.

Generally, special drawings, giving only the dimensions needed, are made for the forge shop.

15.7

Machine Shop. In general, the draftsman is more concerned with machine-shop processes than with the processes in other shops, for all castings and forgings that have been prepared in accordance with his drawings must receive their final machining in the machine shop (see chapter-facing page). Since all machining operations must be considered in the design and then properly specified, a draftsman must be thoroughly familiar with the limitations as well as the possibilities of such common machines as the lathe, drill press, boring

machine, shaper, planer, milling machine, and grinder (Fig. 15.2).

15.8

Standard Stock Forms. Many types of metal shapes, along with other materials that are used in the shops for making parts for structures, are purchased from manufacturers in stock sizes. They are made available from the stock department, where rough stock, such as rods, bars, plates, sheet metal, and so on, is cut into sizes desired by the machine shop.

15.9

Lathe. Many common operations, such as turning, facing, boring, reaming, knurling, threading, and so on, may be performed with this widely used machine. In general, however, it is used principally for machining (roughing-out) cylindrical surfaces to be finished on a grinding machine. Removing metal from the exterior surfaces of cylindrical objects is known as *turning* and is accomplished by a sharp cutting tool that removes a thin layer of metal each time it travels the length of a cylindrical surface on the revolving work (Fig. 15.3). The piece, which is supported in the machine between two aligned centers, known as the *dead center* and the *live center*, is caused to rotate about an axis by power transmitted through a lathe dog, chuck, or faceplate. The work revolves against the cutting tool, held in a tool post, as the tool moves parallel to the longitudinal axis of the piece being turned.

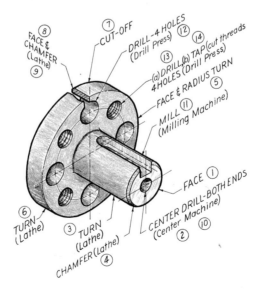

Fig. 15.2 Shop operations required to produce a part from cold-rolled stock. The numbers indicate a possible order in which the operations could be performed in the shops.

Fig. 15.3 Lathe operation—turning.

Cutting an interior surface is known as *boring* (Fig. 15.4). A note is not necessary on a drawing to indicate that a surface is to be turned on a lathe.

When a hole is reamed, it is finished very accurately with a fluted reamer of the exact required diameter. If the operation is performed on a lathe, the work revolves as the nonrotating reamer is fed into the hole by turning the handwheel on the tail stock (see Fig. 15.5).

Screw threads may be cut on a lathe by a cutting tool that has been ground to the shape required for the desired thread. The thread is cut as the tool travels parallel to the axis of the revolving work at a fixed speed (Fig. 15.6).

Knurling is the process of roughening or embossing a cylindrical surface. This is accomplished by means of a knurling tool containing knurl rollers that press into the work as the rollers are fed across the surface (Fig. 15.7).

15.10

Drill Press. A drill press is a necessary piece of equipment in any shop because, although it is used principally for drilling, as the name implies (Fig. 15.8), other operations, such as reaming, counterboring, countersinking, and so on, may be performed on it by merely using the proper type of cutting tool. The cutting tool is held in position in a chuck at the end of a vertical spindle that is made to revolve, through power from a motor, at a particular speed suitable for the type of metal being drilled. The most flexible drill press, especially for large work, is the radial type, which is so designed that the spindle is mounted on a movable arm that can be revolved into any desired position for drilling. With this machine, holes may be drilled at various angles and locations without shifting the work, which may be either clamped to the horizontal table or held in a drill vise or drill jig. The ordinary type of drill press without a movable arm is usually found in most shops along with the radial type. A multispindle drill is used for drilling a number of holes at the same time.

Figure 15.9 shows a setup on a drill press for performing the operation of counterboring. A counterbore is used to enlarge a hole to a depth that will allow the head of a fastener, such as a fillister-head cap screw, to be brought to the level of the surface of the piece through which it passes. A counterbore has a piloted end having approximately the same diameter as the drilled hole.

Figure 15.10 shows a setup for the op-

Fig. 15.4 Boring on a lathe.

Fig. 15.5 Reaming on a lathe.

eration of countersinking. A countersink is used to form a tapering depression that will fit the head of a flathead machine screw or cap screw and allow it to be brought to the level of the surface of the piece through which it passes.

A plug tap is used to cut threads in a drilled hole (Fig. 15.11).

A spotfacer is used to finish a round spot that will provide a good seat for the head of a screw or bolt on the unfinished surface of a casting. Figure 15.12 shows various cutting tools commonly used for forming holes and cutting threads.

Fig. 15.6 Cutting threads on a lathe.

Fig. 15.7 Knurling.

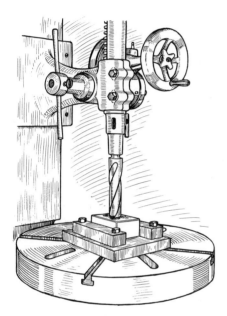

Fig. 15.8 Drill press.

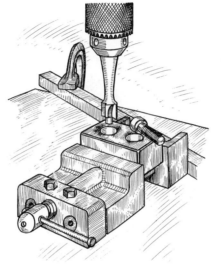

Fig. 15.9 Counterboring on a drill press.

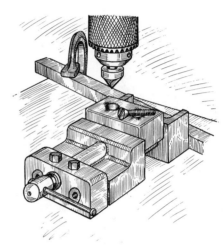

Fig. 15.10 Countersinking on a drill press.

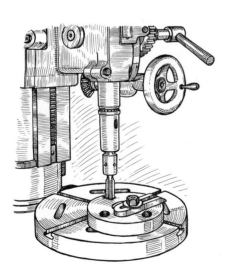

Fig. 15.11 Tapping on a drill press.

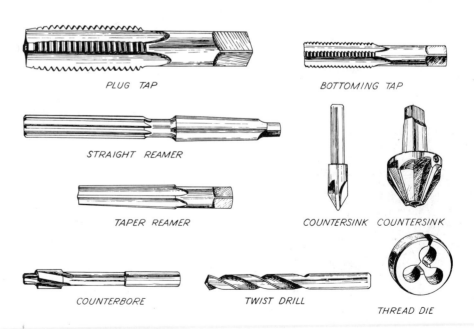

PLUG TAP

BOTTOMING TAP

STRAIGHT REAMER

TAPER REAMER

COUNTERSINK   COUNTERSINK

COUNTERBORE

TWIST DRILL

THREAD DIE

Fig. 15.12 Various cutting tools.

15.11

Hand Reaming. A hole may be finished to an accurate size by hand reaming, as shown in Fig. 15.13. The reamer in this illustration is of a special type known as a *line reamer.*

15.12

Boring (Fig. 15.14). Boring is the operation of enlarging a circular hole for accuracy in roundness or straightness and may be accomplished on a lathe, drill press, milling machine, or boring mill. When the hole is small and of considerable length, the operation may be performed on a lathe. If the hole is large, the

work is usually done on a boring mill, of which there are two types—the vertical and the horizontal. On a vertical boring machine, the work is fastened on a horizontal revolving table, and the cutting tool or tools, which are stationary, advance vertically into it as the table revolves. On a horizontal boring machine, the tool revolves and the work is stationary.

15.13

Milling Machine. A milling machine is used for finishing plane surfaces and for milling gear teeth, slots, keyways, and so on. In finishing a plane surface, a rotating circular cutter removes the metal for a desired cut as the work, fastened to a moving horizontal bed, is automatically fed against it. Several types of milling cutters are shown in Fig. 15.16.

Figure 15.15 shows a setup for milling gear teeth in a gear blank. Note the form of this particular type of cutter.

15.14

Shaper (Fig. 15.17). A shaper is used for finishing small plane surfaces and for cutting slots and grooves. In action, a fast-moving reciprocating ram carries a tool across the surface of the work, which is fastened to an adjustable horizontal table. The tool cuts only on the forward stroke.

15.15

Planer. The planer is a machine particularly designed for cutting down and finishing large flat surfaces. The work is fastened to a long horizontal table that moves back and forth under the cutting

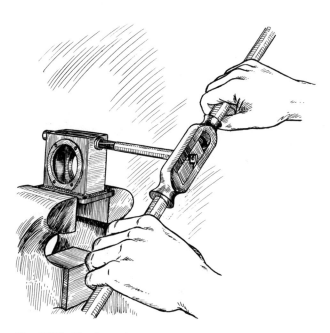

Fig. 15.13 Hand reaming.

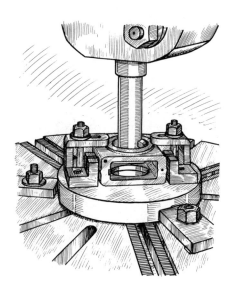

Fig. 15.14 Boring on a boring mill.

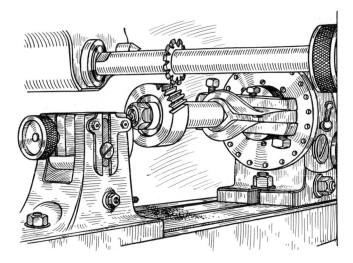

Fig. 15.15 Cutting gear teeth on a milling machine.

tool. In action, the tool cuts as the table moves the surface against it. Unlike the cutter on the shaper, it is stationary except for a slight movement laterally for successive cuts.

15.16
Grinding Machine. A grinding machine has a rotating grinding wheel that, ordinarily, is either an emery wheel (fine or coarse) or some type of high-speed wheel made of carborundum (Fig. 15.18). Grinding consists of bringing the surface to be ground into contact with the wheel. Although grinding machines are often used for "roughing" and for grinding down projections and surfaces on castings, their principal use, as far as a draftsman is concerned, is for the final finishing operation that will bring the cylindrical surface of a piece of work down to accurate dimensions. Very fine surface finishes with tolerances as close as 1 or 2 microinches (0.025 or 0.050 micrometer) may be obtained by grinding. Grinding wheels of special form are used for grinding threads when close fits are desired and for forming other shapes.

When a flat surface is to be brought to a super finish, a surface grinder is used. The work, clamped to either a reciprocating or rotating worktable, passes under a rapidly rotating abrasive wheel, as illustrated in Fig. 15.18(a).

In grinding the external surface of a shaft, as illustrated in (d), the work, mounted between centers, rotates slowly while in contact with the rapidly rotating grinding wheel moving along the work as indicated by the arrows. The grinding wheel is mounted so that it may be moved up to the work or away from it. Normally,

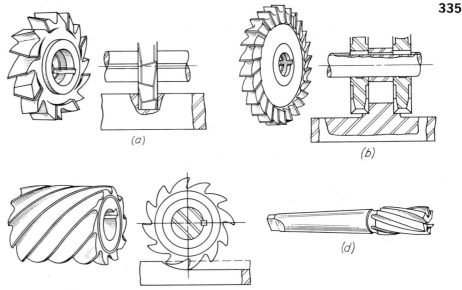

Fig. 15.16 Milling cutters and operations.

the depth of cut is about .002 in. (0.05 mm) per pass. However, finishing cuts may be made as fine as .0001 in. (0.003 mm) per pass to bring the finished size within the dimensional limits given on a detail drawing.

As might be expected with centerless grinding, the work is not held between

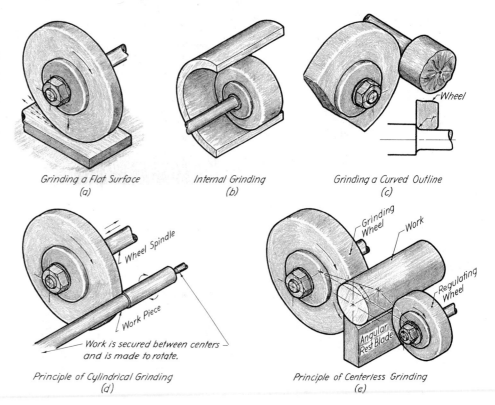

Grinding a Flat Surface
(a)

Internal Grinding
(b)

Grinding a Curved Outline
(c)

Principle of Cylindrical Grinding
(d)

Principle of Centerless Grinding
(e)

Fig. 15.17 Shaper.

Fig. 15.18 Grinding surfaces.

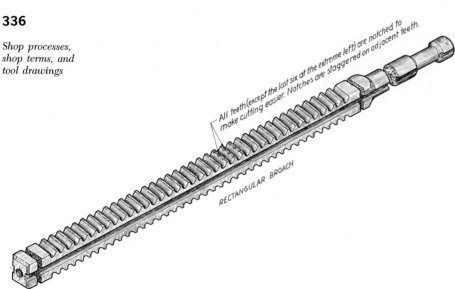

All teeth (except the last six at the extreme left) are notched to make cutting easier. Notches are staggered on adjacent teeth.

RECTANGULAR BROACH

Fig. 15.19   Broach for forming a rectangular hole.

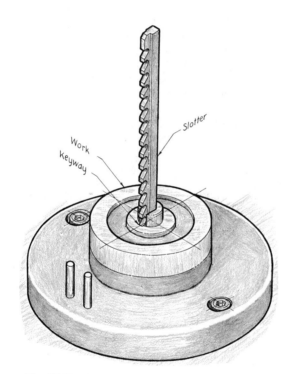

Slotter

Work
Keyway

Fig. 15.20   Cutting a keyway.

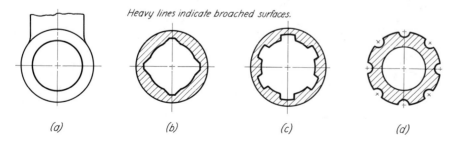

Heavy lines indicate broached surfaces.

(a)   (b)   (c)   (d)

Fig. 15.21   Some typical broached contours.

centers, as in the case of cylindrical grinding. Rather, it may be said to be fed along against the grinding wheel by a regulating wheel while being supported on a rest blade. See Fig. 15.18(*e*). It should be noted that, with both the grinding wheel and regulating wheel rotating in a clockwise direction, the work is forced to rotate counterclockwise. Pressure from the grinding wheel forces the work against both the work rest and the regulating wheel. A centerless grinder produces work that is round and of constant diameter.

Internal grinding for fine finish and close tolerances is illustrated in (*b*). Form grinding is shown in (*c*).

15.17
Broaching.   A broach is a tool used to cut keyways and to form square, rectangular, hexagonal, or irregular-shaped holes. It is a hard, tempered cutting tool with serrated cutting edges that enlarge a drilled, punched, or cored hole to a required shape. In operation, each tooth removes a chip of material from the surface except the last few teeth, which are to size (Fig. 15.19). A broach produces a fine finish with the work accurately sized. This is accomplished with a single pass. Broaches, whether used to finish an internal (hole) or an external surface, are either pushed or pulled. A special broaching machine is used for pulling broaches. Some form of press, hydraulic or otherwise, is required for push broaches.

The cutting of a keyway using a slotter (broach) is illustrated in Fig. 15.20. A guide bushing has been inserted in the hole in the piece to hold the tool in position.

Some typical broached contours are shown in Fig. 15.21.

15.18
Superfinishing: Polishing, Honing, and Lapping.   Polishing consists of bringing a ground surface into contact with a revolving disc of leather or cloth, thus producing a lustrous smoothness that would be impossible to obtain by using even the finest grinding wheel. The operation is specified on a drawing by a note: ''polish'' or ''grind and polish.''

Honing and lapping are methods of producing superfinishes, after grinding, through the use of abrasives. For honing cylindrical bores, fine-grained abrasive sticks which are available in many styles and grades, are used to minimize scratches and to finish the bore to precision limits.

Lapping is a final stock-removing operation that is performed by rubbing the surfaces of parts to be lapped over a lap of soft metal to which has been applied a powdered abrasive mixed with a lubricant.

15.19

Jigs and Fixtures. Often, when an operation must be performed many times in making a part in quantities on a general machine, one of two devices, a *jig* or a *fixture*, may be used to facilitate production and ensure accuracy without making repeated measurements. In general, it might be said that jigs and fixtures are devices for holding work while various machining operations are being performed. Such devices play an important role in the present quantity production of interchangeable parts, for their use greatly reduces the amount of labor required in performing accurate machining operations. However, the advantage that may be gained in the form of lower labor cost in the assembly of units having parts that are interchangeable may be just as important.

Although there are some people who use the terms denoting these devices interchangeably, there is a definite difference between a jig and a fixture that is quite generally recognized. Specifically, a jig is a work-holding device that is capable of controlling the path of the cutting tool. Since it is not fastened to the machine, a jig may be moved around so that several holes may be drilled, tapped, or reamed in their proper locations as established by the position of the drill bushings. Although it is usual for the work to be held by the jig, there are cases where the jig may be clamped on the piece. A well-designed jig should permit the work to be quickly inserted and removed.

A fixture, as the name implies, is fixed or fastened to the table of the machine to locate and hold the work securely in a definite position. Either the cutting tool is moved into position for the operation or the table is moved under the tool. A fixture does not guide the cutting tool. Fixtures are used for such operations as milling, honing, broaching, grinding, and welding.

The three factors that must be considered and evaluated when determining whether or not the cost of making a jig or a fixture will be justified are (1) the number of parts to be produced, (2) the saving in labor cost, and (3) the possibility that the device may be needed for making replacement parts at a later time. Using simple arithmetic, one can compute and compare the cost of producing one unit in the conventional manner against making the part using a jig or fixture. Knowing the cost of the jig or fixture, the number of units to be produced, and the savings per unit, it will not be difficult to arrive at a decision. Generally, the use of the device should result in a definite saving of time and labor.

More could be given here relating to the economic problems surrounding the use and design of jigs and fixtures but it would be beyond the scope of this chapter.

15.20

Jig Borers. Jig borers are provided with accurate lead screws and special measuring devices that make possible the accurate location and boring of holes. Their use greatly reduces the cost of making jigs since the usual layout work with scriber and punch is not required. It is a common practice in many shops to use a jig borer or its equivalent to machine pieces in small lots rather than go to the expense of making jigs.

15.21

Types of Jigs. There are a number of types of jigs and fixtures that have been developed and are suitable for a wide range of work. The drawing for a simple jig for drilling a jacket tube is shown in Fig. 15.22. The work, shown in phantom line, is usually drawn in red, while the jig itself is drawn in black. In preparing the drawing, the visibility of the lines of the jig itself are as they would be if the work were not in place. This treatment of visibility along with showing the piece to be drilled in red is universal practice.

15.22

Fixtures. Fixtures are designed for varied uses. They may be classified as vise, boring, milling machine, tapping, lathe, grinding, broaching, honing, welding, or checking fixtures. Most readers should be familiar with the principle of the vise. Checking fixtures, as the name implies, are used to check accuracy of production. Figure 15.23 shows a special milling machine fixture that is designed to hold a cover while a keyway is being cut.

15.23

Tool Drawings. When the product drawings for a structure or mechanism have

DET.	"A"	MARK
8	1.870	2-1.870-H945737-8
8A	1.803	2-1.813-H945737-8A

NO.	REQ	NAME	SPECIFICATIONS
1	1	BASE	SAE 1020-CARB. HDN.
2	1	VEE BLOCK	SAE 1020-CARB. HDN.
3	1	BUSHING PLATE	C.R.S.
4	1	SLIDING BLOCK	SAE 1020-CARB. HDN.
5	2	X-L-O BUSHING	#O-11 DR. SIZE #29
6	1	X-L-O BUSHING	#X-14 DR. SIZE 11/16
7	1	SCREW	5/16-18UNC-C.R.S.
8	1	STOP BLOCK	SAE 1020-CARB. HDN.
	1	LINER BUSHING	X-L-O #O-41
	2	X-L-O LOCK SC.	#1
	1	X-L-O LOCK SC.	#2
	1	HAND KNOB	STD. 1 1/2 DIA.
	1	PIN	1/8 DIA ×
	1	HEX. NUT	5/8-11UNC-2B
	1	T-BOLT	3/8-16UNC
	1	HEX. NUT	3/8-16UNC-2B
	1	PLAIN WASHER	3/8 I.D.
	2	SOC. HD. SC.	5/16-18UNC
	2	SOC. HD. SC.	3/8-16UNC
	2	DOWEL	5/16 × 1 1/2
	2	DOWEL	1/4 × 1 1/2

Fig. 15.22 Jig for drilling a jacket tube. (*Courtesy Ross Gear Division of TRW, Inc.*)

NO.	REQ	NAME	SPECIFICATIONS
1	1	BASE	SAE 1020 CARB. HDN.
2	2	KEY	C.R.S.
3	1	CLAMP STRAP	C.R.S.
4	1	STOP	C.R.S.
5	1	LOCATING PIN	DR. ROD. HDN.
6	1	PLUG	TOOL STL.
	1	AETNA BEARING	#E-1
	1	STUD	1/2-13UNC × 2 3/8
	1	HEX. NUT	1/2-13UNC
	1	HEX. SOC. HD. CAP SC.	5/16-18UNC × 1
	1	HEX. SOC. HD. CAP SC.	1/4-20UNC × 7/8
	1	FLAT PT. SET SC.	1/2-13UNC × 5/8
	1	FLAT PT. SET SC.	5/16-18UNC × 5/16

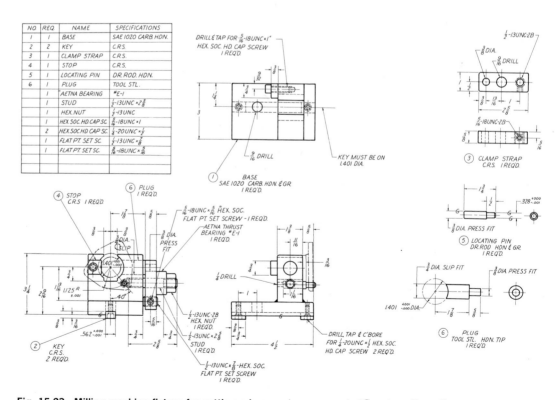

Fig. 15.23 Milling machine fixture for cutting a keyway (upper cover). (*Courtesy Ross Gear Division of TRW, Inc.*)

been completed by the engineering department, they are released to the production division, where a tool routing sheet and a worksheet are prepared by a process engineer. A routing sheet lists needed manufacturing operations in proper sequence and specifies which tools and machines are to be used. It also gives information for checking production accuracy and may specify the type of inspection instruments. The worksheets contain detailed information concerning the manufacturing processes.

In some cases, clearance specifications between mating parts may be given when critical situations are known to exist. These sheets furnish information needed by the tool designer and tool draftsman and ensure that current machine-shop and production practices will be followed.

The drawings for making the holding devices and special tools needed for the various operations are known as *tool drawings*, and they are prepared by the tool engineering branch of the production division. Along with the routing sheet and worksheet, the tool designer and tool draftsman must have catalogs at hand that furnish information about standard parts and devices that may be purchased ready-made in standard arrangements.

In the planning stage of the development of a jig or fixture, freehand sketches may be used by the designer to integrate his thoughts and record information on standard parts. Such sketches assist the designer in preparing the accurate design layouts that are needed, particularly for complicated situations and where different tool draftsmen will be assigned the task of preparing the working drawings.

The tool draftsman starts with an outline of the workpiece traced in red using phantom lines. Then he draws the outline assembly around the workpiece using black lines (Fig. 15.22). Where a black line of the assembly drawing coincides with one of the red lines of the workpiece, the black line is drawn over the phantom red line. In other words, on tool drawings the black lines are given precedence over the red ones. The practice of showing the workpiece in red and the outline of the jig or fixture in black permits easy reading of the drawing by enabling the reader to distinguish quickly between the workpiece and the jig or fixture holding it. An assembly layout may show locating dimensions and toleranced dimensions for critical conditions.

After the design assembly layout has been completed and approved, prints are made for the use of the tool draftsman, who is to make the detail drawings. The draftsman or group of draftsmen under a squad leader prepares drawings for all nonstandard and altered standard parts, placing several detail drawings on each detail sheet, as space will permit without overcrowding. Each detail drawing is identified by the same part number shown for it on the assembly layout. In addition to showing the part number in a small circle, it is customary to letter, just below the drawing, the part name, part material, and number required for one unit (Fig. 15.23). In the case of relatively simple jigs and fixtures, the parts are dimensioned, where possible, directly on the layout so that the layout becomes a working drawing. Those parts that cannot be dimensioned conveniently on the layout are detailed in an open space adjacent to the layout (Fig. 15.23). Of course, additional sheets may be used if necessary.

15.24
Dies and Die Drawings. Since standard die-sets may be found in manufacturing catalogs in different styles and sizes, it is the usual practice of tool designers to analyze the workpiece carefully and then select a suitable die-set on which the required punches and dies can be mounted.

A standard die-set has three principal parts, namely, the punch holder (upper shoe), die holder (lower shoe), and guideposts. Other parts of a standard combination are the shank and guidepost bushings. The shank provides for fastening the punch holder to the ram of the press. The guideposts ensure accurate alignment of the punch with the die under all conditions (Fig. 15.24).

Specific information on die design may be found in the several books listed in the bibliography. It is recommended that one consult the handy reference book *Practical Design of Manufacturing Tools, Dies, and Fixtures*, that was prepared under the direction of the Society of Manufacturing Engineers.

The fundamental press-work operations involving the use of punches and dies are (1) plain blanking, (2) piercing, (3) bending or forming, (4) drawing, (5) coining. and (6) assembling. Die assembly drawings usually show three or four views of the assembled die and the same general procedure is followed in their preparation as for jig and fixture drawings except for a modification of the theoretical arrange-

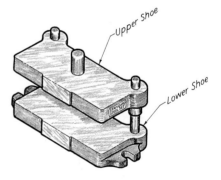

Fig. 15.24 Typical die set.

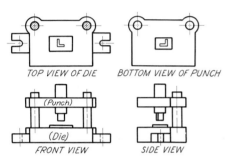

Fig. 15.25 Arrangement of views for a punch and die drawing.

ment of the views as shown in Fig. 15.25. In this illustration, the front view (front elevation) shows both of the main members (punch and die) in working position. The plan view of the lower shoe (die holder) is in the position of a top view directly above the front view. Aligned horizontally with this latter view is the plan view of the upper shoe (punch holder) shown as it would be seen looking upward from below. This is done so that the diemaker has a plan of the actual layout from which to work and to conserve space. The front view, always in section, shows the die as it would be seen from the front of the press and the views of the upper and lower shoes show all parts attached as they would be when in operation. Small part-section views and auxiliary views are used when needed. If a side view is necessary, it should be drawn in section. The assembly drawing should give general information and show the important dimensions needed by the diemaker, such as the shut height of the die, shrank diameter, the type and size of press, and the press stroke. On the die assembly drawing, each component part is numbered as shown in Fig. 15.26.

When it is necessary to detail a designed die, it is the usual practice to place the details on sheets separate from the assembly drawing.

A material list should be given on the die assembly drawing. This should show the number of each part and specify the rough material size.

15.25

Automatic Machines. In large industrial concerns, most mechanical parts are made on either semiautomatic or fully automatic machines by semiskilled operators. A discussion of even a few of these, however, is beyond the scope of a general drawing text in which each subject is limited to a few pages. Since most specialized mass-production machines, having mechanisms that control the movement of cutting tools, operate on the same general principles as the general-purpose machines, a young engineering draftsman should be able to determine their limitations and capabilities through observation, if he has a general knowledge of such machines as the lathe, shaper, drill press, milling machine, and so on. No prospective designer or draftsman should ever

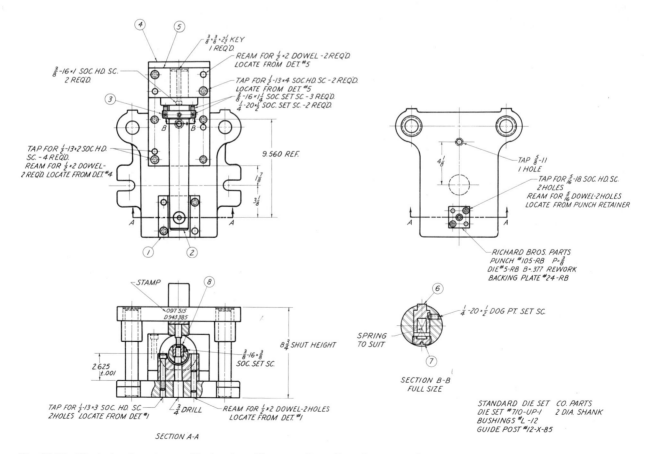

Fig. 15.26 Die design layout assembly drawing. (*Courtesy Ross Gear Division of TRW, Inc.*)

forgo an opportunity to observe special production machines for he must have a thorough understanding of all shop machines and methods, if his drawings are to be satisfactory for the shops.

In use at the present time are production machines that follow directions given on punched tapes and gages that measure electrically to millionths of an inch. The use of tapes and computers by industry does not mean that the draftsmen, technologists, and engineers will have less work to do and that there will be fewer of them. It does mean, however, that the men assigned to both areas must keep themselves informed and up-to-date on late developments. The draftsman should learn how to dimension the drawing of a part to meet the requirement for programming the machine or machines that will perform the shop operations. He must have a more thorough understanding of basic fundamentals than in the past and at the same time be willing to accept change. Numerically controlled machine tools are discussed in Chapter 18.

The fact must be recognized that automation is here and the only question the technically trained man of today can ask himself is whether or not he, as an individual, can adjust to new knowledge and different requirements.

15.26

Manufacturing Processes and the Detail Drawing. In preparing the detail drawing that is needed for the production of a part, the draftsman must give considerable thought to the manufacturing processes that will be required to make the conception a reality. Machined surfaces must be indicated and the dimensions selected and placed with the manufacturing methods in mind. Shop notes, as they may be needed, must be prepared in accordance with the recommendations given in the appropriate ANSI standards.

The detail working drawing of the inlet cover shown on this chapter's facing page has been prepared for use by both the pattern shop and the machine shop since these two shops will be involved in the production of the part. It should and does contain all of the dimensions and notes needed, first, to make the pattern for the casting and, second, to machine the cast part to obtain the finished inlet cover ready for use.

The pictorial drawing gives an interpretation, in shop terms, of the machining operations required by the detail drawing.

The sequence of these operations is indicated by the numbers enclosed within circles.

15.27

Measuring Tools. Figures 15.27, 15.28, and 15.29 show a few of the measuring tools commonly available in shops. When great accuracy is not required, calipers are used (Fig. 15.28). The outside calipers are suited for taking external measurements, as, for example, from a shaft. They are adjusted to fit the piece, and then the setting is applied to a rule to make a reading. The inside calipers have outturned toes, which fit them for taking internal measurements, as, for example, in measuring either a cylindrical or a rectangular hole. When extreme accuracy is required, some form of micrometer calipers may be used (Fig. 15.29).

Quick Reading with Decimal Equivalents

Fig. 15.27 Steel rule.

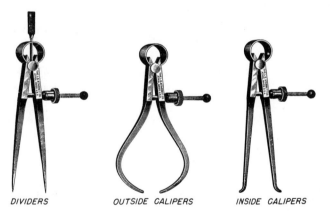

DIVIDERS   OUTSIDE CALIPERS   INSIDE CALIPERS

Fig. 15.28 Calipers.

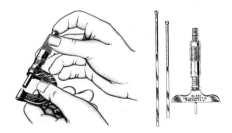

Fig. 15.29 Micrometers.

Shop processes, shop terms, and tool drawings

Glossary
of common shop terms

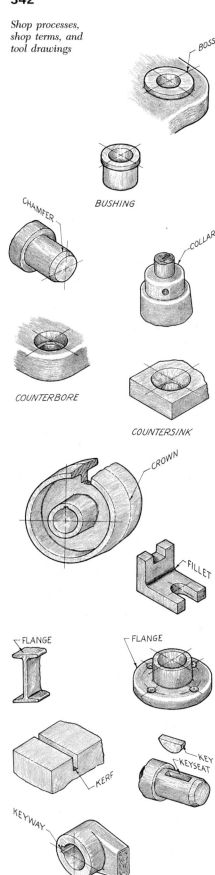

Anneal (*v*). To heat a piece of metal to a particular temperature and then allow it to cool slowly for the purpose of removing internal stresses.

Bore (*v*). To enlarge a hole using a boring bar in order to make it smooth, round, and coaxial. Boring is usually done on a lathe or boring mill.

Boss (*n*). A circular projection, which is raised above a principal surface of a casting or forging.

Braze (*v*). To join two pieces of metal by the use of hard solder. The solder is usually a copper–zinc alloy.

Broach (*v*). To machine a hole to a desired shape, usually other than round. The cutting tool, known as a broach, is pushed or pulled through the rough-finished hole. It has transverse cutting edges.

Burnish (*v*). To smooth or apply a brilliant finish.

Bushing (*n*). A removable cylindrical sleeve, which is used to provide a bearing surface.

Carburize (*v*). To harden the surface of a piece of low-grade steel by heating in a carbonizing material to increase the carbon content and then quenching.

Case-harden (*v*). To harden a surface as described above or through the use of potassium cyanide.

Chamfer (*v*). To bevel an external edge or corner.

Chase (*v*). To cut screw threads on a lathe using a chaser, a tool shaped to the profile of a thread.

Chill (*v*). To cool the surface of a casting suddenly so that the surface will be white and hard.

Chip (*v*). To cut away or remove surface defects with a chisel.

Collar (*n*). A cylindrical part fitted on a shaft to prevent a sliding movement.

Color-harden (*v*). (*See* Case-harden.) A piece is color-hardened mainly for the sake of appearance.

Core (*v*). To form a hole or hollow cavity in a casting through the use of a core.

Counterbore (*v*). To enlarge the end of a cylindrical hole to a certain depth, as is often done to accommodate the head of a fillister-head screw. (*n*) The name of the tool used to produce the enlargement.

Countersink (*v*). To form a conical enlargement at the end of a cylindrical hole to accommodate the head of a screw or rivet. (*n*) The name of the tool used to form a conical-shaped enlargement.

Crown (*n*). The angular or curved contour of the outer surface of a part, such as on a pulley.

Die (*n*). A metal block used for forming or stamping operations. A thread-cutting tool for producing external threads.

Die Casting (*n*). A casting that has been produced by forcing a molten alloy having an aluminum, copper, zinc, tin, or lead base into a metal mold composed of two halves.

Die Stamping (*n*). A piece that has been cut or formed from sheet metal through the use of a die.

Draw (*v*). To form metal, which may be either cold or hot, by a distorting or stretching process. To temper steel by gradual or intermittent quenching.

Drill (*v*). To form a cylindrical hole in metal. (*n*) A revolving cutting tool designed for cutting at the point.

Drop Forging (*n*). A piece formed while hot between two dies under a drop hammer.

Face (*v*). To machine on a lathe a flat face, which is perpendicular to the axis of rotation of the piece.

Feather (*n*). A rectangular sliding key, which permits a pulley to move along the shaft parallel to its axis.

File (*v*). To shape, finish, or trim with a fine-toothed metal cutting tool, which is used with the hands.

Fillet (*n*). A rounded filling, which increases the strength at the junction of two surfaces that form an internal angle.

Fit (*n*). The tightness of adjustment between the contacting surfaces of mating parts.

Flange (*n*). The top and bottom member of a beam. A projecting rim added on the end of a pipe or fitting for making a connection.

Forge (*v*). To shape hot metals by hammering, using a hand-hammer or machine.

Galvanize (*v*). To coat steel or iron by immersion in a bath of zinc.

Graduate (*v*). To mark off or divide a scale into intervals.

Grind (*v*). To finish a surface through the action of a revolving abrasive wheel.

Kerf (*n*). A groove or channel cut by a saw or some other tool.

Key (*n*). A piece used between a shaft and a hub to prevent the movement of one relative to the other.

Keyway or Keyseat (*n*). A longitudinal groove cut in a shaft or a hub to receive a key. A key rests in a keyseat and slides in a keyway.

Knurl (*v*). To roughen a cylindrical surface to produce a better grip for the fingers.

Lap (*v*). To finish or polish with a piece of soft metal, wood, or leather impregnated with an abrasive.

Lug (*n*). A projection or ear, which has been cast or forged as a portion of a piece to provide a support or to allow the attachment of another part.

Malleable Casting (*n*). A casting that has been annealed to toughen it.

Mill (*v*). To machine a piece on a milling machine by means of a rotating toothed cutter.

Neck (*v*). To cut a circumferential groove around a shaft.

Pack-harden (*v*). To case-carburize and harden.

Pad (*n*). A low, projecting surface, usually rectangular.

Peen (*v*). To stretch or bend over metal using the peen end (ball end) of a hammer.

Pickle (*v*). To remove scale and rust from a casting or forging by immersing it in an acid bath.

Plane (*v*). To machine a flat surface on a planer, a machine having a fixed tool and a reciprocating bed.

Polish (*v*). To make a surface smooth and lustrous through the use of a fine abrasive.

Punch (*v*). To perforate a thin piece of metal by shearing out a circular wad with a nonrotating tool under pressure.

Ream (*v*). To finish a hole to an exact size using a rotating fluted cutting tool known as a reamer.

Rib (*n*). A thin component of a part that acts as a brace or support.

Rivet (*n*). A headed shank, which more or less permanently unites two pieces. (*v*) To fasten steel plates with rivets.

Round (*n*). A rounded external corner on a casting.

Sandblast (*v*). To clean the surface of castings or forgings by means of sand forced from a nozzle at a high velocity.

Shear (*v*). To cut off sheet or bar metal through the shearing action of two blades.

Shim (*n*). A thin metal plate, which is inserted between two surfaces for the purpose of adjustment.

Spline (*n*). A keyway, usually for a feather key. (See Feather.)

Spotface (*v*). To finish a round spot on the rough surface of a casting at a drilled hole for the purpose of providing a smooth seat for a bolt or screw head.

Spot Weld (*v*). To weld two overlapping metal sheets in spots by means of the heat of resistance to an electric current between a pair of electrodes.

Steel Casting (*n*). A casting made of cast iron to which scrap steel has been added.

Swage (*v*). To form metal with a "swage block," a tool so constructed that through hammering or pressure the work may be made to take a desired shape.

Sweat (*v*). To solder together by clamping the pieces in contact with soft solder between and then heating.

Tack Weld (*n*). A weld of short intermittent sections.

Tap (*v*). To cut an internal thread, by hand or with power, by screwing into the hole a fluted tapered tool having thread-cutting edges.

Taper (*v*). To make gradually smaller toward one end. (*n*) Gradual diminution of diameter or thickness of an elongated object.

Taper Pin (*n*). A tapered pin used for fastening hubs or collars to shafts.

Temper (*v*). To reduce the hardness of a piece of hardened steel through reheating and sudden quenching.

Template (*n*). A pattern cut to a desired shape, which is used in layout work to establish shearing lines, to locate holes, etc.

Tumble (*v*). To clean and smooth castings and forgings through contact in a revolving barrel. To further the results, small pieces of scrap are added.

Turn (*v*). To turn-down or machine a cylindrical surface on a lathe.

Undercut (*n*). A recessed cut.

Upset (*v*). To increase the diameter or form a shoulder on a piece during forging.

Weld (*v*). To join two pieces of metal by pressure or hammering after heating to the fusion point.

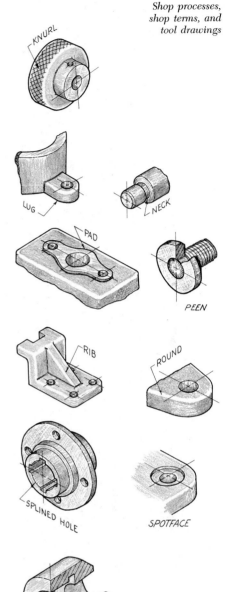

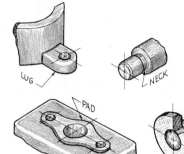

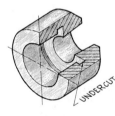

The scene shows the Boeing 747 final assembly line. At the Boeing Company's plant in Everett, Washington, three 747s in final stages of completion can be seen in the line. From the factory, completed airliners are towed to the preflight line, where they are prepared for production test flights and delivery to airlines. Before these giant planes could be produced, many detail and assembly drawings were prepared. (*Courtesy The Boeing Company*)

16

Production drawings: preparation and duplication

A □ Production (shop) drawings

16.1

Communication Drawings. These varied types of engineering drawings, ranging from design drawings to exploded pictorial drawings, have one thing in common, and that is that they are prepared to convey needed ideas and facts to others. Since all serve the same purpose they may be classed together as "communication drawings," this term being almost all-inclusive.

In this chapter we will be concerned mainly with the types of drawings that are prepared by draftsmen under an engineer's supervision and that are to serve as communications to others beyond the engineering department. The preparation of idea sketches, both multiview and pictorial, has been discussed in detail in Chapters 6 and 12. Charts and graphs, which may also be thought of as communication drawings, are presented in Chapter 19. Charts and graphs are used by engineers to supplement written reports and technical papers.

16.2

Sketches and Design Drawings. The first stage in the development of an idea for a structure or machine is to prepare freehand sketches and to make the calculations required to determine the feasibility of the design. From these sketches the designer prepares a layout, on which an accurate analysis of the design is worked out. It is usually drawn full-size and is executed with instruments in pencil (Fig. 12.28). The layout should be complete enough to allow a survey of the location of parts (to avoid interference), the accessibility for maintenance, the requirements for lubrication, and the method of assembly. See Chapter 12.

Usually, only center distances and certain fixed dimensions are given. The general dimensioning, as well as the determination of material and degree of finish of individual parts, is left for the draftsman, who makes the detail drawings while using the layout drawing as a guide.

Design layouts require both empirical and scientific design. Empirical design involves the use of charts, formulas, tables, and so forth, which have been derived from experimental studies and scientific computations. Scientific design, which requires a broad knowledge of the allied fields, such as mechanics, metalurgy, and mathematics, is used when a new machine is designed to operate under special specified conditions for which data are not available in any handbook.

16.3

Classes of Machine Drawings. There are two recognized classes of machine drawings: detail drawings and assembly drawings.

16.4

Set of Working Drawings. A complete set of working drawings for a machine consists of detail sheets giving all necessary shop information for the production of individual pieces and an assembly drawing showing the location of each piece in the finished machine. In addition, the set may inclue drawings showing a foundation plan, piping diagram, oiling diagram, and so on.

16.5

Detail Drawing. A detail drawing should give complete information for the manufacture of a part, describing with adequate dimensions the part's size. Finished surfaces should be indicated and all necessary shop operations shown. The title should give the material of which the part is to be made and should state the number of the parts that are required for the production of an assembled unit of which the part is a member. Commercial examples of detail drawings are shown in Figs. 16.1–16.4.

Since a machinist will ordinarily make one part at a time, it is advisable to detail each piece, regardless of its size, on a separate individual sheet. In some shops,

however, custom dictates that related parts be grouped on the same sheet, particularly when the parts form a unit in themselves. Other concerns sometimes group small parts of the same material together thus: castings on one sheet, forgings on another, special fasteners on still another, and so on.

16.6

Making a Detail Drawing. With a design layout or original sketches as a guide, the procedure for making a detail drawing is as follows:

1. Select the views, remembering that, aside from the view showing the characteristic shape of the object, there should be as many additional views as are necessary to complete the shape description. These may be sectional views that reveal a complicated interior construction, or auxiliary views of surfaces not fully described in any of the principal views.

2. Decide on a scale that will allow, without crowding, a balanced arrangement of all necessary views and the location of dimensions and notes. Although very small parts should be drawn double-size or larger, to show detail and to allow for dimensions, a full-size scale should be used when possible. In general, the same scale should be used for pieces of the same size.

3. Draw the main center lines and block in the general outline of the views with light, sharp 6H pencil lines.

4. Draw main circles and arcs in finished weight.

5. Starting with the characteristic view, work back and forth from view to view until the shape of the object is completed. Lines whose definite location and length are known may be drawn in their finished weight.

6. Put in fillets and rounds.

7. Complete the view by darkening the object lines.

8. Draw extension and dimension lines.

9. Add arrowheads, dimensions, and notes.

10. Complete the title.

11. Check the entire drawing carefully.

16.7

One-view Drawings. Many parts, such as shafts, bolts, studs, and washers, may

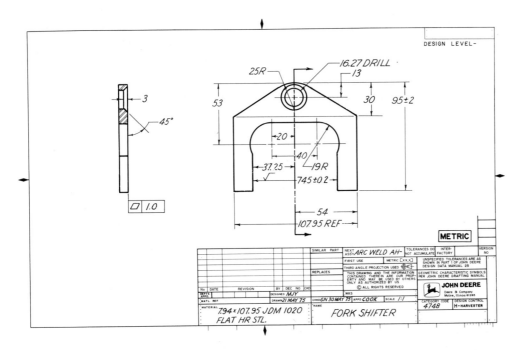

Fig. 16.1 Detail drawing. (*Courtesy John Deere and Company*)

require only one properly dimensioned view. In the case of each of these parts, a note can imply the complete shape of the piece without sacrificing clearness. Most engineering departments, however, deem it better practice to show two views.

16.8
Detail Titles. Every detail drawing must give information not conveyed by the notes and dimensions, such as the name of the part, part number, material, number required, and so on. The method of recording and the location of this information on the drawing varies somewhat in different drafting rooms. It may be lettered either in the record strip or directly below the views (Figs. 16.1 and 16.2).

If all surfaces on a part are machined, finish marks are omitted and a title note, "FINISH ALL OVER," is added to the detail title.

16.9
Title Blocks and Record Strips. The purpose of a title or record strip is to present in an orderly manner the name of the machine, name of the manufacturer, date, scale, drawing number, and other drafting-room information.

Every commercial drafting room has developed its own standard title forms, whose features depend on the processes of manufacture, the peculiarities of the plant organization, and the established customs of particular types of manufacturing. In large organizations, the blank form, along with the borderline, is printed on standard sizes of tracing paper and/or Mylar.

A record strip is a form of title extending almost the entire distance across the bottom of the sheet. In addition to the usual title information, it may contain a

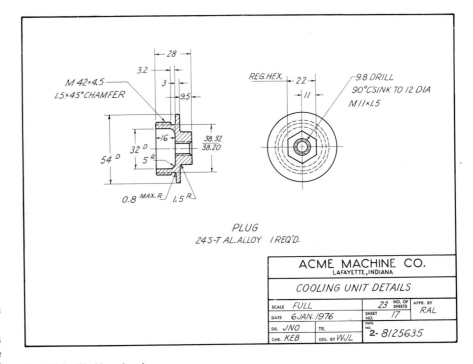

Fig. 16.2 Working drawing.

section for recording revisions, changes, and so on, with the dates on which they were adopted (Fig. 16.3).

16.10
Contents of the Title (Figs. 16.1 and 16.2).
The title on a machine drawing generally contains the following information:

1. Name of the part.

2. Name of the machine or structure. (This is given in the main title and is usually followed by one of two words: *details* or *assembly*.) See Fig. 16.7.

3. Name and location of the manufacturing firm.

4. Name and address of the purchasing firm, if the structure has been designed for a particular company.

5. Scale.

6. Date. (Often spaces are provided for the date of completion of each operation in the preparation of the drawing. If only one date is given, it is usually the date of completion of the drawing.)

7. Initials or name of the draftsman who made the pencil drawing.

8. Initials of the checker.

9. Initials or signature of the chief draftsman, chief engineer, or another in authority who approved the drawing.

10. Initials of the tracer (if drawing has been traced).

11. Drawing number. This generally serves as a filing number and may furnish information in code form. Letters and numbers may be so combined to indicate departments, plants, model, type, order number, filing number, and so on. The drawing number is sometime repeated in the upper-left-hand corner (in an upside-down position), so that the drawing may be quickly identified if it should become reversed in the file.

Some titles furnish information such as material, part number, pattern number,

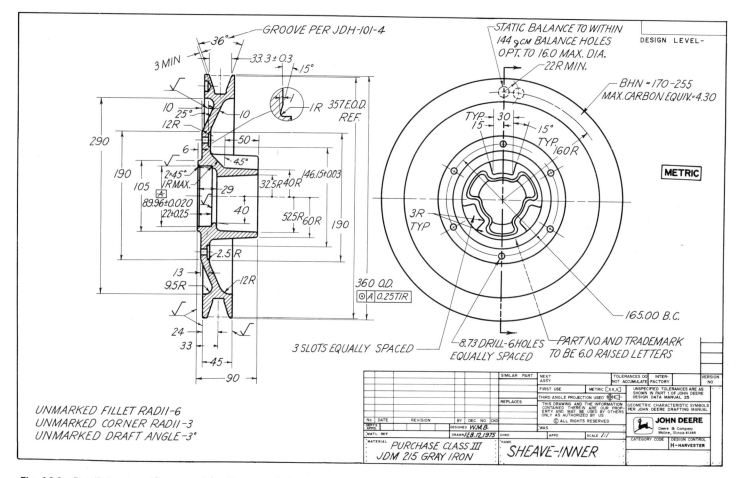

Fig. 16.3 **Detail drawing.** (*Courtesy John Deere and Company*)

finish, treatment, estimated weight, superseded drawing number, and so on.

16.11
Corrections and Alterations. Alterations on working drawings are made either by cancellation or by erasure. Cancellations are indicated by parallel inclined lines drawn through the views, lines, notes, or dimensions to be changed.

Superseding dimensions should be placed above or near the original ones. If alterations are made by erasure, the changed dimensions are often underlined.

All changes on a completed or approved drawing should be recorded in a revision record that may be located either adjacent to the title block (Fig. 16.3) or at one corner of the drawing (Fig. 16.4). This note should contain the identification symbol, date, authorization number, character of the revision, and the initials of the draftsman and checker who made the change. The identification symbol is a numeral or letter placed in a small circle near the alteration on the body of the drawing.

If the changes are made by complete erasure, record prints should be made for the file before the original is altered. Many companies make record prints whenever changes are extensive.

Since revisions on completed drawings are usually necessitated by unsatisfactory methods of production or by a customer's request, they should never be made by a draftsman unless an order has been issued with the approval of the chief engineer's office.

16.12
Pattern-shop Drawings. Sometimes special pattern-shop drawings, giving information needed for making a pattern, are required for large and complicated castings. If the patternmaker receives a drawing that shows finished dimensions, he provides for the draft necessary to draw the pattern and for the extra metal for machining. He allows for shrinkage by making the pattern oversize. When, however, the draft and allowances for finish are determined by the engineering department, no finish marks appear on the drawing. The allowances are included in the dimensions.

16.13
Forge-shop Drawings. If a forging is to be machined, separate detail drawings usually are made for the forge and machine

shops. A forging drawing gives all the nominal dimensions required by the forge shop for a completed rough forging.

16.14
Machine-shop Drawings. Rough castings and forgings are sent to the machine shop to be finished. See Fig. 16.4. Since the machinist is not interested in the dimensions and information for the previous stages, a machine-shop drawing frequently gives only the information necessary for machining.

16.15
Assembly Drawings. A drawing that shows the parts of a machine or machine unit assembled in their relative working positions is an assembly drawing. There are several types of such drawings: design assembly drawings, working assembly drawings, unit assembly drawings, installation diagrams, and so on, each of which will be described separately (Figs. 16.5–16.7 and 16.10–16.12).

16.16
Working Assembly Drawings. A working assembly drawing, showing each piece completely dimensioned, is sometimes made for a simple mechanism or unit of related parts. No additional detail drawings of parts are required.

16.17
Subassembly (Unit) Drawings. A unit assembly is an assembly drawing of a group of related parts that form a unit in a more complicated machine. Such a drawing would be made for the tail stock of a lathe, the clutch of an automobile, or the carburetor of an airplane. A set of assembly drawings thus takes the place of a complete assembly of a complex machine (Fig. 16.6).

16.18
Bill of Material or Parts List. A bill of material is a list of parts placed on an assembly drawing just above the title block, or, in the case of quantity production, on a separate sheet. The bill contains the part (item or key) number, descriptive name, material, quantity (number) required, and so on, of each piece. Additional information, such as stock size, pattern number (castings), and so forth, is sometimes listed.

Suggested dimensions for ruling are shown in Fig. 16.9. For $\frac{1}{8}$-in. letters, the lines should never be spaced closer than

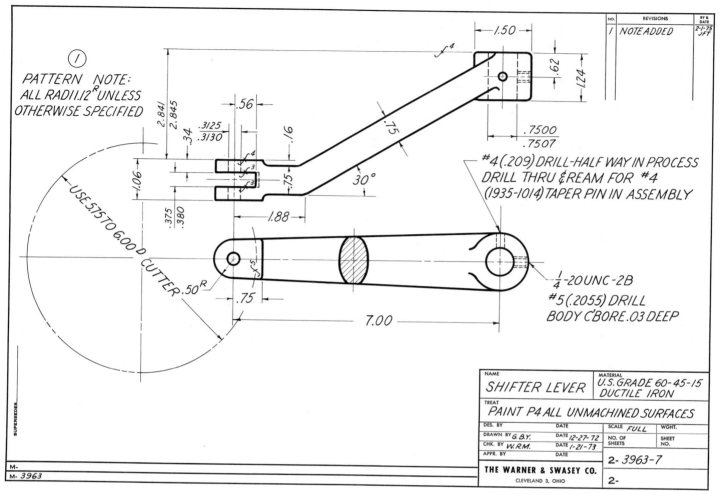

Fig. 16.4 Detail drawing. (*Courtesy Warner & Swasey Co.*)

$5/16$ in. Fractions are made slightly less than full height and are centered between the lines.

When listing standard parts in a bill of material, the general practice is to omit the name of the materials and to use abbreviated descriptive titles. A pattern number may be composed of the commercial job number followed by the assigned number one, two, three, and so on. It is suggested that parts be listed in the following order: (1) castings, (2) forgings, (3) parts made from bar stock, and (4) standard parts.

Sometimes bills of material are first typed on thin paper and then blueprinted. The form may be ruled or printed (Fig. 16.8).

16.19

Title. The title strip on an assembly drawing usually is the same as that used on a detail drawing. It will be noted, when lettering in the block, that the title of the drawing is generally composed of the

name of the machine followed by the word *assembly* (Figs. 16.5 and 16.7).

16.20

Making the Assembly Drawing. The final assembly may be traced from the design assembly drawing, but more often it is redrawn to a smaller scale on a separate sheet. Since the redrawing, being done from both the design and detail drawings, furnishes a check that frequently reveals errors, the assembly always should be drawn before the details are accepted as finished and the blueprints are made. The assembly of a simple machine or unit is sometimes shown on the same sheet with the details.

Accepted practices to be observed on assemblies are as follows:

1. *Sectioning.* Parts should be sectioned using the American Standard symbols shown in Fig. 7.31. The practices of sectioning apply to assemblies.

2. *Views.* The main view, which is usually in full section, should show to the best advantage nearly all the individual parts and their locations. Additional views are shown only when they add necessary information that should be conveyed by the drawing.

3. *Hidden lines.* Hidden lines should be omitted from an assembly drawing, for they tend merely to overload it and create confusion. Complete shape description is unnecessary, since parts are either standard or are shown on detail drawings.

4. *Dimensions.* Overall dimensions and center-to-center distances indicating the relationship of parts in the machine as a whole are sometimes given. Detail dimensions are omitted, except on working-assembly drawings.

5. *Identification of parts.* Parts in a machine or structure are identified on the assembly drawing by numbers that are used on the details and in the bill of material (Fig. 16.5). These should be made at least $\frac{3}{16}$ in. high and enclosed in a $\frac{3}{8}$-in. circle. The centers of the circles are located not less than $\frac{3}{4}$ in. from the nearest line of the drawing. Leaders, terminated by arrowheads touching the parts, are drawn radial with a straightedge. The numbers, in order to be centered in the circles, should be made first and the circles drawn around them. An alternative method used in commercial practice is to letter the name and descriptive information for each part and draw a leader pointing to it in the main view.

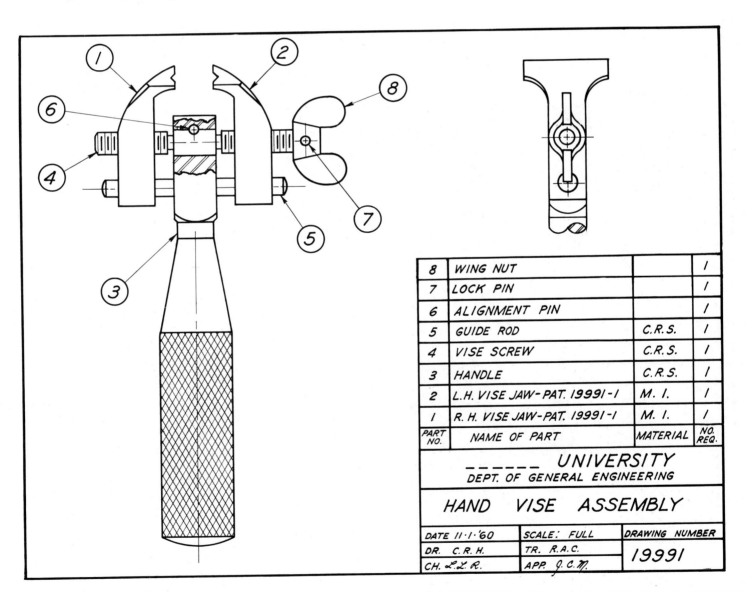

8	WING NUT		1
7	LOCK PIN		1
6	ALIGNMENT PIN		1
5	GUIDE ROD	C.R.S.	1
4	VISE SCREW	C.R.S.	1
3	HANDLE	C.R.S.	1
2	L.H. VISE JAW – PAT. 19991-1	M.I.	1
1	R.H. VISE JAW – PAT. 19991-1	M.I.	1
PART NO.	NAME OF PART	MATERIAL	NO. REQ.

_ _ _ _ _ _ **UNIVERSITY**
DEPT. OF GENERAL ENGINEERING

HAND VISE ASSEMBLY

DATE 11·1·'60	SCALE: FULL	DRAWING NUMBER
DR. C.R.H.	TR. R.A.C.	*19991*
CH. L.L.R.	APP. J.C.M.	

Fig. 16.5 Assembly drawing.

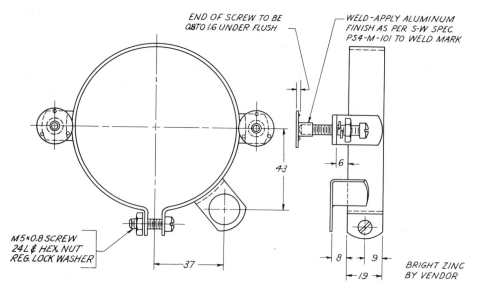

Fig. 16.6 **Tachometer mounting bracket (unit) assembly.** (*Courtesy Studebaker Corp.*)

16.21

Checking Drawings. Checking, the final assurance that the machine is correctly designed, should be done by a person (checker or squad foreman) who has not prepared the drawings but who is thoroughly familiar with the principles of the design. He must have a broad knowledge of shop practices and assembly methods. In commercial drafting rooms, the most experienced persons are assigned to this type of work. The assembly drawing is checked against the detail drawings and corrections are indicated with either a soft or colored pencil. The checker should:

1. Survey the machine as a whole from the standpoint of operation, ease of assembly, and accessibility for repair work. He should consider the type,

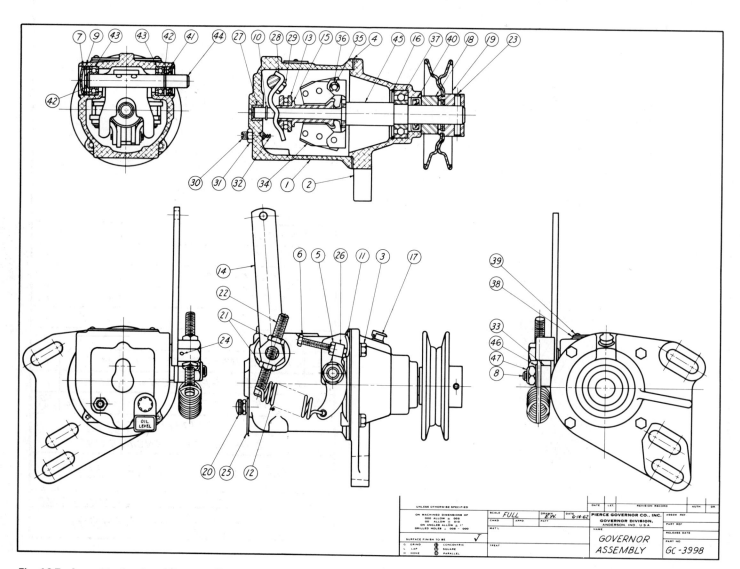

Fig. 16.7 **Assembly drawing.** (*Courtesy Pierce Governor Co., Inc.*)

strength, and suitability of the materials.

2. Check each part with the parts adjacent to it, to make certain that proper clearances are maintained. (To determine whether or not all positions are free of interference, it may be necessary to lay out the extreme travel of moving parts to an enlarged scale.)

3. Study all drawings to see that each piece has been illustrated correctly and that all necessary views, types of views, treatments of views, and scales have been shown.

4. Check dimensions by scaling; calculate and check size and location dimensions that affect mating parts; determine the suitability of dimensions from the standpoint of the various departments' needs, such as pattern, forge, machine, assembly shop, and so on; examine views for proper dimensioning, and mark unnecessary, repeated, or omitted dimensions.

5. Check tolerances, making sure the computations are correct and that proper fits have been used, so that there will be no unnecessary production costs.

6. See that finishes and such operations as drilling, reaming, boring, tapping, and grinding are properly specified.

7. Check specifications for material.

8. Examine notes for correctness and location.

9. See that stock sizes have been used for standard parts, such as bolts, screws, keys, and so on. (Stock sizes may be determined from catalogs.)

10. Add any additional explanatory notes that should supply necessary information.

11. Check the bill of material to see that each part is completely and correctly specified.

12. Check items in the title block.

13. Make a final survey of the drawing in its entirety, making certain there is either a check or correction for each dimension, note, and specification.

16.22

Installation Assembly Drawings. An installation drawing gives useful information for putting a machine or structure together. The names of parts, order of

PIERCE GOVERNOR ASSEMBLY GC-3998 PARTS LIST			
Key No.	Part Name	Part No.	Quantity
1	Governor Body	G-9042-16	1
2	Governor Flange	G-9138-3	1
3	Sems Fastener	X-1784	4
4	Gasket	X-1425	1
5	Hex. Nut	X-977	1
6	Hex. Head Screw	X-890-4	1
7	Welch Plug	X-2019	1
8	Shoulder Stud	G-9799	1
9	Washer	X-2307	2
10	Snap Ring	X-1923	1
11	Stop Bracket	G-9556	1
12	Governor Spring	SN-1304	1
13	Thrust Bearing	X-1336-A	1
14	Throttle Lever Assembly	A-6325	1
15	Thrust Sleeve	G-10813	1
16	Snap Ring—Internal	X-1921	1
17	Oil Cup	X-2053	1
18	Spacer	G-12614-1	1
19	Governor Pulley	G-10908-1	1
20	Oil Lever Check	X-2054	1
21	Hex. Nut	X-1011-1	2
22	Adj. Screw Eye	G-12306	1
23	Roll Pin	X-2620	1
24	Roll Pin	X-2602	1
25	Oil Lever Tag	X-1945	1
26	Spring Adj. Lever	G-5715	1
27	Bushing	X-2721	1
28	Yoke	G-9838	1
29	Sems Fastener	X-1687	2
30	Bumper Screw	G-5113-1	1
31	Hex. Nut	X-246-4	1
32	Bumper Spring	SN-1481	1
33	Spacer	G-11886-1	1
34	Laminated Weight Assembly	A-2446	4
35	Weight Pin	G-14007	4
36	"E" Retaining Ring	X-2996	4
37	Ball Bearing	X-310	1
38	Name Plate	X-581	1
39	Escutcheon Pin	X-455	2
40	Oil Seal	X-652	1
41	Rocker Shaft Oil Seal	A-6118	1
42	External Snap Ring	X-1904	2
43	Ball Bearing	X-328	2
44	Rocker Shaft	G-11698	1
45	Spider and Shaft Assembly	A-6637	1
46	Washer	X-2026-4	1
47	Elastic Stop Nut	X-1845	1

Fig. 16.8 Parts list—governor assembly (Fig. 16.7).

9	$\frac{1}{2} \times 1\frac{1}{16}$ PLAIN WASHER	1	
8	$\frac{3}{8}$-24 x $\frac{1}{2}$ SLOTTED DOG PT. SET SC.	1	
7	#10-24 x $\frac{3}{4}$ FLAT HD. MACH. SC.	6	
6	BALL	2	C.R.S.
5	HANDLE	1	C.R.S.
4	VISE SCREW	1	C.R.S.
3	JAW PLATE	2	C.R.S.
2	JAW PATT. NO. 19742-2	1	C.I.
1	BASE PATT. NO. 19742-1	1	C.I.
ITEM	NAME	NO PER UNIT	MATERIAL

LIMITS, UNLESS OTHERWISE NOTED:
FRACTIONAL ±¹⁄₆₄ DECIMAL ± .010 ANGULAR ±½°

TITLE OF UNIT

VISE ASSEMBLY

SCALE FULL SIZE
APPROVED WJL

LAFAYETTE, INDIANA

TRACED BY J.H.D. CHECKED BY J.H.P. DATE 12-10-61
DRAWN BY DOE, JOHN H. CODE WJL-E-15
DRAWING NO. 19742

Fig. 16.9 Bill of material.

assembling parts, location dimensions, and special instructions for operating may also be shown.

16.23

Outline Assembly Drawings. Outline assembly drawings are most frequently made for illustrative purposes in catalogs. Usually they show merely overall and principal dimensions (Fig. 16.10). Their appearance may be improved by the use of line shading.

16.24

Exploded Pictorial Assembly Drawings for Parts Lists and Instruction Manuals. Exploded pictorial assembly drawings are used frequently in the parts lists sections of company catalogs and in instruction manuals. Drawings of this type are easily understood by those with very little experience in reading multiview drawings. Figure 16.11 shows a commercial example of an exploded pictorial assembly drawing.

16.25

Diagram Assembly Drawings. Diagram

drawings may be grouped into two general classes: (1) those composed of single lines and conventional symbols, such as piping diagrams, wiring diagrams, and so on (Fig. 16.12); and (2) those drawn in regular projection, such as an erection drawing, which may be shown in either orthographic or pictorial projection.

Piping diagrams give the size of pipe, location of fittings, and so on. To draw an assembly of a piping system in true orthographic projection would add no information and merely entail needless work.

A large portion of electrical drawing is composed of diagrammatic sketches using conventional electrical symbols (Fig. 16.13). Electrical engineers therefore need to know the American National (ANSI) Standard wiring symbols given in the Appendix.

16.26

Chemical Engineering Drawings. In general, the chemical engineer is concerned with plant layouts and equipment design. He must be well informed on the types of machinery used in grinding, drying, mixing, evaporation, sedimentation, and dis-

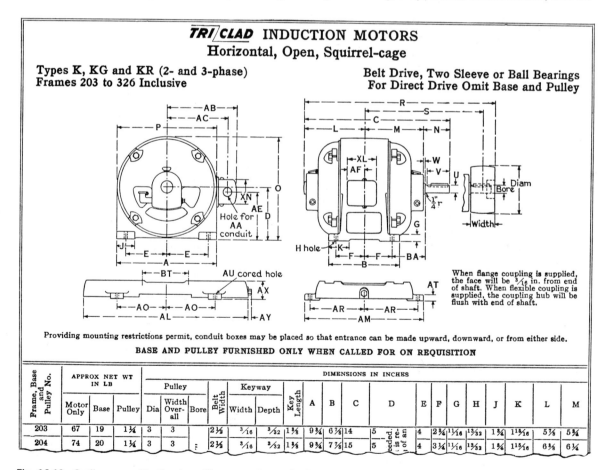

Fig. 16.10 Outline assembly drawing. (*Courtesy General Electric Company*)

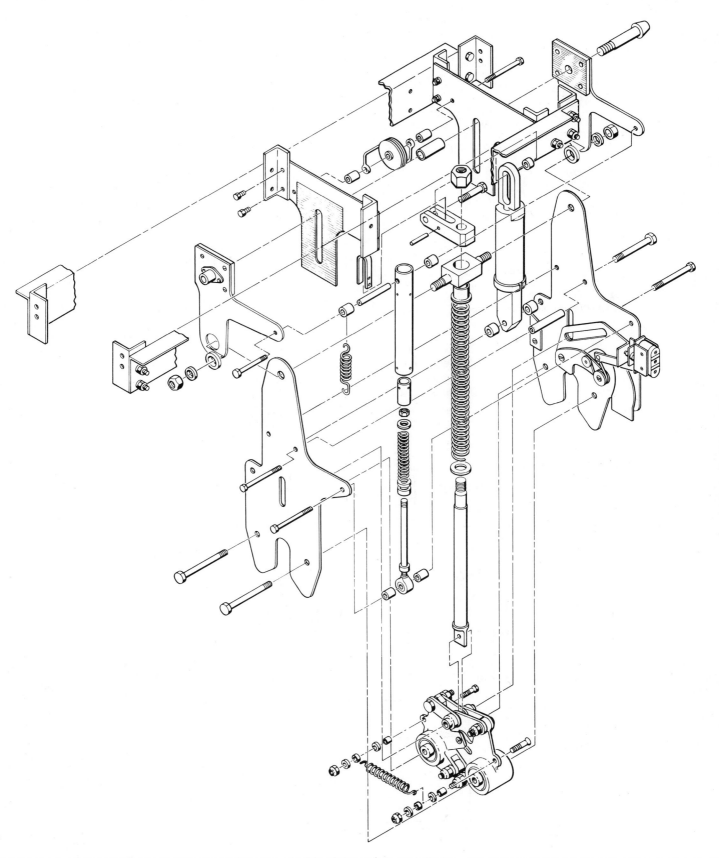

Fig. 16.11 Exploded pictorial assembly drawing. (*Courtesy Lockheed Aircraft Corporation*)

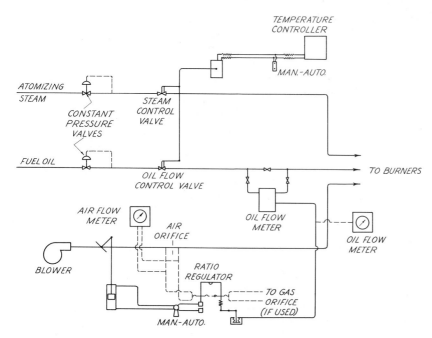

Fig. 16.12 Diagram assembly drawing. (*Courtesy* Instruments Magazine)

neer who can speak the basic language of the mechanical, electrical, or civil engineer with whom he must cooperate. To be able to do this, he must have a thorough knowledge of the principles of engineering drawing.

Plant layout drawings, the satisfactory development of which requires numerous preliminary sketches (layouts, scale diagrams, flow sheets, and so on), show the location of machines, equipment, and the like. Often, if the machinery and apparatus are used in the manufacturing of chemicals and are of a specialized nature, a chemical engineer is called on to do the designing. It even may be necessary for him to build experimental apparatus.

16.27

Electrical Engineering Drawings. Electrical engineering drawings are of two types: machine drawings and diagrammatic assemblies (Fig. 16.13). Working drawings, which are made for electrical machinery, involve all of the principles and conventions of the working drawings of the mechanical engineer. Diagrammatic drawings have been discussed in Sec. 16.25. The practices to be followed in preparing drawings of electrical systems are presented in Chapter 22.

tillation, and must be able to design or select conveying machinery.

It is obvious that the determining of the sequence of operations, selecting of machinery, arranging of piping, and so on, must be done by a trained chemical engi-

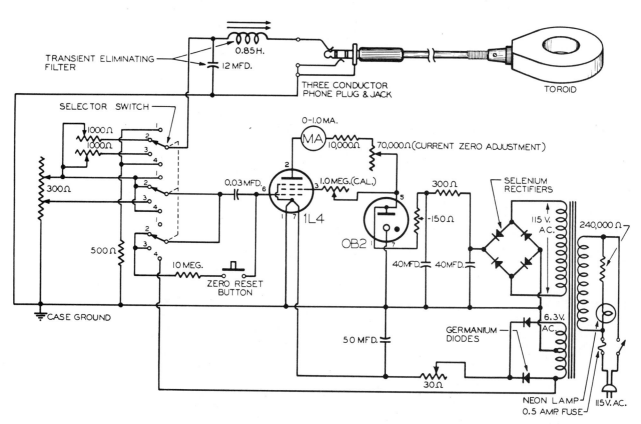

Fig. 16.13 Schematic drawing. (*Courtesy General Motors Corporation*)

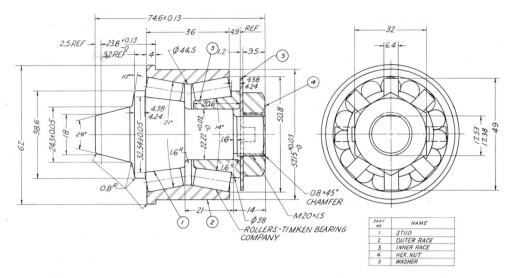

Fig. 16.37 Stud bearing. (*Courtesy Ross Gear Division of TRW, Inc.*)

PART NO.	NAME
1	STUD
2	OUTER RACE
3	INNER RACE
4	HEX. NUT
5	WASHER

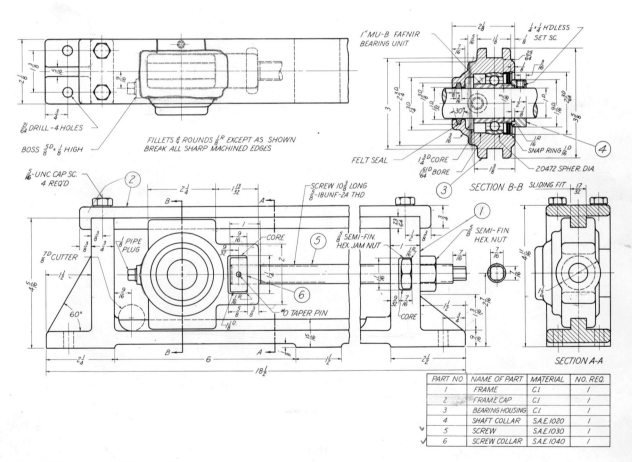

PART NO	NAME OF PART	MATERIAL	NO. REQ.
1	FRAME	C.I.	1
2	FRAME CAP	C.I.	1
3	BEARING HOUSING	C.I.	1
4	SHAFT COLLAR	S.A.E. 1020	1
5	SCREW	S.A.E. 1030	1
6	SCREW COLLAR	S.A.E. 1040	1

Fig. 16.38 Conveyor take-up unit.

PC.NO.	NAME	QUAN.	MATERIAL	PC.NO.	NAME	QUAN.	MATERIAL
1	STANDARD	1	MALL. IRON	5	GROOVE PIN	3	$\frac{7}{32}D \times \frac{5}{8}$ STEEL ROD
2	SCREW	1	S.A.E.1120 FORGING	6	LEVER BAR	1	REROLLED RAIL STK.
3	CAP	1	S.A.E.1045 FORGING	7	$\frac{7}{8}$ DIA. BALL BEARING	1	STD.
4	THRUST WASHER	1	S.A.E. 2315				

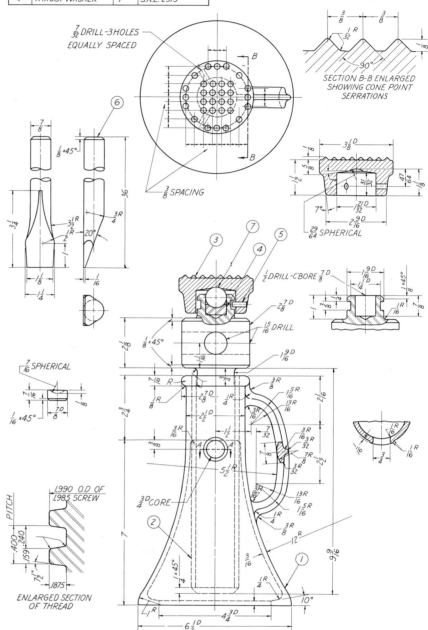

Fig. 16.39 Simplex ball-bearing screw jack. (*Courtesy Templeton, Kenly & Co.*)

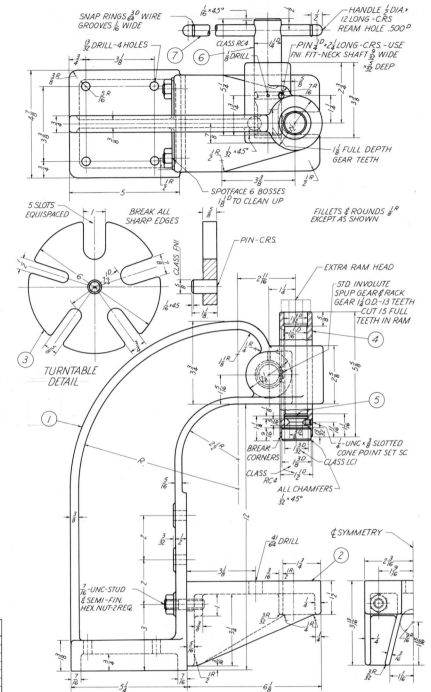

Fig. 16.40 Bench arbor press.

PART NO.	NAME	MATERIAL	NO. REQ.
1	BASE	C.I.	1
2	TABLE	C.I.	1
3	TURNTABLE	C.I.	1
4	RAM	S.A.E. 1045	1
5	RAM HEAD	S.A.E. 1040	1
6	SPINDLE	S.A.E. 1045	1
7	HANDLE	C.R.S.	1

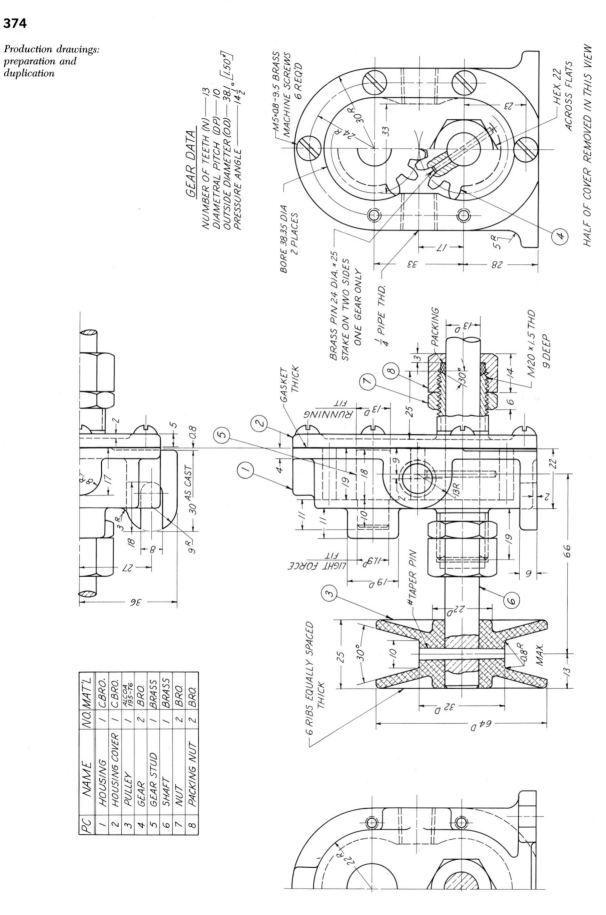

Fig. 16.41 Gear pump.

GEAR DATA

NUMBER OF TEETH (N) — 13
DIAMETRAL PITCH (DP) — 10
OUTSIDE DIAMETER (OD) — 38.1 [.50"]
PRESSURE ANGLE — 14½°

M5×0.8 - 9.5 BRASS
MACHINE SCREWS
6 REQD

HEX 22
ACROSS FLATS

HALF OF COVER REMOVED IN THIS VIEW

BORE 38.35 DIA
2 PLACES

BRASS PIN .24 DIA. × 25
STAKE ON TWO SIDES
ONE GEAR ONLY

¼ PIPE THD.

PACKING

M20 × 1.5 THD
9 DEEP

GASKET
THICK

RUNNING FIT

LIGHT FORCE FIT

30 AS CAST

TAPER PIN

6 RIBS EQUALLY SPACED
THICK

30°

MAX.

PC	NAME	NO.	MAT'L
1	HOUSING	1	C.BRO.
2	HOUSING COVER	1	C.BRO.
3	PULLEY	1	ALCOA 195-T6
4	GEAR	2	BRO.
5	GEAR STUD	1	BRASS
6	SHAFT	1	BRASS
7	NUT	2	BRO.
8	PACKING NUT	2	BRO.

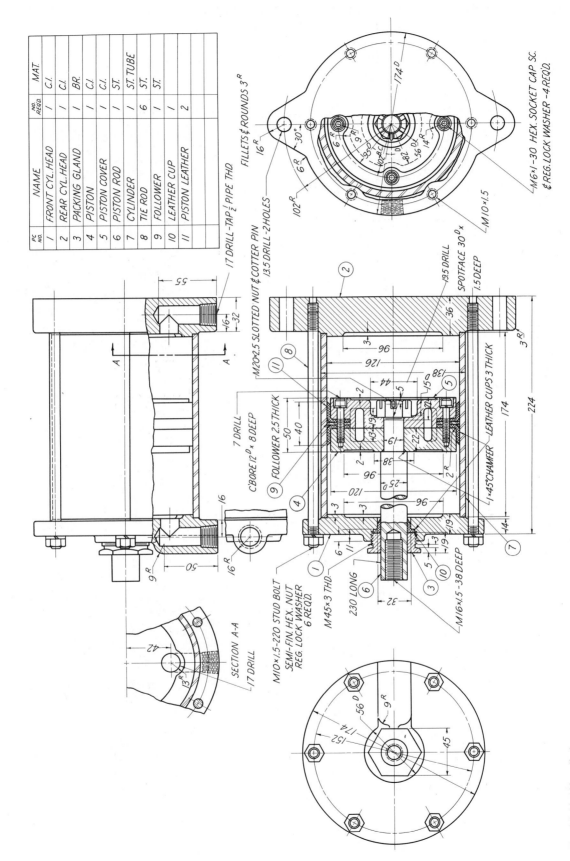

Fig. 16.42 Air cylinder.

PC. NO.	NAME	NO. REQD.	MAT.
1	FRONT CYL. HEAD	1	C.I.
2	REAR CYL. HEAD	1	C.I.
3	PACKING GLAND	1	BR.
4	PISTON	1	C.I.
5	PISTON COVER	1	C.I.
6	PISTON ROD	1	ST.
7	CYLINDER	1	ST. TUBE
8	TIE ROD	6	ST.
9	FOLLOWER	1	ST.
10	LEATHER CUP	1	
11	PISTON LEATHER	2	

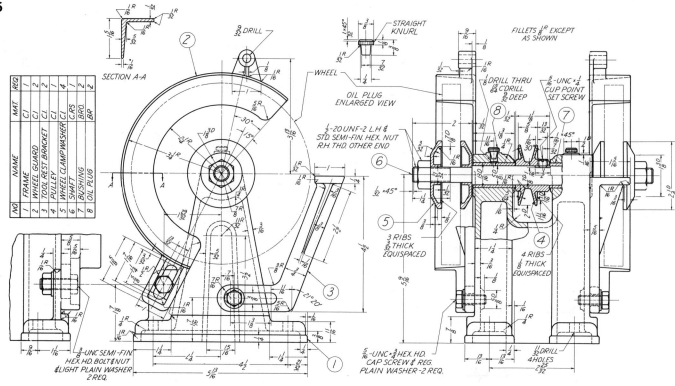

Fig. 16.43 Bench grinder.

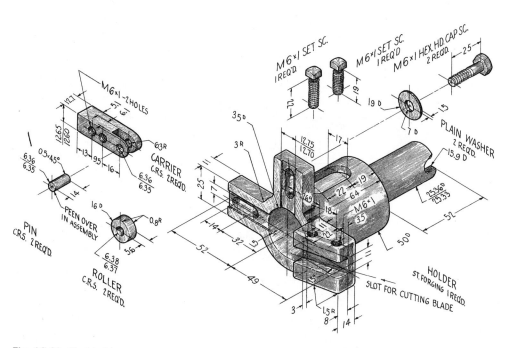

Fig. 16.44 Tool holder.

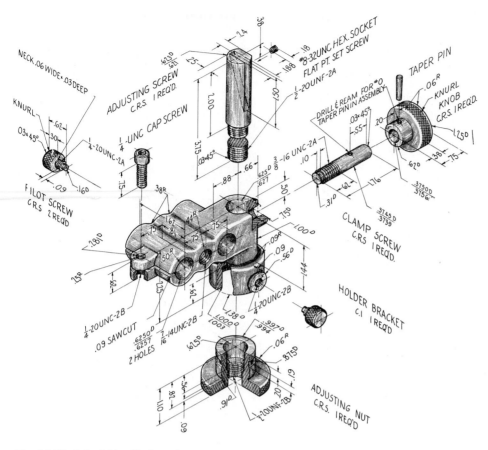

Fig. 16.45 Adjustable attachment.

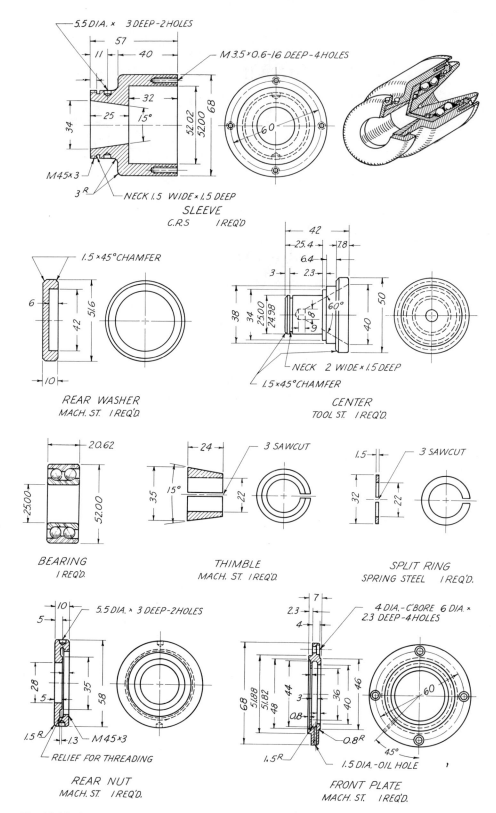

5.5 DIA. × 3 DEEP-2 HOLES

M 3.5 × 0.6-16 DEEP-4 HOLES

57

11 · 40

32

34 · 25 · 15°

52.02
52.00
68

60

M45×3

3 R

NECK 1.5 WIDE × 1.5 DEEP

SLEEVE
C.R.S · 1 REQ'D

1.5 × 45° CHAMFER

6 · 42 · 5.6

10

REAR WASHER
MACH. ST. 1 REQ'D.

42

25.4 · 7.8

6.4

3 · 2.3

38 · 34

25.00
24.98

60°

8

9

40 · 50

NECK 2 WIDE × 1.5 DEEP

1.5 × 45° CHAMFER

CENTER
TOOL ST. 1 REQ'D.

20.62

25.00 · 52.00

BEARING
1 REQ'D.

24 · 3 SAWCUT

35 · 15° · 22

THIMBLE
MACH. ST. 1 REQ'D.

1.5 · 3 SAWCUT

32 · 22

SPLIT RING
SPRING STEEL · 1 REQ'D.

10 · 5.5 DIA. × 3 DEEP-2 HOLES

5

28 · 35 · 58

5

1.5 R · 1.3 · M45×3

RELIEF FOR THREADING

REAR NUT
MACH. ST. 1 REQ'D.

7

2.3

4

4 DIA.-C'BORE 6 DIA. × 2.3 DEEP-4 HOLES

68
51.88
51.82
48 · 44 · 0.8 · 36 · 40 · 46

60

0.8 R

45°

1.5 R · 1.5 DIA.-OIL HOLE

FRONT PLATE
MACH. ST. 1 REQ'D.

Fig. 16.46 Cup center details.

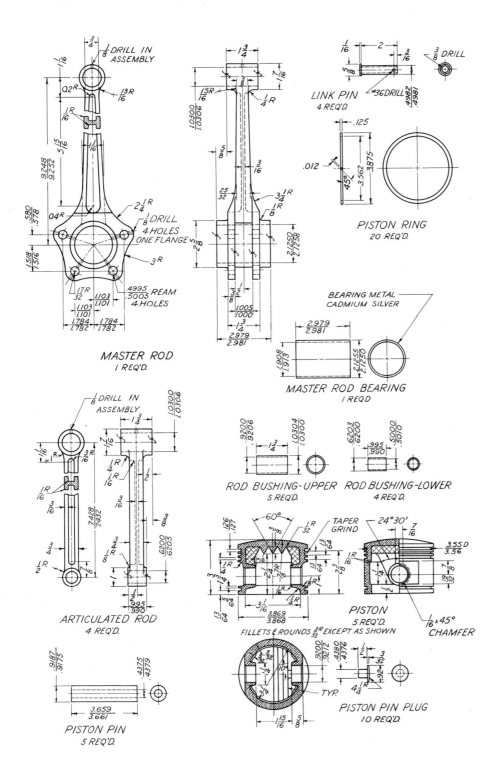

Fig. 16.47 Radial engine details.

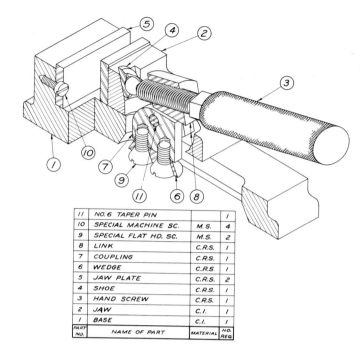

11	NO. 6 TAPER PIN		1
10	SPECIAL MACHINE SC.	M.S.	4
9	SPECIAL FLAT HD. SC.	M.S.	2
8	LINK	C.R.S.	1
7	COUPLING	C.R.S.	1
6	WEDGE	C.R.S.	1
5	JAW PLATE	C.R.S.	2
4	SHOE	C.R.S.	1
3	HAND SCREW	C.R.S.	1
2	JAW	C.I.	1
1	BASE	C.I.	1
PART NO.	NAME OF PART	MATERIAL	NO. REQ.

Fig. 16.48 Hand clamp vise.

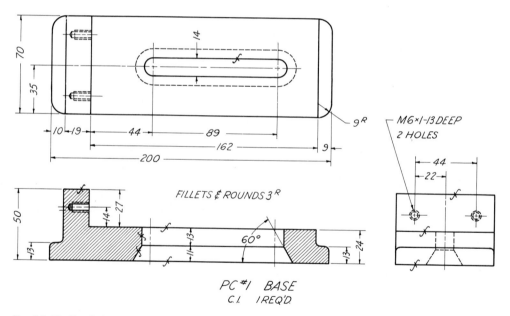

PC #1 BASE
C.I. 1 REQ'D.

Fig. 16.49 Hand-clamp-vise details.

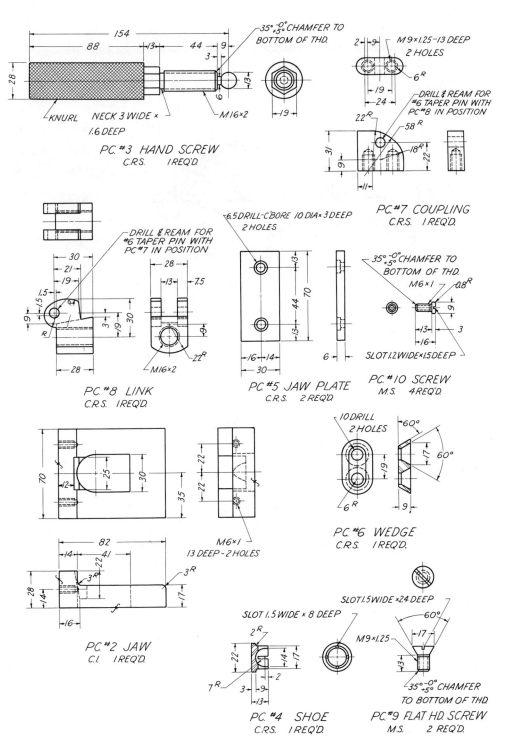

Fig. 16.50 Hand-clamp-vise details.

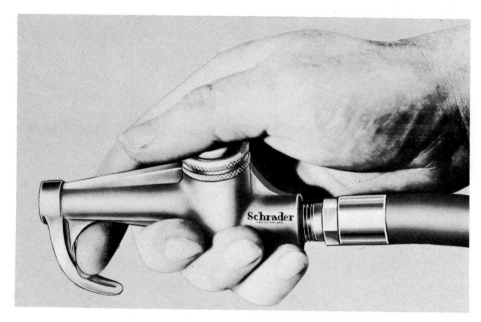

Fig. 16.51 Blowgun. (*Courtesy A. Schrader's Son Mfg. Co.*)

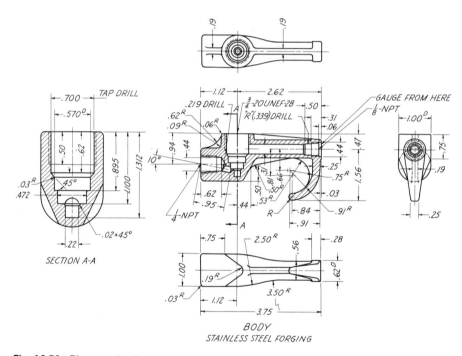

Fig. 16.52 Blowgun details.

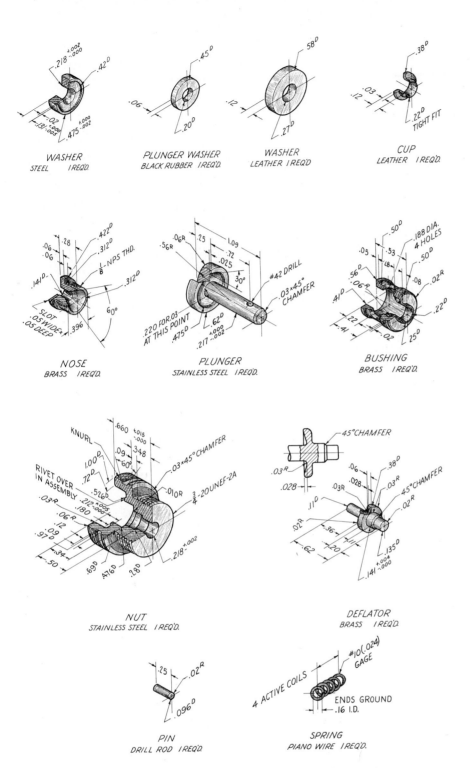

WASHER
STEEL 1 REQ'D.

PLUNGER WASHER
BLACK RUBBER 1 REQ'D.

WASHER
LEATHER 1 REQ'D.

CUP
LEATHER 1 REQ'D.

NOSE
BRASS 1 REQ'D.

PLUNGER
STAINLESS STEEL 1 REQ'D.

BUSHING
BRASS 1 REQ'D.

NUT
STAINLESS STEEL 1 REQ'D.

DEFLATOR
BRASS 1 REQ'D.

PIN
DRILL ROD 1 REQ'D.

SPRING
PIANO WIRE 1 REQ'D.

Fig. 16.53 Blowgun.

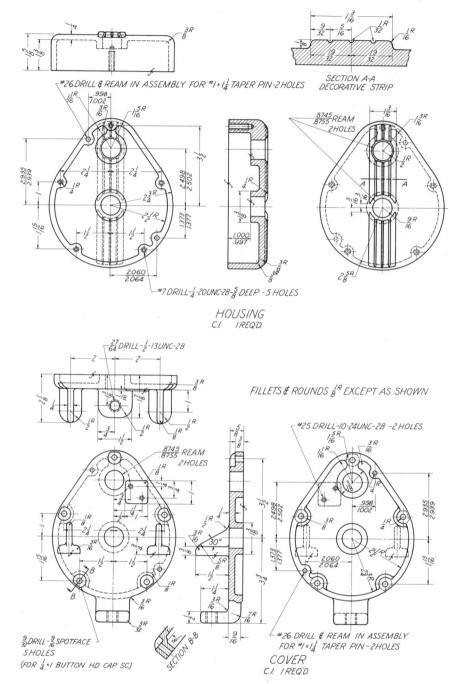

Fig. 16.54 Hand grinder details. (See Fig. 16.56.)

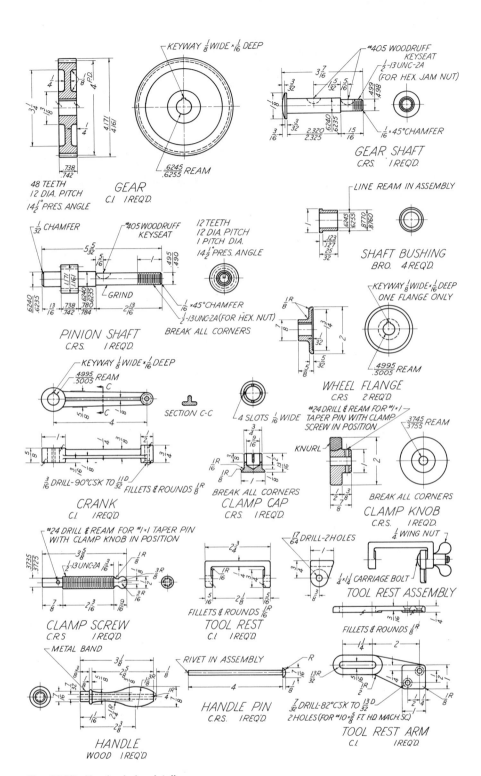

Fig. 16.55 Hand grinder details.

Fig. 16.56 Hand grinder.

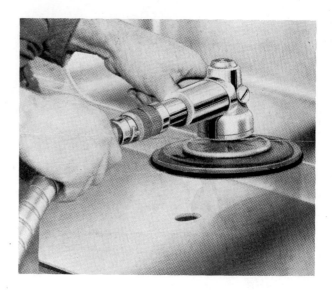

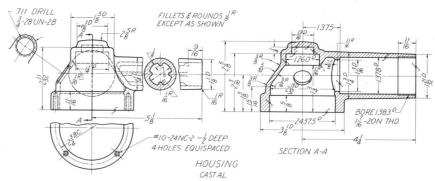

Fig. 16.57 Right-angle-head details. (*Courtesy R. C. Haskins Co.*)

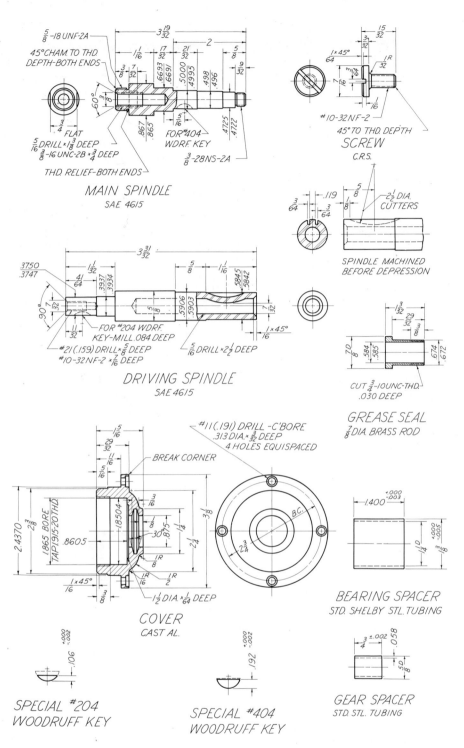

$\frac{5}{8}$-18 UNF-2A

45° CHAM. TO THD.
DEPTH–BOTH ENDS

60°

FLAT

$\frac{5}{16}$ DRILL × $\frac{3}{8}$ DEEP
$\frac{3}{8}$-16 UNC-2B × $\frac{3}{4}$ DEEP

THD. RELIEF–BOTH ENDS

FOR #404
WDRF. KEY

$\frac{3}{8}$-28NS-2A

MAIN SPINDLE
SAE 4615

SCREW
C.R.S.

#10-32NF-2

45° TO THD. DEPTH

SPINDLE MACHINED
BEFORE DEPRESSION

2$\frac{1}{2}$ DIA.
CUTTERS

FOR #204 WDRF.
KEY–MILL .084 DEEP

#21 (.159) DRILL × $\frac{5}{8}$ DEEP
#10-32NF-2 × $\frac{7}{16}$ DEEP

$\frac{5}{16}$ DRILL × 2$\frac{1}{4}$ DEEP

1 × 45°

90°

DRIVING SPINDLE
SAE 4615

CUT $\frac{3}{4}$-10UNC-THD.
.030 DEEP

GREASE SEAL
$\frac{7}{8}$ DIA. BRASS ROD

#11 (.191) DRILL -C'BORE
.313 DIA. × $\frac{3}{32}$ DEEP
4 HOLES EQUISPACED

BREAK CORNER

TAP .1915-20THD.
.1865 BORE

B.C.

1 × 45°

1$\frac{1}{2}$ DIA. × $\frac{1}{64}$ DEEP

COVER
CAST AL.

BEARING SPACER
STD. SHELBY STL. TUBING

1.400

SPECIAL #204
WOODRUFF KEY

SPECIAL #404
WOODRUFF KEY

GEAR SPACER
STD. STL. TUBING

Fig. 16.58 Right-angle-head details.

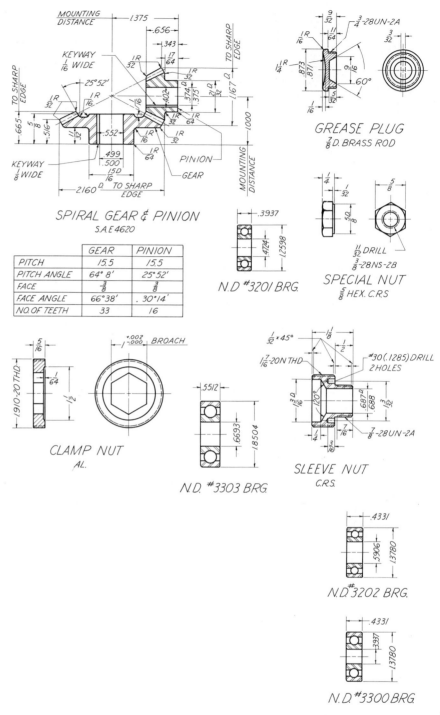

Fig. 16.59 Right-angle-head details.

Part V

The computer

FRONT SUSPENSION-LOWER ARM

PHOTO OF ACTUAL PARTS

**TOP VIEW
DEFORMED AND UNDEFORMED**

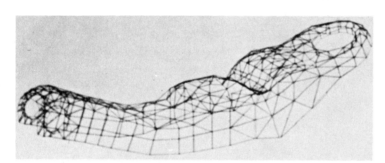

**PLOTTED
PERSPECTIVE VIEW**

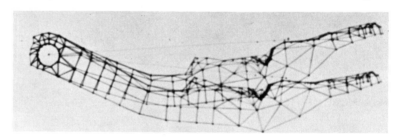

**SIDE VIEW
DEFORMED AND UNDEFORMED**

NASA computer program aids American designers "This photograph of the lower arm in the front suspension of a Ford Motor Company vehicle shows the actual part (upper left) and a grid system in various views of the same part as it appears on a computer plot produced by the NASA-developed NASTRAN program. The picture shows plots of the deformed (stresses) and undeformed (not stressed) equipment. The NASTRAN program has given Ford engineers capability to analyze structures, particularly under dynamic loading, which has not previously available to them. More than 70 industrial firms, universities, laboratories and government agencies are now using NASTRAN, a NASA-developed computer program, to solve their structural engineering problems. The program is presently being used in more than 185 different applications, ranging from suspension units and steering linkages on automobiles to the design of power plants and skyscrapers. At least 55 more uses of NASTRAN are currently planned." (*Photo and quoted statement courtesy National Aeronautics and Space Administration*)

17

Computer-aided design and automated drafting

A □ The computer and computer graphics

17.1

The Computer (Fig. 17.1). A computer is not an electronic brain, as many would have us believe. Rather, it is an electronic calculating machine working in a manner that would be considered inefficient if it were done by a human being, since essentially the machine only adds and subtracts. And yet, by doing what it has been directed to do, with greater than human speed, the computer is capable of masterful performances that are carried out with technical precision. At times, it seems almost to perform miracles—but only almost.

A computer differs from what one ordinarily considers a calculator in that it has components for an internal memory storage system for instructions. It can be said that a computer can calculate, store, compare, correct itself, and then calculate some more *as programmed*.

17.2

Use of the Computer (Fig. 17.1). Computers must be instructed in considerable detail. However, the computer program represents more then just detailed instructions. It includes the problem definition, analysis, and flow charting that are a part of the initial preparation (Fig. 17.2). When the program is written in actual machine coding, the programmer must first analyze the problem in terms of operations that the computer can perform and then write the program supplying tables, formulas, codes, and so forth, as needed

Fig. 17.1 Computer system.
Shown is the IBM System/370 Model 155 with a variety of input and output units. The keyboard/printer operator console, attached to the Model 155's central processing unit (center), can be duplicated in a second key operational area of an installation to increase the efficiency of computer room personnel. The key components shown numbered are: (1) processing unit and main storage, console, and typewriter; (2) magnetic tape units; (3) card read/punch units; (4) magnetic disc drives; (5) printers. (*Courtesy International Business Machines Corporation*)

for the specific application. To do this, the programmer must understand the computer in detail. Since this method of programming is often difficult, even for a qualified person, and at times is found to be impractical, several recognized programming languages have been developed that permit the programmer to give instructions to the computer by using statements and symbols, each statement representing many machine language instructions (Fig. 17.3). Use of these high-level languages, such as FORTRAN (FORmula TRANslation), eliminates much of the painstaking detail of computer programming. Furthermore, since FORTRAN statements closely resemble mathematical terminology, a working knowledge of the language can be acquired with about a week of instruction. With a program called a *compiler*, the computer translates FORTRAN programs into more detailed instructions for its own operation. Other programming languages are COBOL (COmmon Business Oriented Language), and ALGOL (ALGOrithmic Language).

Instructions, fed to the computer by punched cards, paper tape, or magnetic tape (Fig. 17.4), constitute the *program*.

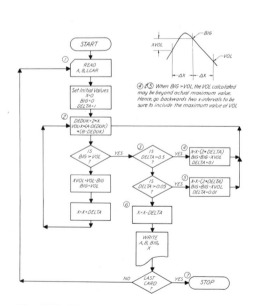

Fig. 17.2 Flow chart for maximum-volume problem (See Fig. 17.3).
A flow chart presents in graphical form the logic of a programming task facing the programmer.

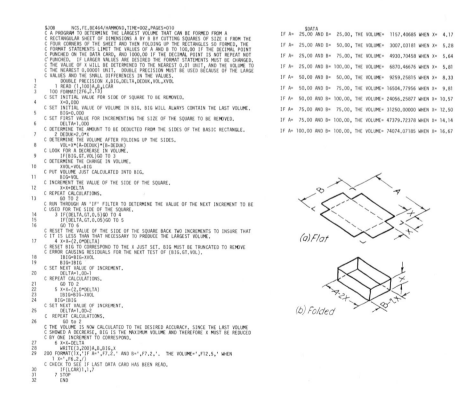

Fig. 17.3 Computer print-out of maximum volume problem (Fig. 17.2).
The program was written by Robert H. Hammond, Director of the Freshman Engineering Division, North Carolina State University.

17.3

Language of the Computer (Input). Punched cards, punched tapes, and magnetic tapes (Fig. 17.4) with information recorded as magnetized spots (called *bits*) are the conventional means for making contact with an electronic computer in the only language that the computer recognizes, the language of electrical impulses. The information supplied by these patterns of holes or magnetic spots are translated by a *sensing element* or *feeler* into the on–off binary language of the computer. Up-to-date punched-tape and punched-card sensing elements use photoelectric cells that transmit current when light reaches them. Paper tapes are punched by an input tape-punching device.

On magnetic tape, data are recorded in continuous parallel tracks. Usually, there are either six or eight data tracks and one checking track, depending on the format. Clocking mechanisms and changes in magnetic polarity indicate separate bits in the tracks. (A bit is a binary digit, either 0 or 1.) In the tape drive, a magnetic read–write head simultaneously reads a bit at a time from each of the tracks. The six or eight bits read represent a complete character. (Eight bits, called a *byte*, can represent two decimal digits or one character.) Using binary-coded decimal coding, the letters of the alphabet as well as decimal numbers may be recorded on magnetic tape. The BCD format also provides for punctuation marks and permits the use of special characters.

Punched cards, such as the one shown in Fig. 17.4, are stacked into a *card reader*, which senses the cards and translates the information furnished by the perforations into the pulse language of the computer.

To permit more efficient use of a computer, the pulses punched into the paper may be transferred first to a magnetic tape and then from it to the computer. Magnetic tapes can deliver magnetized pulses at the rate of 180,000 characters (or 360,000 decimal digits) per second, a stretch of time in which a human can say only "one and two" at the usual word rate of a speaking voice.

17.4

Computer Storages. Every computer must have a memory system; that is, it must have repositories for the storage of information. Input and output storages receive data needed for calculation and

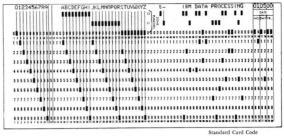

(a) Punched Card

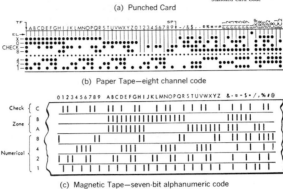

(b) Paper Tape—eight channel code

(c) Magnetic Tape—seven-bit alphanumeric code

Fig. 17.4 Data recording media—punched card, paper tape, and seven-track magnetic tape. (*Courtesy International Business Machines Corporation*)

note the results before these results are typed out by the high-speed printer. There are also intermediate (short) storages that receive results that the computer will need for continued calculation. A special component for intermediate storage, the accumulator, receives totals to be returned immediately to the matrix.

The three categories of computer storage are (1) the main memory; (2) auxiliary storage, which includes direct-access devices such as magnetic disc units and magnetic drums; and (3) bulk storage, involving the use of punched cards and magnetic tape. (See Fig. 17.1)

17.5

Computer Output. Up to this point, we have focused our attention on how information is fed into electronic computers by means of an input unit and to some extent on how the computer digests data and makes the necessary calculations. All of this action would quite naturally be useless, however, if the computer could not express the results of the calculations (Fig. 17.3). The apparatus needed for this purpose is called an *output unit*. Several types of units are available to satisfy different requirements.

One such unit is the tape punch having perforating pins that are actuated by the electrical impulses of the computer. A tape-punch unit is capable of punching 50 or more characters per second in the paper tape. Punched tapes are deciphered by a printer that converts every character into either a letter or a number. Output card punches are also available. Punched-card information may be converted to printed copy through the computer or by using special auxiliary equipment.

When the results obtained represent an intermediate step and there is no need to read the results at the time, the pulses issued by the computer are picked up on magnetic tape to be held in reserve, say, for use a month later. When it becomes necessary to learn what is on a magnetic tape, it may be run through a complicated electronic device that sends the magnetic impulses to a high-speed output printer. This high-speed printing device is almost a miracle in itself because it can print a full line at a time (up to 132 characters) at a rate of 1100 lines per minute.

Information may be transferred, without changing the code, from magnetic tapes to punched tapes or cards and, as also might be expected, from punched cards to magnetic tapes. This is accomplished by devices known as *converters*.

17.6

Communicating with a Computer. Improvement in programming is gradually eliminating the formidable barrier that has existed between the computer and its prospective users. In the years ahead computers will become commonplace and they will have capabilities that will make it possible for them to be used by almost anyone. The day when each of us can consult with his own computer lies just ahead. With this thought in mind, it will be shown how man and computer can work together as partners, with man in control of the creative aspect of the design process. To instruct the computer, the user needs little more than a working knowledge of basic graphics, since the programming barrier is fading away to a great extent in this area.

17.7

Computer Graphics. In addition to making extensive use of conventional computer techniques for engineering and scientific analysis, the engineering team can, with the aid of one of several graphic data processing systems, bring a large-scale digital computer to bear on those problems that can best be analyzed and finally formalized by using and modifying graphical representation. This great step for-

ward in automated graphics makes it possible for an engineer or designer to do at least a part of his creative work on a graphic input–output cathode-ray console that is connected directly to the computer (see Fig. 17.5). The means for man–machine communication is not in the future; it is available now and several practical systems are in daily use for design and production work in the aerospace and automotive fields.

The direct man–computer interaction that this system affords leads to quick answers for even small problems and permits concepts to be evaluated and tested and then accepted or rejected. This relationship in which the computer has become man's immediate partner in creative design became possible only when a way was found to make the computer correctly interpret drawings and, if desired, make a drawing after the intentions of its human partner had been made known. Properly used, these systems can relieve both the engineer and the man at the drawing board of endless calculations and of much of the tedious work connected with making layout and detail drawings. Such systems bring together the talents and creativity of the user and the power and speed of the digital computer in conjunction with digitized plotters or cathode-ray-tube consoles with appropriate supporting equipment.

In general, the term computer-aided design implies that the computer assists the designer in analyzing and modifying a previously created design within new design parameters that he, the designer, has established. Almost any designer or draftsman can use graphic data processing effectively and can analyze, modify, and distribute design information even though he may have only a limited knowledge of computer programming. This is true because the designer at the console and the computer are in two-way communication while using the designer's own graphic language of lines, symbols, pictures, and words.

The computer system described thus does not replace the man, nor does it eliminate the requirement that a designer have a working knowledge of engineering graphics. The graphic language, so fundamental to the design process, has now become a computer language. As such, it has proved to be a reliable means for a conversational type of man–machine communication.

The GM DAC–1 (Fig. 17.5) system that

will be discussed in Secs. 17.8–17.10 is no longer in general use. However, full descriptions of its operations and capabilities have been given here in this text in order to point up the potential of a complete system of this type. Ordinarily, it is much more practical for an industrial organization to assemble a needed graphic system using units now available on the market than to develop a special "one-of-kind" system that may not justify the cost and may soon be out-dated. In a sense, it can be said that DAC–1 was developed as somewhat of a laboratory experiment to determine the potential of a total system of computer graphics in the area of design. As such, it represented a great leap forward at that time.

Fig. 17.5 Man–machine communication.
"At the General Motors Research Laboratories, using the new GM DAC-I system (Design Augmented by Computers), a research engineer checks out a computer program that allows him to modify a design 'drawing.' A touch of the electric 'pencil' to the tube face signals the computer to begin an assigned task, in this case, 'Line Deletion,' where indicated. The man may also instruct the computer using the keyboard at right, the card reader below the keyboard, or the program control buttons below the screen. Hundreds of special computer programs, written by GM computer research programmers, are needed to carry out these studies in man–machine communications."

17.8

Main Elements of the DAC–1 Digital Graphic Input–Output System. The main elements of the GM graphic data processing system shown in Fig. 17.6 are as follows:

1. A large-scale digital computer system (with supporting software) that is capable of storing vast amounts of information and retrieving specifically desired information rapidly.

2. One or more graphic consoles with viewing screens.

3. A film recorder (35 mm) for "hard copy" of graphic information that has been computer generated.

4. A 35 mm film scanner for the input of data in graphical form. This unit of the system converts images recorded on

microfilm to digital form for further processing.

17.9

DAC–1 Display Console. It is at the display console that the man–machine conversation takes place. The principal feature of the console is the cathode-ray-tube

Fig. 17.6 GM DAC–1 graphic processing components. (*Courtesy General Motors Corporation*)

(CRT) screen, on which the computer-programmed graphic and alphanumeric information is displayed (Fig. 17.5). Close at hand on the console are the keyboards and the light pencil, which provide convenient means of entering and modifying rapidly displayed computer-programmed graphic and alphanumeric information. At the console unit the users have direct access to data (previously stored in the system's main or auxiliary storage units in digital form) that they may study and then modify and redisplay.

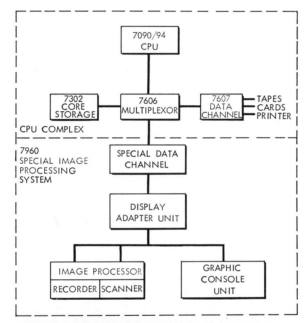

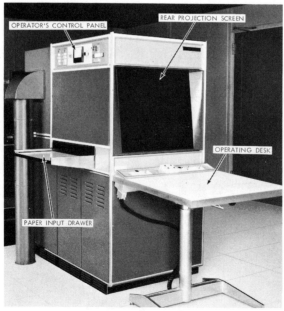

Fig. 17.7 DAC–1 image processing unit. (*Courtesy General Motors Corporation*)

17.10

GM DAC–1 Image Processing Unit. The DAC–1 computerized design and drafting system provides for both the input and output of data in a graphical form. A direct output for engineering and manufacturing use can be in the form of hard copy, a drawing prepared by an on-line plotter, or a programmed tape that can be used to actuate an automated production machine. At this point in our discussion our attention will be directed to the special image processor that enables the computer to read and generate drawings (Fig. 17.7). The use of tapes with plotters adaptable for off-line operation will be discussed later in this chapter.

The image processing unit provides means for the input and output of data in graphical form. Primarily, this component contains a CRT photo-recorder, projectors, a CRT photo-scanner, an input camera, and the rapid film processing equipment needed to develop 35mm film. When the film is developed within the recorder, the image can be viewed on a rear projection screen that is located for easy viewing by the user (Fig. 17.6).

After an input copy has been inserted into the image processor on a tray and then photographed and developed, the 35mm negative is positioned so that the CRT photo-scanner can "read it." The film scanner converts the image furnished by the 35mm negative (light lines on dark background) directly to digital data that is sent either to the computer or to memory storage. Basically, the near miracle is accomplished via an electronic beam that, when it determines whether each of many addressable points is above or below a set level of light intensity, notes the result in computer (digital) language. The digital information thus developed may now be either modified or stored permanently.

From the high-speed recorder of the image processor, on 35mm film, come the permanent records of drawings, charts, and parts lists that are needed by designers and draftsmen. The film, exposed to a high-resolution CRT and developed within the recorder, can be viewed at development on a rear projection screen, as stated previously. Finally, conventional working copies (on paper) may be readily obtained using modern viewer–copier equipment and recorder-produced exposures mounted in aperture cards. It was the development of image processing units, such as the DAC–1 described, that extended the capability of computers to the point where they can accept, interpret, analyze, relate, and finally produce needed output in graphic form.

17.11
IBM 2250 Display Unit (Fig. 17.8). Images are generated on the CRT at the console by computer-programmed positioning and deflection of the CRT's electronic beam to the more than one million possible points (grid format of 1024 by 1024 addressable points) of the 12 in. × 12 in. effective display area. The beam is deflected to each point on the screen that is addressed by the program, either intensifying it or not intensifying it, as directed. The image as we see it on the face of the CRT (screen) is produced by the beam hitting a phosphor coating, causing the coating to glow for just an instant. Operating under this condition, the information normally displayed fades rapidly (within a fraction of a second); therefore, the display must be regenerated approximately 30–40 times every second for a flicker-free representation of high resolution. A buffer is used to regenerate the display independently of the computer.

Alphanumeric symbols may be formed by synthesizing the characters from a series of individually programmed dots or line segments, either horizontal or vertical. To improve computer efficiency, an optional character generator can be used for computer-independent formation of all alphanumeric symbols.

The position-indicating light pencil, when used in conjunction with the keyboards and programs in storage, provides the means whereby the image displayed can be altered (Fig. 17.5). It may be used at will to create, remove (erase), enlarge, or rearrange any part of a displayed image that, as a whole, may consist of points, line segments of any length and orientation, simple and complex geometric shapes, and appropriate graphical symbols and statements. The light pencil (receiving device), having a photocell element that senses a spot of light directly

Fig. 17.8 Example of a CRT display.
The IBM 2250, an advanced graphics display unit, permits users to exchange visual information with IBM's lowest-cost computing system, the desk-sized IBM 1130 (right). This particular unit was developed especially for engineers and designers who need fingertip access to their computers. The display enables them to work with charts, diagrams, drawings, or printed letters and numbers directly on the face of the television-like screen. With the electronic "light pen" the designer can revise images, add or delete lines, or change dimensions. He also can change information and images on the screen through the 1130 and 2250 keyboards (note programmed function and alphanumeric keyboards). A civil engineer, for example, could use the system to help design a building or a bridge, or an electrical engineer could use it to diagram new circuits. As the designer works he can call on the 1130 to perform calculations and update information previously stored in the system. If the user wants a more powerful computer, he can link his combined 1130/2250 system to a larger IBM System/360 through standard communication lines. (*Courtesy International Business Machines Corporation*)